THE HUMAN RIGHTS-BASED APPROACH TO CARBON FINANCE

This book explores the topical and contentious issue of the human rights risks associated with carbon projects, especially in developing countries. It outlines a human rights-based approach to carbon finance as a functional framework for mainstreaming human rights into the design, approval, finance and implementation of carbon projects. It also describes the nature and scope of carbon projects, available legal options for their financing and key human rights issues at stake in their planning and execution. Written in a user-friendly style, the proposal for a rights-based due-diligence framework through which human rights risks can be anticipated and addressed makes this book relevant to all stakeholders in carbon, energy and environmental investments and projects.

DAMILOLA S. OLAWUYI is a senior lecturer in law and director of the Institute for Oil, Gas, Energy, Environment and Sustainable Development at Afe Babalola University, Nigeria. He also provides legal advice and services as an energy lawyer with the leading global law firm Norton Rose Fulbright Canada LLP.

THE HUMAN RIGHTS-BASED APPROACH TO CARBON FINANCE

DAMILOLA S. OLAWUYI

Barrister and Solicitor, Norton Rose Fulbright Canada LLP
Senior Lecturer and Institute Director, Afe Babalola University, Nigeria

CAMBRIDGE
UNIVERSITY PRESS

CAMBRIDGE
UNIVERSITY PRESS

University Printing House, Cambridge CB2 8BS, United Kingdom

Cambridge University Press is part of the University of Cambridge.

It furthers the University's mission by disseminating knowledge in the pursuit of education, learning and research at the highest international levels of excellence.

www.cambridge.org
Information on this title: www.cambridge.org/9781107105515

First published 2016

A catalogue record for this publication is available from the British Library

Library of Congress Cataloguing in Publication data
Olawuyi, Damilola S. (Damilola Sunday), 1983– author.
The human rights-based approach to carbon finance / Damilola S Olawuyi, Barrister and Solicitor, Norton Rose Fulbright Canada LLP, Senior Lecturer and Institute Director, Afe Babalola University, Nigeria.
Cambridge, United Kingdom : Cambridge University Press, 2016. | Based on author's thesis (doctoral – University of Oxford, 2013) issued under title: The human rights based approach to climate change mitigation.
LCCN 2016003558 | ISBN 9781107105515 (hardback)
LCSH: Carbon dioxide mitigation – Law and legislation. | Carbon offsetting – Law and legislation. | Law and economic development. | International law and human rights.
LCC K3593.5.C37 O423 2016 | DDC 344.04/633–dc23
LC record available at https://lccn.loc.gov/2016003558

ISBN 978-1-107-10551-5 Hardback

To the memory of my mother, Victoria Ebunlomo Olawuyi

CONTENTS

FOREWORD

This book is about the big issue in the transition to sustainability – the linking of equity with sustainability. No longer is it tolerable to focus only on environmental impact. Development must be approached and financed through the triple lens of environmental, social and economic sustainability.

The message is one which courts will have to increasingly address, whether they are enforcing laws which governments have enacted, enforcing European Union law against their domestic governments or, more innovatively, finding in general domestic principles a basis for ordering further governmental action.

The UK Supreme Court recently made a mandatory order that the secretary of state prepare new air quality plans under Article 23(1) of Directive 2008/50/EC in accordance with a defined timetable to lead to revised plans being submitted to the European Commission not later than 31 December 2015. This followed the United Kingdom's failure to reduce nitrogen dioxide levels to the levels prescribed by the Directive. A court in The Hague recently held that plans by the Dutch government to reduce greenhouse gas emissions by 2020 by only 14–17 per cent compared to 1990 levels were unlawful under the general law of tort, having regard to the scale of the threat posed by such emissions, and that the Dutch government should instead reduce emissions by at least 25 per cent by the end of 2020. Its reasoning was expressly inspired by general constitutional, international legal and human rights principles and by its conviction that, as an industrialized country, the Netherlands must contribute towards the solution of a worldwide problem.

The essential background to the present book is Dr Damilola S. Olawuyi's doctoral degree in energy and environmental law from the University of Oxford, United Kingdom, and his extensive international experience as an academic and legal practitioner. The work is an innovative and valuable study of the human rights implications of climate change and carbon projects. Its starting point, supported by examples, is that development projects can violate human rights – in some cases even though these may have been conceived and supported in order to reduce carbon emissions. His thesis is that, what is required under international and national law is an approach recognizing human rights considerations at the heart of any development project.

Dr Olawuyi supports the view that all human rights – political, civil, social, cultural and economic – are equal in importance and none can be fully enjoyed without the others: 'without a belly full of food and other vital needs, human rights and fundamental freedoms become meaningless'.[1] But, at the same time, there needs to be 'a holistic approach that mainstreams human rights norms into environmental planning to avoid situations whereby governmental policies or project, for example in the areas of climate change mitigation impact, threaten social, economic and cultural rights'.[2]

The thread running through the book is human rights mainstreaming – the concept of embedding human rights norms in the design, finance and implementation of carbon projects, advocated by the United Nations for nearly twenty years (and echoed recently in the Pope's encyclical 'Laudato Si' – Care of our common home' addressed to 'every person living on this planet'). Rather than an approach involving the recognition or expansion of substantive rights, Dr Damilola Olawuyi focuses on a procedural approach to the achievement of such integration – better access to information, wider public participation, greater attention to the needs to ensure equality and avoid discrimination, improved accountability and monitoring, always backed by access to justice.

The book discusses what this means in practical terms – an expanded role for bodies presently engaged in screening carbon projects, a requirement for a specific human rights impact assessment, review mechanisms, better governance and compliance and complaints committees.

Dr Damilola Olawuyi has written a clear, well-informed and well-documented work, in a field which should and does engage all of us evermore. If the world continues as it is, there may, as a commentator quoted in one footnote observes, come a time when the only habitable place is the Antarctic.[3] This study will stimulate thinking not only about how to avoid that drastic infringement of all our human rights but about

[1] See Chapter 3, where Dr Olawuyi evaluates and reviews the full belly thesis.

[2] Dr Olawuyi discusses the potentials and paradoxes of mainstreaming human rights to climate governance in Chapter 4.

[3] See Chapter 1, note 6.

a whole range of potential infringements which carbon projects threaten in our present everyday existence. The author's skill and initiative in researching and writing it are to be commended. It will be an important stimulus for governments, peoples, industrial, commercial and non-governmental organizations worldwide involved in or affected by carbon and other environmental projects.

The Right Honourable the Lord Jonathan Hugh Mance
Justice of the Supreme Court of the United Kingdom
Chair, International Law Association

PREFACE AND ACKNOWLEDGEMENTS

Over the last decade, the effects of an unprecedented rise in global temperature caused by climate change, on the enjoyment of human rights, especially the right to life, have been subjects of intensive scholarly attention. Many literatures have been espoused on the need for governments, business enterprises and other key stakeholders to adopt policy measures aimed at combating climate change. However, recent findings show that climate change response measures and projects, particularly clean development mechanism (CDM) and multilateral REDD+ (Reducing Emissions from Deforestation and Forest Degradation together with forest conservation, sustainable forest management and enhancement of carbon stocks), are in turn producing, arguably, more serious human rights concerns. While the direct impacts of climate change on human rights have received enormous attention, issues relating to the secondary impacts of climate change responses and projects on human rights, have yet to receive a book-length, exhaustive and rigorous exposition and analysis. This book is an attempt to fill that gap.

This book examines the negative implications and impacts of projects designed to combat climate change ('carbon projects'), on the effective enjoyment of fundamental human rights. It reviews growing evidence that carbon projects currently produce serious human rights concerns, especially in developing countries. The main topics covered lay at the heart of contemporary debates on the impacts of climate change on human rights and larger debates on the need to recognize and reinforce the synergies between human rights and climate change in designing climate change responses.

The ten chapters of the book examine and discuss legal and institutional frameworks required to incorporate human rights norms and safeguards into the design, approval, finance and implementation of carbon projects, specifically the CDM, REDD+ and green climate funds, to address negative human rights impacts and consequences. Despite growing recognition that carbon projects may foster human rights violations, most especially the direct recognition by the 2010 Cancun climate conference that countries 'should fully respect human rights in all climate change related actions', legal and institutional frameworks required to anticipate, mitigate and prevent these human rights impacts of carbon projects have not been coherently developed and communicated.

Overall, the ultimate aim of the book is to answer the question of 'how', that is, how can the international climate change regime be organized to fully integrate and mainstream human rights safeguards and considerations into the design, approval, finance and implementation of carbon projects to protect the rights of affected peoples and communities? The book discusses the guiding principles of a human rights-based approach to carbon finance as a legal and institutional framework for mainstreaming human rights principles into the design, approval, finance and implementation of carbon projects. The book also takes a comprehensive, highly analytical approach to discuss how lessons learned from the implementation of select climate finance mechanisms, specifically CDM and REDD+, may inform thoughts on practical and institutional requirements for mainstreaming human rights principles into the international climate change regime in general.

This book is prepared in a user-friendly style to enhance its utility among its primary audience, namely environmental and human rights administrators, students, lawyers, bankers, auditors, government representatives, judges, analysts, pundits, international negotiators and policy leaders. It provides environmental and human rights stakeholders from all backgrounds with a toolkit and operational framework to understand, digest and appreciate the legal and institutional framework for reflecting human rights norms in climate actions and projects.

In examining these issues, this book takes a comparative approach that benefits immensely from the author's research, professional and teaching experience acquired over the years while working on environmental issues in Nigeria, Kenya, Canada, Australia, India, China, United States, United Kingdom, Switzerland, Spain, Denmark, France and Italy. It offers discussions and analyses of concepts, instruments and principles on climate change and human rights, that are key to a rights-based approach to the design, approval, finance and implementation of carbon projects. A discussion of diverse institutions that could aid the systemic integration and stronger legal partnerships between human rights and climate change, under a rights-based approach, is also provided in the book.

It remains for me to register my profound gratitude to many helping hands, without whom the publication of this book would not have been possible. First and foremost, my thanks and appreciation go to God the Almighty for the successful publication of this book. Second, my profound gratitude goes to Dr Andrew Shacknove and Professor Alastair R Lucas Q.C., who supervised my doctoral thesis at the University of Oxford – the research work that culminated in the publication of this book. I am highly indebted to them for their scholarly guidance, mentoring and intellectual generosity in the course of my research. Furthermore, I owe a debt of gratitude to Professor Dapo Akande and Professor Edwin Peel for their constructive comments and encouragements during my academic voyage at Oxford. I am also indebted to Professor Catherine Redgwell and Professor Phillipe Cullet for providing insightful and helpful comments during my doctoral thesis examination. Their constructive criticisms and suggestions greatly enhanced the framing of this book.

I also appreciate the kind motivation and support of the president and founder of Afe Babalola University, Aare Afe Babalola, OFR, CON, SAN, FNIALS, FCIArb., LLD, who has taken unparalleled personal interests in my professional development. Aare Afe Babalola's journey and global achievements provide reinforcing pillars and instructive pathways that one could only aspire to follow.

Furthermore, I am especially indebted to everyone at Norton Rose Fulbright Canada LLP, for their exceptional support and kind mentoring over the years. I would also like to extend a word of gratitude to Orieji Onuma and Omolola Olarinde for providing careful and thoughtful research assistance, and to my legal assistants, Calvin Jim, Curtis Snow, Moronkeji Kolawole and Tosin Adewunmi, who have all worked tirelessly for the success of this book.

Finally, I would like to thank my dear wife Oluwabunmi, and my twin girls, Titilayo and Oluwatoni for their love, dedication and inspiration. Thanks are also due to my exceptional students at the College of Law, Afe Babalola University, whose class contributions and brilliant ideas served as timely seeds that blossomed into this book. I thank you all for your support and kindness, and I commit this book to your scholastic minds.

I have endeavoured to state the position of the law as of 30 June 2015, although I have been able to take into account subsequent developments in one or two instances.

Damilola S. Olawuyi
Ado Ekiti, June 2015

ABBREVIATIONS

ACHR	American Convention on Human Rights
AfCHPR	African Charter on Human and Peoples Rights
ATCA	United States Aliens Torts Act
ATI	Access to Information
BAP	Bali Action Plan
CDM	Clean Development Mechanism
CDM EB	Executive Board of the Clean Development Mechanism
CEDAW	Convention on the Elimination of All Forms of Discrimination against Women
CER	Certified Emissions Reductions
CERD	Convention on Elimination of Racial Discrimination
CESCR	Committee on Economic, Social and Cultural Rights
CFCs	Chlorofluorocarbons
CFU	Carbon Finance Unit
COP	Conference of the Parties
CSR	Corporate social responsibility
DfNS	Debt for Nature Swaps
DNA	Designated National Authority
DOE	Designated Operational Entity
ECHR	European Convention on Human Rights
EIA	Environmental Impact Assessment
EP	Equator Principles
ERPA	Emissions Reduction Purchase Agreement
ERU	Emission Reduction Unit
ET	Emissions Trading
FAO	Food and Agriculture Organization of the United Nations
FCPF	Forest Carbon Partnership Facility
FPIC	Free Prior Informed Consent
GCF	Green Climate Fund
GEF	Global Environment Facility
GHGs	Green House Gases
GWP	Global Warming Potential
HFCs	Hydrofluorocarbons

HRBA	Human Rights-Based Approach
HRCU	Human Rights And Climate Change Unit
HRE	Human Rights Education
HRIA	Human Rights Impact Assessment
ICC	Inuit Circumpolar Conference
ICCPR	International Covenant on Civil and Political Rights
ICERD	International Convention on the Elimination of All Forms of Racial Discrimination
ICESCR	International Covenant on Economic, Social and Cultural Rights
ICJ	International Court of Justice
IFC	International Finance Corporation
ILC	International Law Commission
ILO	International Labour Organization
IMF	International Monetary Fund
IPCC	Intergovernmental Panel on Climate Change
IUCN	International Union for Conservation of Nature
JI	Joint Implementation
KPCS	Kimberley Process Certification Scheme
LDC	Least Developed Countries
LDCF	Least Developed Country Fund
MOP	Meeting of the Parties
NGOs	Non Governmental Organizations
NIMBY	Not-in-My-Backyard
ODA	Official Development Assistance
OHCHR	Office of the High Commissioner for Human Rights
PCB	Public Complaints Branch
PDD	Project Design Document
PDT	Participation Development Theory
PFCs	Perfluorocarbons
PRTR	Protocol UNECE Kiev Protocol on Pollutant Release and Transfer Registers
REDD	Reducing Emissions from Deforestation and Forest Degradation
REDD+	REDD together with sustainable forest management and enhancement of carbon stocks
RTD	Right to Development
SCCF	Special Climate Change Fund
SF6	Sulphur Hexafluoride
UDHR	Universal Declaration of Human Rights
UNDG	United Nations Development Group
UNDRIP	United Nations Declaration on the Rights of Indigenous Peoples
UNECE	United Nations Economic Commission for Europe
UNEP	United Nations Environmental Programme

UNESCO	United Nations Educational, Scientific and Cultural Organization
UNFCCC	United Nations Framework Convention on Climate Change
UNHRC	United Nations Human Rights Council
WBIP	World Bank Inspection Panel
WWF	World Wildlife Fund

PART I

Carbon projects and human rights: introductory context and principles

1

Introduction

> There is virtually no aspect of our work that does not have a human rights dimension. Whether we are talking about peace and security, development, humanitarian action, the struggle against terrorism, climate change, none of these challenges can be addressed in isolation from human rights.
>
> Ban Ki-moon, Secretary-General of the United Nations.[1]

The aim of this book is to examine and analyse the impacts of climate change mitigation projects on human rights; that is, the potential infringement of fundamental human rights by projects undertaken to cut the emission of greenhouse gases (GHGs) that contribute to climate change ('carbon projects').[2] Drawing examples from CDM and REDD+ projects, this book discusses the need to mainstream human rights safeguards into the design, approval, finance and implementation of carbon projects to avoid human rights impacts and risks. It also takes a wider look at how lessons learnt from the implementation of CDM and REDD+ projects could inform thoughts on the value of, and requirements for, mainstreaming human rights principles into the design of project-based mechanisms and into the international climate change regime in general. This book examines legal and theoretical prospects and paradoxes of adopting the United Nations Human Rights-Based Approach (HRBA) as a framework through which human rights norms may be mainstreamed into the design, approval, finance and implementation of carbon projects.[3]

[1] See the United Nations Human Rights Mainstreaming Portal, hrbaportal.org/the-un-and-hrba, accessed 12 January 2015.

[2] While mitigation covers reducing greenhouse gas (GHG) emissions to prevent climate change, adaptation means coping with those impacts that cannot be avoided. It includes policy adjustments or changes proposed as means for ameliorating the anticipated adverse consequences associated with climate change. See B. Smith, I. Burton, R. Klein, and J. Wandel, 'An Anatomy of Adaptation to Climate Change and Variability', *Climatic Change*, 45(1) (2000), 233–251.

[3] The HRBA in this book specifically refers to the framework proposed by the United Nations, which encourages all entities of the UN system to mainstream human rights

Questions unpacked and examined in this book are as follows: Firstly, are there sufficient legal supports for claims that carbon projects currently threaten or could threaten the protection and fulfilment of fundamental human rights under international law? Secondly, are there legal and theoretical justifications for emerging proposals on the need to integrate and mainstream human rights standards into the international climate change regime? Specifically, what are the theoretical prospects and problems associated with the emerging 'human rights mainstream paradigm' under international law? Thirdly, could the HRBA be applied as a framework for mainstreaming human rights standards into project design, approval, finance and implementation procedures under the international climate change regime? What is the basis, value and novelty of the HRBA as a human rights mainstreaming approach as compared to traditional human rights mainstreaming approaches? Fourthly, what are the practical and normative implications, problems and paradoxes of mainstreaming human rights norms into the international climate change regime? And finally, what are the legal and institutional strategies and reforms needed to overcome these practical problems?

This chapter provides factual context and background on the linkages between carbon projects and human rights. It begins with an evaluation and review of pertinent human rights questions and challenges that have trailed the implementation of carbon projects under the international climate change regime. It then examines how these challenges have resulted in increased calls for an approach that recognizes the cross-cutting synergies between human rights and carbon projects. This chapter will also lay out the scope and structure of the book.

norms into their areas of activities. The implementation of the HRBA across national and regional levels includes enacting laws and policies that would recognize the rights of citizens to demand participation, information, protection and justice in environmental issues and the duty of states and non-state actors to protect, respect and fulfill the provisions of such laws. At the international level, it includes mainstreaming human rights into multilateral treaties. See United Nations, 'The Human Rights Based Approach to Development Cooperation: Towards a Common Understanding among UN Agencies' (2003) www.undg.org/archive_docs/6959-The_Human_Rights_Based_Approach_to_Development_Cooperation_Towards_a_Common_Understanding_among_UN.pdf, accessed 12 October 2015.

1.1 Climate change, carbon finance and human rights: background and factual context

Climate change is perhaps one of the greatest concerns facing our present generation.[4] Each of the last three decades has been successively warmer than any preceding decade since 1850.[5] With emerging signs of temperature change all over the world, it is now widely accepted that climate change is real; that human emissions of GHGs are a cause; that if left unchecked, climate change may lead to: extreme weather events such as droughts and flooding; food scarcity, ill health and economic decline in nations of the world.[6]

[4] Global warming is often used interchangeably with climate change. Climate change is the generic term used to describe any increase in the average temperature of the atmosphere at an abnormal rate, caused by the anthropogenic emission of gases that trap the sun's heat in the atmosphere. These gases called 'greenhouse gases' (GHGs) act like blankets which absorb heat radiation that should escape to space, thereby heating the atmosphere at a rate far from normal. GHGs that contribute to global warming include carbon dioxide ($C0_2$), methane (CH_4), nitrous oxides, chlorofluorocarbons (CFCs) and halocarbons. J. Houghton, *Global Warming: The Complete Briefing*, 4th edn (Cambridge: Cambridge University Press, 2009); See also United Nations, 'Glossary of Climate Change Acronyms and Jargon', unfccc.int/siteinfo/glossary.html, accessed 08 October 2014; J. Von Doussa, 'Climate Change: Catastrophic Impacts and Human Rights' (2007), www.humanrights.gov.au/about/media/speeches/speeches_president/2007/20071211_Climate_Change.html, accessed 10 October 2014.

[5] This is according to the Intergovernmental Panel on Climate Change (IPCC) - a body of over 2,000 scientists established by the United Nations to provide comprehensive scientific assessments of current scientific, technical and socio-economic information about the risk of climate change, its potential environmental and socio-economic consequences, and possible options for adapting to these consequences or mitigating the effects. According to the IPCC, the current evidence of changing climate includes an increase in night-time temperatures over many land areas at about twice the rate of day-time temperatures; an increase in the length of the freeze-free season in many Northern Hemisphere mid-to-high latitude land areas; more intense rainfall events over many Northern Hemisphere mid-to-high latitude land areas; a near-worldwide decrease in mountain glacier extent and ice mass; a decrease in Northern Hemisphere sea-ice amounts; and a substantial thinning of Arctic sea-ice in late summer. See Intergovernmental Panel on Climate Change (IPCC), 'Summary for Policymakers', in T.F. Stocker, D. Qin, G.-K. Plattner, M. Tignor, S.K. Allen, J. Boschung, A. Nauels, Y. Xia, V. Bex, and P.M. Midgley (eds.), *Climate Change 2013: The Physical Science Basis. Contribution of Working Group I to the Fifth Assessment Report of the Intergovernmental Panel on Climate Change* (Cambridge: Cambridge University Press, 2013) 5–7. See also IPCC, *Climate Change 2014: Impacts, Adaptation, and Vulnerability, Contribution of the Working Group II to the Fifth Assessment Report of the Intergovernmental Panel on Climate Change* (Cambridge University Press 2014) 232–275.

[6] J. Hansen, 'Defusing the Global Warming Time Bomb', *Scientific American*, 3 (2004), 68–77; see also M. Wallstrom, 'Meeting the Long Term Challenge of Global Warming: A European Perspective', in D. Michel (ed.), *Climate Policy for the 21st Century: Meeting the Long-term Challenge of Global Warming* (Washington, DC: John Hopkins University,

It was in order to avert these disasters that the global community agreed to the Kyoto Protocol in 1997.[7] The Kyoto Protocol gained recognition as one of the most important global agreements of the late twentieth century, not only for fixing GHG emission limits to be achieved by industrialized nations by 2012, but also for providing three flexible, project-based mechanisms through which industrialized countries can achieve their emission reduction objectives.[8] Apart from encouraging countries to design domestic policy measures aimed at climate change mitigation and adaptation, the flexible mechanisms of the Kyoto Protocol presented industrialized countries with additional opportunities to earn emission reduction credits anywhere in the world at the lowest cost possible by investing in projects that lead to emission reduction and sustainable development.

For example, studies confirm that while it requires US$50 to mitigate one tonne of carbon dioxide equivalent (CO_2 eq.) in developed countries, the same reduction can be accomplished in developing countries at US$15 per tonne of CO_2 eq.[9] The mechanisms are therefore designed to deliver cheap emission reductions to the investing country and sustainable development gains to the host country. According to the United Nations, through carbon projects, a country could: attract capital for infrastructural development that could assist in a shift to a more prosperous but less carbon-intensive

Center for Transatlantic Relations 2003), 17–25. Professor David King was perhaps making this same point when he said that 'Antarctica is likely going to be the world's only habitable continent by the end of this century, if global warming remains unchecked.' See G. Lean, 'Global Warming Could Soon Make Antarctica the Only Place to Live, Says Chief British Scientist', *The Independent on Sunday* (London, 2 May 2004).

7 The Kyoto Protocol is a legally binding set of obligations ratified by thirty-eight industrialized countries and eleven countries in Central and Eastern Europe to return their emissions of GHGs to an average of approximately 5.2 per cent below their 1990 levels over the commitment period of 2008–2012. Conference of the Parties to the Framework Convention on Climate Change, *Kyoto Protocol*, 10 December 1997, UN Doc FCCC/CP/1997/L.7/add.1, 37 ILM 22 (1998) [*Kyoto Protocol*].

8 The three flexible mechanisms: Joint Implementation (JI), Emission Trading (ET) and the Clean Development Mechanism (CDM) allow industrialized countries to meet their emission reduction targets by investing in projects abroad rather than through domestic actions alone. For a detailed and excellent discussion of these mechanisms, see F. Yamin and J. Depledge, *The International Climate Change Regime* (Cambridge: Cambridge University Press, 2005), 25.

9 See F. Ackerman, 'Financing the Climate Mitigation and Adaptation Measures in Developing Countries' (2009), sei-us.org/Publications_PDF/SEI-WorkingPaperUS-0910.pdf, accessed 02 June 2015. See also United Nations Framework Convention on Climate Change, 'Investment and Financial Flows to Address Climate Change: An Update' (2008) FCCC/TP/2008/7.

economy; foster the active participation of both its private and public sectors; provide a tool for technology transfer through projects that replace old, dirty and inefficient fossil fuel technology with cleaner ones; create new industries using environmentally sustainable technologies; and define investment priorities in projects that meet sustainable development goals.[10] Specifically, project-based emission reduction measures could assist developing countries to achieve progress in environmental issues, such as cleaner air and water, reduced deforestation, soil conservation and biodiversity protection. They could also result in social benefits such as rural development, employment and poverty alleviation.[11] Project-based mitigation efforts therefore provide significant social and economic benefits for poorer societies. The magnitude of financial flows to major developing countries generated by the GHG allowance trading market could be substantial, rising from approximately US$10 billion to over US$100 billion per year in future decades.[12]

The CDM has been particularly important over the last decade because it is the only Kyoto mechanism that allows project investments in developing countries.[13] It allows developing countries to host emission reduction projects sponsored by industrialized countries. The CDM was designed to assist developing countries in achieving sustainable

[10] United Nations Environmental Programme, *CDM Information and Guide Book* (3rd Edition, Roskilde: UNEP 2011) 14; K. Olsen and J. Fenhann, 'Sustainable Development Benefits of Clean Development Mechanism Projects: A New Methodology for Sustainability Assessment Based on Text Analysis of the Project Design Documents Submitted for Validation', *Energy Policy*, 36 (2008), 2819–2830.

[11] Olsen and Fenhann, *ibid.*

[12] See J. Wiener, 'Global Environmental Regulation: Instrument Choice in Legal Context', *Yale Law Journal*, 108 (1999), 677, 691–692, also the 2012 Report of the UNFCCC, 'The Benefits of the Clean Development Mechanism 2012', cdm.unfccc.int/about/dev_ben/ABC_2012.pdf, accessed 12 January 2015.

[13] Joint Implementation (JI) and Emission Trading (ET) take place between two industrialized countries with emission reduction targets. Since developing countries do not have emission reduction targets under the Kyoto Protocol, they are only eligible to take part in the CDM, which is a cooperative mechanism that allows developed countries to invest in developing countries in exchange for emission reduction credits. There are also human rights issues in emission trading and joint implementation; however, this book focuses on the CDM because it has attracted the most protests and court cases. Secondly, it is also the only mechanism with global coverage as it allows for collaborative project development between any developed country and a developing country party to the Protocol. This global reach has generated more concerns, especially in developing countries. Despite the focus on the CDM however, proposals in this book could be applied to address human rights issues in project-based mechanisms in general.

development by boosting their economies and by promoting environmentally friendly investment from governments and businesses of industrialized countries. CDM projects have provided enormous opportunities for developing countries to reach some of their economic, social, environmental and sustainable development objectives, such as cleaner air and water, improved land use, accompanied by social benefits such as rural development, employment, and poverty alleviation and reduced dependence on fossil fuels.[14] Estimates indicate that average foreign investments in participating developing countries through the CDM could be as high as US$4752 million annually.[15]

Studies show that over the past decade, the CDM alone has helped nations to mitigate approximately one billion tonnes of GHGs while saving developed countries a total of US$3.6 billion in mitigation costs.[16] With over 7,597 registered CDM projects in over eighty countries, the CDM has mobilized more than US$215.4 billion in investments in developing countries, thereby providing opportunities for socio-economic growth and poverty alleviation in many developing countries.[17]

However, implementations of project-based mechanisms under the Kyoto Protocol have been fraught with challenges.[18] For example, the

[14] C. Figueres, *Establishing National Authorities for the CDM: A Guide for Developing Countries* (Washington DC: International Institute for Sustainable Development, 2004), 21. See also D. Behn, 'Linking Climate Change Mitigation and Poverty Reduction: Using Kyoto's Clean Development Mechanism to Promote Energy Development on the African Continent', in Y. Le Bouthillier *et al.* (eds.), *Poverty Alleviation and Environmental Law* (Cheltenham, UK: Edward Elgar, 2012), chp. 13.

[15] See the 2012 Report of the UNFCCC, 'The Benefits of the Clean Development Mechanism 2012', (n 12).

[16] See Benefits of the Clean Development Mechanism, *ibid.* According to the report, the total investment in registered or soon-to-be-registered CDM projects as of June 2012 is estimated at US$ 215.4 billion. Investments in projects that are known to be operating is US$ 92.2 billion. The annual investment peaked in 2008 at US$ 13.9 billion (operating projects) and US$ 40.4 billion (all projects), but the large number of projects undergoing validation could lead to a new, much higher, peak in 2012 or thereafter. Over 750 million CERs had been transferred from the CDM registry by the end of 2011. The revenue generated by the sale of these CERs is estimated to be at least US$ 9.5 billion and possibly as much as US$ 13.5 billion.

[17] See Behn, 'Linking Climate Change Mitigation and Poverty Reduction', (n 14), chp. 13; also UNEP/ Danish Technical University, CDM/JI Pipeline Analysis and Database, www.cdmpipeline.org, accessed 08 February 2015.

[18] See generally N. Roht-Arriaza, 'Human Rights in the Climate Change Regime', *Journal of Human Rights and the Environment*, 1(2) (2010) (where the author identifies areas where current climate change regimes may cause human rights violations in local communities. These include some projects under the Clean Development Mechanism, large hydropower and biomass projects, use of biofuels, choices on energy and adaptation, and

CDM has been criticized for not delivering on its sustainable development promises.[19] Furthermore, a number of CDM projects approved by the CDM Executive Board (CDM EB) have been criticized for resulting in the violation of fundamental human rights in developing countries.[20] It is estimated that climate change mitigation and adaptation projects already displace several million people a year.[21] These tend to be the poorest and the most vulnerable citizens in developing countries where these projects

REDD+ projects); see also K. Umamaheswaran and A. Michaelowa, 'Additionality and Sustainable Development Issues Regarding CDM Projects in Energy Efficiency Sector' (2006) HWWA Discussion Paper 346; I. Rowlands, 'The Kyoto Protocol's Clean Development Mechanism: A Sustainability Assessment', *Third World Quarterly*, 22 (2001), 795.

19 See L. Schneider, 'Is the CDM Fulfilling its Environmental and Sustainable Development Objectives? An Evaluation of the CDM and Options for Improvement', a report prepared for the World Wildlife Fund (Berlin: Öko-Institut, 2007); C. Sutter, 'Does the Current Clean Development Mechanism Deliver Its Sustainable Development Claims' (2005) *HWWA Report* 1; R. Saner and A. Neiderberger, 'Hype or Reality: Can the CDM Trigger FDI?', *European Climate Platform*, 2 (2005), 12; B. Haya, *Damming the CDM: Why Big Hydro Is Ruining the Clean Development Mechanism* (Berkeley, CA: International Rivers Network, 2002) 1; H. Kolshus, 'Can the Clean Development Mechanism Attain Both Cost effectiveness and Sustainable Development Objectives?', CICERO Working Paper, 8 (2001), 1; Umamaheswaran and Michaelowa, 'Additionality and Sustainable Development Issues', *ibid.*

20 See C. McDowell, 'Climate-Change Adaptation and Mitigation: Implications for Land Acquisition and Population Relocation', *Development and Policy Review*, 31(6) (2013), 677–679; T. Griffiths and F. Martone, 'Seeing "REDD"? Forests, Climate Change Mitigation and the Rights of Indigenous Peoples and Local Communities' (May 2009), www.rightsandresources.org/documents/files/doc_923.pdf, accessed 12 January 2015.

21 See M.A. Orellana, 'A Human Rights-Based Approach to Climate Change', in J. Parra (ed.), *The Human Rights-Based Approach: A Field of Action for Human Rights Education* (Cifedhop, 2012) 58, www.ciel.org/Publications/HR_Approach_Climate_2012.pdf, accessed 12 December 2015 (noting that 'a large-scale hydroelectric project in a developing country under the CDM could force local communities to relocate from their traditional lands and livelihoods resulting in infringement of their rights to self-determination, property and life'); also A. de Sherbinin *et al.*, 'Preparing for Resettlement Associated with Climate Change', *Science*, 334(6055) (2011), 456–457; R. Bronen, 'Climate-Induced Community Relocations: Creating an Adaptive Governance Framework Based in Human Rights Doctrine', *New York University Review of Law & Social Change*, 35 (2011), 357–407; K. Hoshour and J. Kalafut, *A Growing Global Crisis: Development-Induced Displacement and Resettlement* (International Accountability Project Issue Paper 2010); J. McAdam, *Climate Change, Forced Migration, and International Law* (Oxford: Oxford University Press, 2012); B. Mayer, 'International Legal Challenges of Climate-Induced Migration: Proposal for an International Legal Framework', *Colorado Journal of International Environmental Law and Policy*, 22 (2011), 357–416; J. McAdam (ed.), *Climate Change and Displacement: Multidisciplinary Perspectives* (Oxford and Portland, OR: Hart Publishing Ltd, 2010); M. Couldrey and M. Herson (eds.), *Forced Migration Review: Climate Change and Displacements* (Oxford: Oxford Refugees Studies Center, 2008), 6.

are located.[22] In addition, there are concerns related to pollution caused by the transfer of outdated and inefficient technologies for emission credits.[23] Other human rights concerns include lack of opportunities for participation by stakeholders in project planning and implementation,[24] siting of projects in poor and vulnerable communities, lack of governmental accountability on projects and the absence of judicial and quasi-judicial remedies for victims of the above-mentioned problems.[25] Furthermore, considering that many communities live in and are dependent on forests for their lives and livelihoods, there is great concern around potential negative impacts of REDD+ projects on traditional land rights.

Due to these problems, the credibility and integrity of project-based mechanisms under the Kyoto Protocol have come under intensive scrutiny.[26] Projects that carry significant potential for climate change

[22] As the United Nations Human Rights Council recognized in its resolution 16/11, environmental damage is felt most acutely by those segments of the population already in vulnerable situations. See United Nations Human Rights Council, *Resolution adopted by the Human Rights Council 16/11, Human rights and the environment*, www2.ohchr.org/english/bodies/hrcouncil/docs/16session/A.HRC.RES.16.11_en.pdf, accessed 12 December, 2015.

[23] See generally P. Lucas and T. Patzek, 'The Disastrous Local and Global Impacts of Tropical Biofuel Production', *Energy Tribune* (March 2007), 19. See also Carbon Trade Watch, 'Groups Slam Nigeria's Submission of Gas Flare Reductions for Carbon Credits' (2006), www.carbontradewatch.org/index.php?option=com_content&task=view&id=171&Itemid=36, accessed 12 October 2014.

[24] See F. Seymour, 'Forests, Climate Change, and Human Rights: Managing Risks and Trade-offs', in S. Humphreys, (ed.), *Human Rights and Climate Change* (Cambridge: Cambridge University Press, 2010), 207; S. Jodoin, 'From Copenhagen to Cancun: A Changing Climate for Human Rights in the UNFCCC?' CISDL & IDLO Sustainable Development Law on Climate Change Working Paper Series (January 2011), www.idlo.int/Download.aspx?Id=282&LinkUrl=Publications/3_JodoinSébastien%20_ChangingClimateforHumanRights.pdf&FileName=3_JodoinSébastien%20_ChangingClimateforHumanRights.pdf, accessed 03 August 2014.

[25] See Petition to the Inter-American Commission on Human Rights Seeking Relief from Violations Resulting from Global Warming caused by Acts and Omissions of the United States, by the Inuit people of the Arctic Regions of the United States and Canada, 7 December 2005. The IACHR informed the petitioners that it would not consider the petition because the information it provided was not sufficient for making a determination and that no legally enforceable right has been violated. Available at inuitcircumpolar.com/files/uploads/iccfiles/ FINALPetitionICC.pdf, accessed 12 July 2014.

[26] See L. Pottinger, *The CDM's Hydro Hall of Shame, Bad Deal for the Planet: Why Carbon Offsets Aren't Working. . .And How to Create a Fair Global Climate Accord* (Dams, Rivers and People Report 2008, International Rivers, 2008); S. McInerney-Lankford, 'Climate Change and Human Rights: An Introduction to Legal Issues', *Harvard Environmental Law Review*, 33(2) (2009), 431–437; Oxfam, Climate Wrongs and Human Rights: Putting People at the Heart of Climate-Change Policy (2008), Oxfam Briefing Paper on Climate Change and Human Rights, www.oxfam.org.uk/resources/

mitigation have been met with resistance, criticisms and protests due to alleged impacts on human rights of local communities.[27] These concerns have led to questions on whether solutions to climate change are not in fact more threatening to human rights and livelihood than climate change itself.[28] In some cases, human rights and accountability concerns associated with carbon projects have been so significant that advocacy groups have condemned the CDM as a 'Cheap and Corrupt Development Mechanism', while REDD+ projects have also been openly denounced by indigenous communities.[29]

A good example is the level of protests and international condemnation that trailed the approval of the Aguan biogas CDM project in the Bajo Aguan region of Honduras.[30] This project, sponsored by the United Kingdom (UK), has been heavily linked with human rights violations.[31] According to a report of the international human rights mission submitted

policy/climate_change/downloads/bp117_climatewrongs.pdf, accessed 12 July 2014; see also C. Aminzadeh 'A Moral Imperative: The Human Rights Implications of Climate Change', *Hastings International and Comparative Law Review*, 30 (2007), 259; M. Limon, 'Human Rights and Climate Change: Constructing a Case for Political Action', *Harvard Environmental Law Review*, 33(2) (2009), 451.

27 See German Energy Transition, 'Carbon Credits Tarnished by Human Rights Disgrace', www.euractiv.com/climate-environment/carbon-credits-tarnished-human-r-news-508068, accessed 02 August 2014.

28 S. Long, E. Roberts, and J. Dehm, 'Climate Justice Inside and Outside the UNFCCC: The Example of REDD', *Journal of Australian Political Economy*, 66 (December 2010), 229; H. Morten Haugen, 'What Role for Human Rights in Clean Development Mechanism, REDD+, and Green Climate Fund Projects?' *Nordic Environmental Law Journal*, 5 (2013), 51.

29 See Down to Earth Group, 'Issues: Flexibility Mechanisms', *Down to Earth Magazine* (15 November 2005); see also Centre for Science and Environment, 'Current CDM Design Corrupt and "Unclean"', www.cseindia.org/node/3031, accessed 21 March 2015. This book disagrees with this rather sentimental condemnation of the CDM. While it acknowledges the flaws in the implementation of the CDM, this book argues that project-based mechanisms, such as the CDM, could stimulate real economic, social and environmental growth in developing countries if properly restructured. This book calls for a reform of project-based mechanisms to make them more transparent, accountable and rights-based.

30 See Carbon Market Watch, 'Open Letter to UK: UK Government must withdraw authorization for Aguan and Lean CDM projects linked to assassinations and other human rights abuses in Honduras' (2011), carbonmarketwatch.org/open-letter-uk-government-must-withdraw-authorisation-for-aguan-and-lean-cdm-projects-linked-to-assassinations-and-other-human-rights-abuses-in-honduras/ accessed 21 January 2015.

31 See BIOMASS Hub, 'Human Rights Violations Linked to CDM Biogas Project in Honduras' (2011), biomasshub.com/human-rights-violations-linked-cdm-biogas-honduras/accessed 12 July 2014. See also CDM Watch, 'Press Release: United Nations under Pressure to denounce Human Rights Abuses in Carbon Offsetting Scheme' (2011), www.cdm-watch.org/?p=1872, accessed 02 August 2014.

on 25 March 2011 to the Inter-American Commission on Human Rights, the local project developer was alleged to have been at the centre of violent conflicts with local people, who were deprived of their land by the project; about twenty-three local farmers were allegedly killed.[32] A coalition of over seventy international human rights groups called on the United Kingdom to withdraw sponsorship for the project and for the CDM EB not to approve or register the project.[33] In response, the UK Secretary of State for Energy promised to follow the findings of the CDM EB on the project.[34] Despite these protests, the CDM EB approved this project. As in most cases, the CDM EB argued that it had no mandate to investigate human rights abuses and that any matter related to the sustainable development of a CDM project or its human rights impacts is determined by the government that hosts the project.[35]

The absence of a comprehensive project review mechanism that could allow an international review panel to evaluate and address human rights complaints emanating from the execution of CDM projects has perennially presented complex dilemmas to human rights protection in

[32] For comprehensive details of human rights violations by this project, see CDM Watch, *Petition to the CDM Executive Board on Aguan Gas project* (2011), www.cdm-watch.org/wordpress/wp-content/uploads/2011/02/unsolicited_letter_cdmproject_application_3197_honduras.pdf., accessed 02 August 2014; See also FIAN, Human Rights Violation in Bajo Arguan (2011), www.fian.org/resources/documents/others/honduras-human-rights-violations-in-bajo-aguan/pdf, accessed 02 August 2014.

[33] CDM Watch, 'Open Letter: UK Government must withdraw authorisation for Aguan and Lean CDM projects linked to assassinations and other human rights abuses in Honduras' (2011), www.cdm-watch.org/?p=1648, accessed 02 August 2014. As a response to protests by several international human rights groups, German public development bank DEG (Deutsche Entwicklungsgesellschaft) declared that it will not pay out an already approved loan of US$20 million for the project. Similarly, EDF Trading, a wholly owned subsidiary of Electricité de France SA's and one of the biggest CDM investors, pulled out from a contract to buy carbon credits from the project. See CDM Watch, 'German Bank Won't Lend to Honduran CO_2 Project' (April, 2011), www.bloomberg.com/news/2011-04-18/german-bank-won-t-lend-to-honduran-co2-project-cdm-watch-says.html, accessed 1 August 2014.

[34] See Letter by Rt. Hon Chris Huhne M.P., (April 2011), www.cdm-watch.org/wordpress /wp-content/uploads/2011/02/UK_Gov_reponse_on_aguan_130411.pdf, accessed 31 July 2014.

[35] According to the spokesman for the CDM Executive Board, 'The allegations are deplorable. If human life has been taken, or human rights violated in any other way, it is a flagrant violation of the most fundamental principles of the United Nations... However the CDM board has no mandate to investigate human rights abuses. Any matters related to the sustainable development of the project are determined by the government that hosts the project, in this case the de-facto government of Honduras.' See the Response of the CDM Executive Board to the Petition (2011), www.cdm-watch.org/wordpress/wp-content/uploads/2011/02/EB59-14_CDM-Watch_Concerns-on-CDM-project-in-Honduras-Ref.-3197_Response.pdf, accessed 02 August 2014.

developing countries where CDM projects are sited.[36] Furthermore, placing the decision to host (or discontinue) a project in the hands of national governments that are financially interested in such projects has arguably resulted in the approval of questionable projects, including projects that have fostered monumental loss of lives and the violation of human rights. Arguably, national authorities in developing countries have been more concerned with hosting carbon projects at all costs, irrespective of the human rights consequences of such projects.[37] Due to this scramble by developing countries to host CDM projects, there have been increased tendencies to lower sustainability standards and to encourage foreign CDM projects despite their potential short- and long-term detrimental effects on human rights.[38] These concerns are most serious in developing countries with already existing abysmal human rights records.[39]

[36] See Roht-Arriaza, 'Human Rights in the Climate Change Regime', (n 18), 211; see also D. Olawuyi, 'Towards a Transparent and Accountable Clean Development Mechanism: Legal and Institutional Imperatives', *Nordic Environmental Law Journal*, 2 (2) (2012), 33–52.

[37] As the CDM Watch Group rightly noted: 'Current CDM Rules rely on the CDM host country government to assess whether a project contributes to sustainable development. This places the assessment of sustainable development in the hands of governments that would like to see more investment in their respective countries. As a consequence it is quite logical that essential criteria to assess sustainable development, which are chosen by host country governments themselves, are already deemed to be fulfilled with the mere requirement that the projects "increase GDP per capita". As a consequence, no CDM project has ever been rejected on the basis that it did not contribute to sustainable development. Allegations of serious human rights abuses related to CDM project applications in Honduras and Panama have caused an outcry amongst civil society organizations and widespread dismay that human rights are not being taken seriously under the CDM'. See CDM Watch Group News Letter (February 2011), www.cdm-watch.org/wordpress/wp-content/uploads/2011/02/cdm_watch_newsletter_15_february_2011.pdf, accessed 23 July 2014.

[38] M. Jung, 'Host Country Attractiveness for CDM Non-sink Projects', *Energy Policy* 34 (15) (2006), 2174; A. Silayan, 'Equitable Distribution of CDM Projects Among Developing Countries', *HWWA-Institut für Wirtschaftsforschung-Hamburg (HWWA) Report*, 255 (2006), 1; K. Olsen, 'The Clean Development Mechanism's Contribution to Sustainable Development: A Review of the Literature', cd4cdm.org/Publication/, accessed 21 January 2014.

[39] In Nigeria, for example, six different emission reduction projects have been subjects of petitions and criticisms over alleged human rights violations. See K. Adeyemo, 'Nigerians Oppose Climate Development Projects', *The Tribune* (Ibadan, 12 September 2010), 3; 'REDD Under Fire in Nigeria', uk.oneworld.net/article/view/165950/1/246, accessed 23 July 2014; 'Don't Sell Forests: Group Urge Nigerian Government', www.scoop.co.nz/stories/WO1008/S00467/dont-sell-forests-groups-urge-nigerian-govts.htm, accessed 23 July 2014; See Carbon Trade Watch, 'Groups Slam Nigeria's Submission of Gas Flare Reductions for Carbon Credits' (2006), (n 23). Bank Information Center, 'Local Groups Say Project Will Not End Gas Flaring, Could Exacerbate Conflicts in the Niger Delta', www.bicusa.org/en/Project.39.aspx, accessed 14 January 2015.

1.1.1 *Human rights risks in carbon projects: calls for change*

Human rights risks in carbon projects manifest in several ways. For residents of communities where projects are located, ineffectively designed carbon projects may result in loss of lives, homes, and means of livelihood. For business enterprises, host governments and project proponents, defective carbon investments or projects may trigger financial, reputational or investment loss, legal liability, or significant dispute settlement costs, all of which may ultimately erode the short and long term value of such projects. Similarly, clients and beneficiaries of investment firms might have a legal case to bring against their investors, who stand idly by as human rights litigation erode the value of their carbon investments.

Concerns about the high incidence of human rights violations and risks from carbon projects have led to increased expectation and calls for a more transparent, accountable and human rights-based approach to the design, approval and execution of carbon projects. In 2012, a coalition of developing countries petitioned the United Nations to seek powers to withdraw approvals they have given to emission reduction projects if evidence emerges that they breach human rights or harm the environment.[40] The petition emphasizes how carbon projects could have serious negative impacts on fundamental human rights.

Considering the increased importance of market-based mechanisms and carbon projects to climate change mitigation efforts, it is imperative to take stock and examine how lessons learnt from the design, approval, finance and implementation of CDM projects could inform thoughts on how to address human rights impacts and risks in climate actions and projects. The sustained execution of emission reduction schemes that affect human health, lead to loss of property or the denial of the right to water, or lead to loss of life must be avoided to prevent trade-offs, legal liability and investment risks. Improperly executed carbon projects might be counterproductive to government efforts aimed at addressing climate change, and for business enterprises involved at all stages of design, approval, finance and implementation.[41] At international governance levels, excluding carbon

[40] Point Carbon, 'CDM Host Nations Ask UN for Power to Ban Projects', www.pointcarbon.com/news/1.1802068, accessed 30 March 2014.

[41] See P. Cullet, 'Rethinking the Legal Regime for Climate Change: The Human Rights and Equity Imperative', in B. Steffen and D. Siddhartha (eds.), *Upsetting the Offset – The Political Economy of Carbon Markets* (London: MayFly Books, 2009), 292–306; see also P. Cullet, 'The Kyoto Protocol and Vulnerability: Human Rights and Equity Dimensions', in S. Humphreys

projects that fund human rights abuses and violate international human rights norms from global carbon markets would be a logical move to prevent such trade-offs and risks.[42]

In order to revitalize and restore faith in project-based mechanisms and global carbon markets generally, it is necessary to examine how human rights principles could be mainstreamed into the international climate change regime to make it more transparent, accountable and rights-based.[43] Rights-based frameworks are needed to ensure that carbon projects respect, protect and fulfil human rights throughout their various stages, including planning, funding, implementation, monitoring and evaluation. Rights-based frameworks and approaches would not only ensure that carbon projects comply with existing international human rights obligations and principles, they also have the potential to strengthen their acceptability, effectiveness and long-term efficiency.[44] As the United Nations Human Rights Commissioner rightly noted:

(ed.), *Human Rights and Climate Change*, 183–206; Olawuyi, 'Towards a Transparent and Accountable Clean Development Mechanism', (n 36), 33–52.

[42] See A. Filmozer, 'CDM Projects Affect Human Rights' (February 2011), www.cdm-watch.org/wordpress/wp-content/, accessed 2 August 2014; also Roht-Arriaza, 'Human Rights in the Climate Change Regime', (n 18), 211 in Humphreys (ed.), *Human Rights and Climate Change*, 23. See also S. Lankford, M. Darrow and L. Rajamani, *Human Rights and Climate Change: A Review of the International Legal Dimensions* (Washington, DC: World Bank, 2011), 56–57, where the authors argued that policy measures designed to address climate change may impact the realization of human rights. They argued that human rights principles, both substantive and procedural, would be relevant to the design and implementation of effective responses to climate change, particularly in relation to adaptation and to some extent also to mitigation.

[43] See Lankford, et al., *Human Rights and Climate Change: A Review, ibid.*; M. Macchi, *Indigenous and Traditional Peoples and Climate Change* (Gland: International Union for Conservation of Nature, 2008). IUCN expressed concern about plans to establish a large-scale biofuel plantation and the threat it constituted to the rights of indigenous peoples to own their lands and enjoy their culture (CERD/C/IDN/CO/3, para. 17); E. Meijer and J. Werksman, 'Keeping It Clean: Safeguarding the Environmental Integrity of the Clean Development Mechanism', in D. Freestone and C. Streck (eds.), *Legal Aspects of Implementing the Kyoto Protocol* (Oxford: Oxford University Press, 2005) 191, 201; Oxfam International, *Climate Wrongs and Human Rights: Putting People at the Heart of Climate-Change Policy*, Oxfam Briefing Paper 117 (Oxford: Oxfam International, 2008); P. Baer, T. Athanasiou, and S. Kartha, *The Right to Development in a Climate Constrained World: The Greenhouse Development Rights Framework* (Berlin: Heinrich Böll Foundation, 2007); E. Page, *Climate Change, Justice and Future Generations* (Cheltenham: Edward Elgar 2006), 132–160.

[44] The United Nations High Commissioner for Human Rights (OHCHR) emphasized this when it noted that 'human rights standards and principles should inform and strengthen policy measures in the area of climate change…a human rights framework underlines the critical importance of effective participation of individuals and communities in

> Without explicit human rights safeguards, policies intended to advance environmental or development goals can have serious negative impacts on those rights. Thus, technocratic processes have excluded women from decision-making, economic and social inequalities have been exacerbated (and, with them, societal tensions), indigenous peoples have seen threats to their lands and livelihoods from some emission reduction schemes, scarce food-growing lands have sometimes been diverted for the production of biofuels, and massive infrastructure projects have resulted in the forced eviction and relocations of entire communities. Simply put, participatory, accountable, non-discriminatory and empowering development is more effective, more just, and, ultimately, more sustainable… Member States should commit to ensuring full coherence between efforts to advance the green economy, on the one hand, and their solemn human rights obligations on the other. They should recognize that all policies and measures adopted to advance sustainable development must be firmly grounded in, and respectful of, all internationally agreed human rights and fundamental freedoms, including the right to development… States should resolve to work to advance a *human rights-based approach* to the green economy, based on the principles of participation, accountability (at the national and international levels), non-discrimination, empowerment, and the rule of law in green economy efforts, and to pursue a model of economic growth that is socially and environmentally sustainable, just and equitable, and respectful of all human rights.[45]

The above views adequately sum up the importance of mainstreaming human rights principles into the approval and implementation of carbon projects. A human rights-based approach to carbon finance will integrate human rights safeguards into international climate finance architecture and could help ensure that countries, and project participants, avoid or minimize the human rights impacts of carbon projects at the design,

decision-making processes affecting their lives, stresses the importance of accountability mechanisms in the implementation of measures and policies in the area of climate change, requiring access to administrative and judicial remedies in cases of human rights violations'. See OHCHR Report on Climate Change and Human Rights, para. 95. See also S. Foot, 'An Evaluation of the Present Clean Development Mechanism', *Environmental Law Management*, 16(3) (2004), 3; A. Najama, 'Climate Negotiations Beyond Kyoto: Developing Countries Concerns and Interests', Climate Policy, 3 (2009), 221–231; also see S. Jodoin, 'Rights-Based Frameworks for Climate Finance' (Center for International Sustainable Development Law, Legal Working Paper Series 2011), 1–2; see also S. Jodoin, 'Human Rights and Climate Change Governance' (Center for International Sustainable Development Law, Legal Working Paper, December 2010).

[45] Emphasis added. See N. Pillay, 'Open Letter by the United Nations Human Rights Commissioner' (30 March 2012), www.ohchr.org/Documents/Issues/Development/OpenLetterHC.pdf, accessed 14 April 2015; see also OHCHR, 'Pillay urges states to inject human rights into Rio+20' (18 April 2012), www.ohchr.org/en/NewsEvents/Pages/DisplayNews.aspx?NewsID=12070&LangID=E, accessed 21 April 2015.

approval, finance and implementation stages. Recognizing and adopting human rights principles of Participation, Accountability, Non-discrimination, Transparency, Human Dignity, Empowerment and Rule of Law (PANTHER Principles) in carbon projects may address human rights concerns relating to land grabs, exclusion and discrimination, inadequate dissemination of project information, lack of accountability, lack of participation and inadequate access to review mechanisms, that have derailed the sustainable development goals and intrinsic value of carbon projects and mechanisms such as the CDM.[46]

While project-based mechanisms remain very important to global efforts to combat climate change, robust legal frameworks and governance structures must be put in place to avoid human rights concerns that have, arguably, eroded the relevance and value of the CDM. This book examines how lessons learnt from the CDM governance structure could inform thoughts on the need to mainstream human rights safeguards into the international legal regime on climate change in general, in order to make it more accountable, right-based and sustainable.

1.1.2 Emergence of rights-based approaches to climate change

Over the last decade, there have been increased discussions at the international level on how human rights obligations can be invoked to inform and strengthen climate change governance, specifically the design, approval, finance and implementation of carbon projects.[47]

For example, the United Nations has recognized the importance of addressing the linkages between human rights and climate change through several official resolutions and reports. On 28 March 2008, the United Nations Human Rights Council (UNHRC) adopted Resolution 7/23, which expressed that 'climate change poses an immediate and far-reaching threat to people and communities around the world and has implications for the full enjoyment of human rights', and directed the Office of the

[46] See Food and Agricultural Organization of the United Nations, 'Human Right Principles: PANTHER' stating that 'a rights-based perspective requires decision-making processes (from policy formulation to law-making down to administrative acts) to comply with the seven important PANTHER principles: participation, accountability, non-discrimination, transparency, human dignity, empowerment and rule of law', www.fao.org/righttofood/about-right-to-food/human-right-principles-panther/en/, accessed 12 February 2015.

[47] For a review of these efforts, see L. Rajamani, 'The Increasing Currency and Relevance of Rights-Based Perspectives in the International Negotiations on Climate Change', *Journal of Environmental Law* 22(3) (2010), 391–429.

United Nations High Commissioner for Human Rights (OHCHR) to conduct a 'detailed analytical study of the relationship between climate change and human rights'.[48] Following its study, the OHCHR released a report in 2009, which concluded that 'there is broad agreement that climate change has generally negative effects on the realization of human rights'.[49]

Since the 2009 report of the OHCHR, the UNHRC has issued additional resolutions recognizing the linkages between climate change and human rights. On 25 March 2009, the UNHRC adopted its second resolution – Resolution 10/4 – which noted that 'climate change-related impacts have a range of implications, both direct and indirect, for the effective enjoyment of human rights'.[50]

In recognition of these indirect effects of climate change measures on human rights, the Conference of Parties (COP 16) to the United Nations Framework Convention on Climate Change (UNFCCC) in Cancun, Mexico, released a decision stating that 'Parties should, in all climate change-related actions, fully respect human rights.'[51] Similarly, in September 2011, the UNHRC adopted Resolution 18/22, its third resolution on 'human rights and climate change'.[52] This resolution affirmed that human rights obligations, standards and principles have the potential to inform and strengthen international and national policymaking in the area of climate change, promoting policy coherence, legitimacy and sustainable outcomes. Resolution 18/22 also called for further study and dialogue on how to address the adverse impacts of climate change on the full enjoyment of human rights.

[48] UN Human Rights Council Res. 7/23, in UN Human Rights Council, Report of the Human Rights Council on Its Seventh Session, at 65, UN Doc A/HRC/7/78 (14 July 2008).

[49] OHCHR, Report of the Office of the United Nations High Commissioner for Human Rights on the Relationship between Climate Change and Human Rights, A/HRC/10/61 (2009), 69–70.

[50] UN Human Rights Council Res. 10/4, in U.N. Human Rights Council, Report of the Human Rights Council on its Tenth Session 13, at 15, UN Doc A/HRC/10/L.11 (31 March 2009). www.google.com/search?hl=en&rlz=1T4RNWN_enUS302US303&ei =lXnES8fgCIH9_AbZsPG8Dw&sa=X&oi=spell&resnum=0&ct=result&cd=1&ved=0CAUQBSgA&q=UNHCR+Res+10/4+(2009)+UN+Doc+E/CN.4/RES/2009&spell=1

[51] International Shared Vision on Long-Term Cooperative Action, Cancun LCA Outcome, para. 8. UNFCCC, unfccc.int/files/meetings/cop_16/application/pdf/cop16_lca.pdf, accessed 15 April 2014.

[52] UN Human Rights Council Res. 18/22, in UN Human Rights Council, Report of the Human Rights Council on Its eighteenth Session, at A/HRC/RES/18/22 (30 September 2011).

Furthermore, during the 2011 Durban Conference (COP18), the UNFCCC Executive Secretary Christiana Figueres launched a CDM Policy Dialogue – a panel of independent experts – to take stock of some of the concerns facing the implementation of the CDM, notably the human rights criticisms against CDM projects.[53] This panel was mandated to spearhead widespread dialogue on the framework for improving the transparency, effectiveness and accountability of the CDM.[54]

The outcome document of the CDM Policy Dialogue identified the several human rights impacts of CDM projects on local communities and noted that the CDM EB must strengthen and restructure its governance in order to become a more accountable and efficient organization.[55] It called on the UNFCCC to establish robust guidelines for adequate local consultation procedures to ensure that local community stakeholders are properly notified and consulted on proposed project activities. The Panel also called for a redesign of governance structures under the UNFCCC by establishing a grievance mechanism for local stakeholders to address environmental and social concerns of projects and to facilitate the resolution of issues emerging after the registration of a project.[56] The Panel was however silent on legal and institutional prerequisites for achieving such robust reform of governance structures under the UNFCCC.

The UNHRC released its fourth resolution in 2014 – Resolution 26/27 (2014), which noted the 'urgent importance of continuing to address, as they relate to States' human rights obligations, the adverse consequences of climate change for all, particularly in developing countries and its

[53] UNFCCC, 'Input to the High-Level Panel for the CDM Policy Dialogue: Background Paper by the Secretariat' (22 December 2011), www.cdmpolicydialogue.org/background/CDM_policy_background.pdf, accessed 10 May 2015.

[54] As noted by the UNFCCC: 'While a condition of registering each CDM project is the receipt of a letter from the host country government confirming that the project assists it to achieve sustainable development, there are frequently voiced concerns that this stage in the process is a formality, and that national governments frequently issue such letters without proper investigation and/or appropriate regard to concerns raised by stakeholders. In extreme cases, it has been alleged that CDM projects, despite having been issued with such letters, have been the scene of environmental damages or human rights abuses', UNFCCC, 'Terms of Reference for the Policy Dialogue on the Clean Development Mechanism' (2012), cdm.unfccc.int/filestorage/S/D/9/SD94R1COMGNYTP3BIW7QAZ5UH026KJ/eb64_repan01.pdf?t=VVN8bTN0ZXg5fDDM3LBP1KZGmNsZihzMWYmh, accessed 10 May 2015.

[55] CDM Policy Dialogue, 'Climate Change, Carbon Markets and the CDM: A Call to Action: Recommendations of the High Level Panel on the CDM Policy Dialogue' (2012), www.cdmpolicydialogue.org/report/ues_en.pdf, accessed 10 January 2015.

[56] *Ibid.*

people whose situation is most vulnerable to climate change, especially those in a situation of extreme poverty, and deteriorating livelihood conditions'.[57] It called upon all states to continue to enhance international dialogue and cooperation in relation to the adverse impacts of climate change on the enjoyment of human rights, including the right to development, particularly in developing countries, especially least developed countries, small island developing states and African countries.

In July 2015, the UNHRC released its fifth resolution on human rights and climate change. This resolution stresses the importance of coherence and legal partnerships, between human rights and climate change communities, in order to holistically address the problem of climate change. Apart from affirming that human rights obligations, standards and principles have the potential to inform and strengthen international, regional and national policymaking in the area of climate change, it expressly noted the importance of facilitating meaningful interaction between the human rights and climate change communities in order to build capacity to deliver responses to climate change.[58]

The foregoing recognitions and linkages emphasize the increasing relevance and importance of human rights in the design, finance and implementation of carbon projects.[59] Before now, most of the focus had been on the direct impacts of climate change on human rights, thereby obscuring discussions on how policy measures and projects aimed at combating climate change are currently producing human rights violations.[60]

Discussions on the linkages between climate change and human rights must now move beyond simply warning policymakers about policy

[57] UNHRC Res. 26/27, *Human Rights and Climate Change*, A/HRC/Res/26/27 (15 July 2014), daccess-dds-ny.un.org/doc/UNDOC/LTD/G14/065/81/PDF/G1406581.pdf?OpenElement, accessed 12 December, 2015.

[58] UNHRC Res. 29/15, *Human Rights and Climate Change*. A/HRC/Res/29/15 (2 July 2015), daccess-dds-ny.un.org/doc/UNDOC/LTD/G15/137/48/PDF/G1513748.pdf?OpenElement, accessed 12 December, 2015.

[59] In addition, many regional bodies have explicitly recognized the linkages between climate change and human rights. See African Commission on Human and Peoples' Rights (ACHPR), *Resolution on Climate Change and Human Rights and the Need to Study its Impact in Africa*, ACHPR/Res.153(XLVI)09 (2009), which requested the Working Group on Extractive Industries, Environment and Human Rights Violations in Africa to undertake an in-depth study on the impact of climate change on human rights in Africa. See also Organization of American States General Assembly Resolution No. 2429 (XXXVIII–O/08) 'Human Rights and Climate Change in the Americas', recognizing the 'adverse effects of climate change might have a negative impact on the enjoyment of human rights'.

[60] Humphreys (ed.), *Human Rights and Climate Change*, (n 24), 16.

measures that might violate human rights.[61] It is important to consider how human rights may strengthen and inform emission reduction efforts and how carbon projects may contribute to the protection and fulfilment of human rights. For example, properly implemented carbon projects would result in the reduction of GHGs that cause climate change, and could provide social, environmental and economic solutions to several local problems and concerns. Unfortunately, policy discussions at the intersection of human rights and climate change rarely consider the possibilities for win-win scenarios of this sort.[62]

In order to move discussions on the interactions between carbon projects and human rights beyond mere rhetoric, questions on the legal and institutional frameworks required to incorporate human rights issues into climate actions and projects must be addressed in greater detail.[63] As the International Union for Conservation of Nature (IUCN) rightly notes, one reason for the limited literature on a rights-based approach in the climate change field 'is the current lack of an operational framework that would guide participants through such an approach'.[64] Furthermore, proponents of the human rights approach to climate change have not clearly articulated the conceptual promises, normative basis and practical implications of this approach.[65] Consequently, human rights issues only get onto the agenda when there is a crisis or a carbon project that affects a wide sector of the general public. While emphasizing the need to move beyond mere preambular recognition of the linkages between human rights and climate change, the United Nations Environment Programme recently noted that:

[61] S. Jodoin, 'Human Rights and the Environment: Key Challenges and Best Practices' (2011), Center for International Sustainable Development Law Brief presented to the United Nations Office of the High Commissioner for Human Rights, 10–11, www.cisdl.org/publications/collaborative005.html, accessed 03 August 2014.

[62] *Ibid.*

[63] These issues were given some thoughts in Lankford, Darrow and Rajamani, *Human Rights and Climate Change: A Review*, (n 42), which considers procedural and substantive human rights as possible frames for addressing these secondary impacts of climate change policy measures and projects. In the two pages they devoted to this subject, the authors did not provide a detailed examination of the theoretical implications of both approaches and the legal framework required under these frames. See also S. Kas, 'Integrated Justice: Human Rights, Climate Change, and Poverty', *Transnational Law & Contemporary Problems*, 18(1) (2009), 115–138; J. Carlson, 'Reflections on a Problem of Climate Justice: Climate Change and the Rights of States in a Minimalist International Legal Order', *Transnational Law & Contemporary Problems*, 18(1) (2009), 45–68.

[64] See IUCN, *Conservation with Justice: A Rights-Based Approach* (International Union for Conservation of Nature (IUCN), (2009), 3.

[65] See Lankford, Darrow and Rajamani, *Human Rights and Climate Change: A Review*, (n 42), 27–28.

> The UNFCCC COP also formally recognized the linkages between climate change and human rights in the 2010 Cancun Agreements, where the Parties agreed to emphasize that countries 'should, in all climate change-related actions, fully respect human rights'. But the COP has done very little in the past five years to elaborate on *how* countries should fulfill that goal.[66]

This book is an attempt to fill this gap. Its ultimate contribution is to answer the question of *how* – that is, how could the international climate change regime be reformed to efficiently address the impacts of carbon projects on human rights. In other words, how could the international climate regime be enhanced to become more rights-based, accountable and transparent such that carbon actions and projects will not foster or exacerbate human rights violations and risks? This book proposes a legal and institutional framework that could serve as a process-based guideline or template for mainstreaming human rights principles into project design, approval and execution under the international climate change regime. It takes a lessons-learnt approach in reviewing the web of legal and institutional loopholes within the CDM project approval framework that arguably resulted in a culture of 'approvals', human rights violations and procedural impunity in the execution of CDM projects.

The book then takes a broader look at the interactions between carbon projects and human rights generally and how ineffectively designed carbon projects may result in human rights violations. It discusses the importance of integrating human rights safeguards into the approval, implementation and funding of carbon projects. It argues for a process-based approach that entrenches procedural human rights such as access to information, participation, accountability, non-discrimination, empowerment and access to justice as cornerstones for the design, approval, finance and implementation of carbon projects.

In order to achieve the above goals, first, the book examines the theoretical basis and value of the human rights mainstream paradigm under international law. It starts by evaluating the moral, philosophical, ethical, legal or economic value that human rights norms would bring to, or take away from, the international legal regime on carbon finance. The book then explores some of the contested and unresolved theoretical issues on the prospects and

[66] Emphasis added. United Nations Environment Program (UNEP), *Climate Change and Human Rights* (UNEP, 2015), 14.

paradoxes of mainstreaming human rights into the international climate change regime. It then considers the potentials and pitfalls of the HRBA as a policy framework through which human rights standards could be mainstreamed into international climate change instruments and governance structures. In discussing the HRBA, the book examines and analyses the legal status of the key normative elements of the HRBA with a view of situating them under international human rights and international environmental law.

1.2 Scope, definitions and structure

1.2.1 Scope

This book contributes to emerging literature on the human rights responsibilities of business actors in the implementation of carbon projects. It discusses how the international climate change regime may better integrate and mainstream human rights safeguards into project-based instruments on climate change, to address human rights risks.[67] It proposes and develops a legal and institutional framework for mainstreaming human rights protection into the international climate finance regime – both the approval of carbon projects and distribution of funds in support of such projects. As such, the main concern in this book is not to contribute to debates on the direct impacts of climate change on human rights but to examine emerging debates on the indirect effects of carbon projects on the effective enjoyment of fundamental human rights.

The main topics covered lay at the heart of contemporary debates on the theoretical justifications, basis and value of the human rights mainstream paradigm. The human rights mainstream paradigm encompasses proposals that human rights norms should be integrated or 'mainstreamed' into international environmental regimes to address the intersections between human rights and environmental issues such as climate change. The key questions are: what are the justifications – moral, philosophical, ethical, legal or economic – prospects, practical implications, paradoxes and

[67] One of the more recent and robust contributions to these debates is the International Bar Association Climate Change Justice and Human Rights Task Force Report, *Achieving Justice and Human Rights in an Era of Climate Disruption* (International Bar Association, 2014), 49, which notes that although the UNFCCC has recognized that 'parties to the [UNFCCC] should, in all climate change-related actions, fully respect human rights', there are as yet no practical mechanisms to ensure accountability.

frameworks for mainstreaming human rights obligations into the international climate change regime?

Even though the linkages between human rights and carbon finance has gained increased attention and recognition, legal and institutional frameworks and guidelines on how policymakers could achieve this aim have not been espoused in sufficient detail. This book discusses legal frameworks and governance structures that could bring together and address the critical linkages between human rights and climate change actions, to ensure that carbon projects do not result in human rights violations.[68]

Establishing strong international governance structures to monitor human rights risks in carbon projects is very important to address these risks in a transparent, democratic and accountable way. As examples in this book show, many national governments in developing countries have either turned away from preventing human rights violations emanating from carbon projects or have, in extreme cases, aided such violations. This calls for concerted international action and leadership if this concern must be addressed. A rights-based reform at the international level could spur action at national and regional levels to ensure that climate actions and projects do not threaten the enjoyment and protection of fundamental human rights.

In order for 'parties to fully respect human rights in all climate change-related actions', a starting point would be to recognize the linkages between human rights and climate change in the wider context in the international climate change regime. While specifically discussing the idea of mainstreaming human rights principles into the approval and execution of carbon projects, the book discusses legal and institutional reforms that are pertinent to reflect human rights principles in the broader context of the international climate change regime. It provides insights on how international human rights thresholds could be adopted

[68] As the UN Secretary General noted: 'There still remains a need for a body that brings together the key developed and developing countries to address the critical inter linkages between human rights and the environment, trade, finance, the handling of pandemic diseases and economic and social development. To be effective, such a body must operate at the level of national leaders. . .'. Report of the UN Secretary General's High-Level Panel on Threats, Challenges and Climate Change (2004), www.un.org/secureworld/report.pdf, accessed 12 May 2014; see also H. Shue, 'Human Rights, Climate Change, and the Trillionth Ton', in D. Arnold (ed.), *The Ethics of Global Climate Change* (Cambridge: Cambridge University Press, 2011), 292, where it was noted that any institutions to protect the rights threatened by climate change must be international.

to stem the tide of gross human rights violations in the design, approval, finance and implementation of carbon projects.

1.2.2 Definition of key concepts

Some key terms and concepts run across this book. Readers will find terms such as 'human rights', 'environment', 'climate change regime', 'sustainable development', 'soft law', 'mitigation' and 'carbon projects' in almost every page of the book. Many of these concepts have acquired varied, and most times highly contested, meanings. To avoid any confusion, it is pertinent to clarify their contextual utilization and connotations within this book as early as possible. The definitions provided in this section are far from exhaustive, they are however intended to provide appropriate signposts and background for subsequent and more detailed discussions in the book.

1.2.2.1 Human rights

In the literature on human rights mainstreaming, 'human rights' itself is a contested concept.[69] The different interpretations of human rights influence human rights mainstreaming efforts. On the one hand, for example, many scholars fault attempts to mainstream a human right to food or right to water into decision-making processes, pointing to their contested legal statuses under international law.[70] On the other hand, theorists point to the interdependent and indivisible nature of all human rights and argue that some human rights are prerequisites for enjoying other rights and as such all rights must be mainstreamed into policymaking. These are debates on the sources and hierarchy of rights. It is pertinent to

69 A. Langlois, 'The Elusive Ontology of Human Rights', *Global Society*, 18(3) (2004), 245; See also D. Kennedy, 'The International Human Rights Movement: Part of the Problem', *European Human Rights Law Review*, 3 (2001), 245–267, where Kennedy argues that one of the problems of the human rights movement is that the concept of human rights 'generalizes too much'. See also J. Nickel, *Making Sense of Human Rights: Philosophical Reflections on the Universal Declaration of Human Rights* (Berkeley, CA: University of California Press, 1987), 561–562; See J. Shestack, 'The Philosophical Foundations of Human Rights', in J. Symonides (ed.), *Human Rights Concepts and Standards* (England and Burlington, VT: Ashgate, 2000), 33–35.

70 See N. Aryeh, 'Social and Economic Rights: A Critique', *Human Rights Brief*, 13(2) (2006), 1; see also P. Alston, 'Resisting the Merger and Acquisition of Human Rights by Trade Law: A Reply to Petersmann', *European Journal of International Law*, 13(8) (2002), 815.

clarify the meaning we attach to the idea of human rights in general in order to put human rights mainstreaming into proper context.

Human rights are those inalienable legal entitlements recognized in core international human rights instruments, conventions and legal instruments under international law.[71] This conception of rights denies *a priori* sources of rights and posits that all human rights stem from what legal institutions prescribe. It contrasts sharply with natural law theories of human rights.[72] As Kamenka rightly notes, rights are claims that have achieved a special kind of endorsement or legal success, by a legal system, by widespread sentiment or an international order.[73] This book will not delve into debates on the theories of rights or on what human rights ought to be. Instead, it will work within a human rights framework which assumes that human rights are entitlements that have achieved legal recognition under existing international treaties, conventions, courts, and legal instruments under international law. As such, analysis on human rights mainstreaming in this book focus on human rights that

[71] There are ten core international human rights instruments: International Covenant on Civil and Political Rights, 19 December 1966, 999 UNTS 171 (hereinafter ICCPR); the International Covenant on Economic, Social and Cultural Rights, 16 December 1966, 993 UNTS 3(hereinafter ICESCR); the International Convention on the Elimination of All Forms of Racial Discrimination, 7 March 1966, 660 UNTS 195 (hereinafter CERD); Convention on the Elimination of All Forms of Discrimination against Women, 18 December 1979, 1249 UNTS 13 (hereinafter CEDAW); Convention against Torture and Other Cruel, Inhuman or Degrading Treatment or Punishment, 10 December 1984, 1465 UNTS 85; Convention on the Rights of the Child, 20 November 1989, 1577 UNTS 3; International Convention on the Protection of the Rights of All Migrant Workers and Members of Their Families, 18 December 1990, 2220 UNTS 3; the Convention on the Rights of Persons with Disabilities, 13 December 2006, UN Doc A/61/611. See OHCHR, 'Core International Human Rights Instruments', www.ohchr.org/EN/ProfessionalInterest/Pages/CoreInstruments.aspx, accessed 12 March 2014.

[72] See J. Finnis, *Natural Law and Natural Rights* (Oxford: Oxford University Press, 1980), 280. The doctrine of natural law was popularized and developed by a number of classical Greek philosophers, including Aristotle. According to Hettinger, 'the best description of natural law is that it provides a name for the point of intersection between law and morals. The principal claim of natural law is that what naturally is, ought to be'. See R. Hittinger, 'Natural Law and Virtue: Theories at Cross Purposes', in R. George (ed.), Natural *Law Theory Contemporary Essays* (Oxford: Oxford University Press, 1994), 63; See also R. Wacks, *Understanding Jurisprudence: An Introduction to Legal Theory* (Oxford: Oxford University Press, 2005), 15; R. George, 'Natural Law', *Harvard Journal of Law and Public Policy*, 31(1) (2008), 171–196; N. Ferreira, 'The Expanding Realm of Human Rights', *Res Publica*, 14 (2008), 57; M. Cranston, *What are Human Rights?* (Taplinger Publishing Co., 1973), 36; Shestack 'The Philosophical Foundations of Human Rights', (n 69), 33–35.

[73] E. Kamenka, 'Human Rights, Peoples' Rights', in J. Crawford (ed.), *The Rights of Peoples* (Oxford: Oxford University Press, 1988), 127.

have been endorsed and enshrined in existing legal instruments under international law.

Notions of 'legal instruments under international law' discussed in this book are based on the broad understanding of the sources of international law. Article 38(1) of the International Court of Justice (ICJ) statute is generally recognized as an authoritative statement of the sources of international law.[74] It requires the ICJ to apply, among other things, international conventions, whether general or particular, establishing rules expressly recognized by the contesting states; international custom, as evidence of a general practice accepted as law; the general principles of law recognized by civilized nations; and subject to the provisions of Article 59, judicial decisions and the teachings of the most highly qualified publicists of the various nations, as subsidiary means for the determination of rules of law. This includes international treaties, conventions, customs, general principles and rules that have achieved legal recognition under existing legal regimes on international human rights. Discussions of the human rights-based approach in this book shall specifically rely on human rights identified in core international human rights treaties and the Universal Declaration of Human Rights (UDHR).[75]

This further raises questions on the relative importance of different types of rights under international law. For example, speaking of mainstreaming a human right to food or right to participation may sound irrational to scholars who question the legal status of such rights in the first place, despite their broad recognition in the International Covenant on Economic, Social and Cultural Rights (ICESCR) and other core international human rights instruments.[76] These are questions that go

74 Statute of the International Court of Justice, 3 Bevans 1179; 59 Stat. 1031; T.S. 993; 39 AJIL Supp. 215 (1945).

75 Though not binding, the UDHR is arguably an important foundation for many binding international human rights instruments and is considered as providing evidence of customary international human rights law. See J. Morsink, *The Universal Declaration of Human Rights: Origins, Drafting, and Intent* (Philadelphia, PA: University of Pennsylvania Press, 1999), 2.

76 Article 11 of the International Covenant on Economic, Social and Cultural Rights recognizes the right of everyone to an adequate standard of living. This includes, but is not limited to, the right to adequate food, clothing, housing, and the continuous improvement of living conditions. It also creates an obligation on parties to work together to eliminate world hunger. It is also enshrined in the 1948 Universal Declaration of Human Rights (Article 25). Similarly, the 2009 Optional Protocol to the International Covenant on Economic, Social and Cultural Rights establishes a right to food justiciable at the international level. The right to participation is recognized in Art. 25 of the ICCPR, (n 71).

back to debates on the hierarchy of rights.[77] This book shares the view that all human rights – political, civil, social, cultural and economic – are equal in importance and none can be fully enjoyed without the other. All human rights are important, indivisible and interdependent whether they relate to civil, political or economic and social issues and cannot be positioned in a hierarchical order. Denial of one right invariably impedes enjoyment of other rights. This book therefore argues for an integrated understanding and implementation of all human rights under international law.

In this book, human rights mainstreaming connotes integrating those human rights that have been recognized in existing core international human rights instruments into the international legal regime on climate change. The human rights mainstream approach, as espoused in this book, focuses on integrating endorsed human rights under international law into climate actions and projects.

1.2.2.2 Environment

The term 'environment' has acquired different meanings over time. While some scholars define it from an eco-centric and nature standpoint, others assign an anthropocentric/human-related connotation to the term. For example, the environment has been defined as the sum of all external conditions affecting the life, development and survival of an organism.[78] It has also been defined as all the elements of the physical and biological world (including humans) and the interactions between them.[79] From an anthropocentric or human point of view, the

[77] The term 'hierarchy of rights' flows from the idea that certain categories of human rights are more important than others. While some scholars stress the importance of civil and political rights others mostly scholars from the South tend to attach primary importance to economic, social and cultural rights. See H. Quane, 'Further Dimension to the Interdependence and Indivisibility of Human Rights? Recent Developments Concerning the Rights of Indigenous People', *Harvard Human Rights Journal*, 25(1) (2012); J. Donnelly, *Universal Human Rights in Theory and Practice*, 2nd edn. (Ithaca, NY: Cornell University Press, 2003), 27–33; G. Teeple, *The Riddle of Human Rights* (Toronto: University of Toronto Press, 2004), 24; J. Nickel, 'Rethinking Indivisibility: Towards a Theory of Supporting Relations between Human Rights', *Human Rights Quarterly*, 30(4) (2008), 984, 985.

[78] United States Environmental Protection Agency, Terms of Environment: Glossary, Abbreviations, and Acronyms, EPA,ofmpub.epa.gov/sor_internet/registry/termreg/searchandretrieve/glossariesandkeywordlists/search.do;jsessionid=2cwMQCcdsnSchlrQXyXqGfhcYt2yWm4hCsYltCL22QFzlpFNd6p1!709404430, accessed 12 January 2015.

[79] See A. Ballance, 'What is the State of the Environment?' *Poverty Times* 1, UNEP & GRID Arendal, GRIDA, www.grida.no/publications/et/ep1/page/2509.aspx, accessed 12 January 2014.

environment has been defined as the relationship of human beings with water, air, land and all biological forms – the combination of elements whose inter-relationships make up the settings, surroundings and conditions of life of the individuals and the society.[80] The Stockholm Declaration of the United Nations Conference on the Environment (the Stockholm Declaration) recognizes both aspects of the environment, the natural and the man-made, as essential to well-being and to the enjoyment of basic human rights, the right to life itself.[81] It describes attributes that make up the environment to include the natural resources of the earth, such as air, water, land, flora and fauna and wildlife.[82] This definition in the Stockholm Declaration is central to the recognition of the linkages between elements of the environment and the enjoyment of basic human rights and that environmental protection is a *sine qua non* for the enjoyment of other human rights. This book adopts this cross-cutting understanding of the environment to mean the interactions between natural ecosystems and humans. Most of the discussions here centre on how failures to protect the environment from climate change and its direct and indirect impacts could affect human life and the enjoyment of fundamental human rights. The book examines how efforts to combat climate change, if not properly designed and implemented, may threaten the environment and the fundamental human rights of its inhabitants.

1.2.2.3 Climate change and carbon finance mechanisms

Climate change is an increase in the average temperature of the atmosphere caused by the anthropogenic emission of gases (known as 'greenhouse gases') that trap the sun's heat in the atmosphere. Despite overwhelming scientific predictions on climate change, its certainty and the accuracy of scientific predictions themselves are not without debates.[83]

80 See D. Bodansky, *The Art and Craft of International Environmental Law* (Cambridge, MA: Harvard University Press, 2010), 9–11.

81 The *Stockholm Declaration on the Human Environment*, UN Doc A/CONF48/14/Rev.1, (1972), para. 1.

82 *Ibid.*

83 This book proceeds on the notion that there is now sufficient scientific consensus on the fact that climate change is real and that anthropogenic emissions are a cause. See Intergovernmental Panel on Climate Change (IPCC), *Climate Change 2007*, the Fourth Assessment Report (AR4), in which the IPCC stated that the likelihood was 90–99 per cent that emissions of heat-trapping greenhouse gases such as carbon dioxide were the dominant cause of the observed warming of the last 50 years. According to the Panel, it is 'very likely' that human activity is the main driver of the rise in temperatures since 1950.

GHGs that contribute to climate change have been identified to include carbon dioxide (CO_2), methane (CH_4), nitrous oxides (N_2O), water vapour (H_2O), ozone (O_3) and high global warming potential (GWP) gases such as chlorofluorocarbons (CFCs), hydrofluorocarbons (HFCs), perfluorocarbons (PFCs) and sulphur hexafluoride (SF6).[84] Climate change mitigation therefore refers to efforts designed to reduce or prevent the emission of GHGs. Although CO_2 is not the only GHG, projects designed to reduce the volume of GHGs emitted into the atmosphere are often referred to as 'carbon projects' due to scientific recognitions of CO_2 as the principal GHG.

Carbon projects are not mere foreign development investments (FDI) or infrastructural projects; they are projects undertaken and financed specifically for the purpose of emission reduction as part of efforts to mitigate climate change.[85] They are designed and undertaken to meet specific, and most times, binding emission reduction targets. Carbon finance therefore refers to funding, investments and instruments related to the development of carbon projects and the acquisition of carbon credits. Carbon finance mechanisms treat GHG reductions as a commodities to be bought, sold or traded in accordance with specified valuations, measurable as CO_2 equivalents. In essence, a carbon credit

See also W. Stevens, 'On the Climate Change Beat, Doubt Gives Way to Certainty', *New York Times* (2007); E. Rosenthal and A. Revkin, 'Science Panel Calls Global Warming "Unequivocal"', *New York Times* (02 March 2007). See also C. Tickell, *Climatic Change and World Affairs* (Boston, MA: Harvard University Press, 1986).

84 See Intergovernmental Panel on Climate Change (IPCC), *Climate Change 2007*, the Fourth Assessment Report (AR4), *ibid.*

85 For example, one of the eligibility requirements for CDM projects is that the project must be 'additional'. Additionality refers to an effort that is supplemental to the business-as-usual (BAU) scenario in at least two areas: (i) the additionality of financial contributions of developed countries to mitigate climate change in developing countries; and (ii) the additionality of greenhouse gas (GHG) emissions generated by mitigation activities. According to Paragraph 43 of the CDM Modalities and Rules, a CDM Project activity is additional if anthropogenic emissions of greenhouse gases by sources are reduced below those that would have occurred in the absence of the registered CDM Project activity. FDI projects, infrastructural projects or other projects that do not result in real and measureable emission reductions are not eligible under the CDM. However, the problem of defining and measuring additionality remains one of the main challenges of the CDM. See J. McFarland, 'Carbon Reduction Projects and the Concept of Additionality', *Sustainable Development Law & Policy*, 11(2) (2011), 15–18; C. Streck, 'The Concept of Additionality under the UNFCCC and the Kyoto Protocol: Implications for Environmental Integrity and Equity' (University College London, 2010); Umamaheswaran and Michaelowa, 'Additionality and Sustainable Development Issues', (n 18).

is a financial instrument that represents the removal of one ton of carbon dioxide equivalent (tCO_2e) from the atmosphere from a carbon project.[86]

Generally, carbon trade is a transaction whereby a buyer purchases, and a seller sells, carbon credits, while carbon markets are virtual financial marketplaces where sale and exchange of carbon credits occur. Participants in the carbon market, mostly governments and business enterprises, divide carbon credits into commodity units, which are then tracked, priced and traded, depending on the participants' relative capacity and needs vis-à-vis their targets.[87] Participants who have excess carbon credits receive payments, while those who need carbon credits are able to buy carbon credits to offset their emissions. An international transaction log, a software-based accounting system, ensures secure transfer of emission reduction units (ERU) between countries. To find a common unit for this commodity, all GHGs are converted to CO_2 equivalents (CO_2-eq).[88] CO_2-eqs are traded on carbon markets.

The last two decades have seen the emergence of hundreds of carbon finance instruments, mechanisms and projects, across the world, worth billions of dollars. Carbon finance instruments are not only within the purview of the UNFCCC, they are also found in regional, multilateral and bilateral agreements, as well as in voluntary financial markets. Participants in the carbon market include private sector corporations with binding or voluntary emission reduction commitments, carbon project developers, investment firms, brokerages, audit firms, law firms, accounting firms, technology developers and public sector organizations such as multilateral development banks, government agencies, UN agencies and non-governmental organizations (NGOs) that finance carbon projects.

86 See Carbon Planet, 'Introduction to Carbon Credits', www.carbonplanet.com/introduction_to_carbon_credits, accessed 10 December 2015.

87 Emission reduction targets may be voluntary or compliance-based. The compliance carbon market is used by companies and governments that have mandatory emission reduction targets under national, regional or international carbon reduction regimes. Voluntary markets on the other hand are utilized by companies and investors who buy and sell carbon credits on a voluntary basis, often to satisfy personal or corporate social responsibility goals. The main difference between the two is that the voluntary market is unregulated. See Gold Standard, 'Carbon Markets FAQ', www.goldstandard.org/resources/faqs, accessed 10 December 2015.

88 See C. Seeberg-Elverfeldt, *Carbon Finance Possibilities for Agriculture, Forestry and Other Land Use Projects in a Smallholder Context* (Food and Agricultural Organization, 2010) 4, www.fao.org/fileadmin/templates/ex_act/pdf/carbone_finance.pdf, accessed 25 May 2015.

International supervisory institutions such as the UNFCCC and the CDM EB oversee the design, approval, finance, implementation and enforcement of carbon finance projects through a wide-range of activities, including setting project approval frameworks, overseeing project execution, compiling market studies and data, advising on financing mechanisms, and developing technical standards and best operational practices.

One of the key concerns facing the implementation of carbon projects is that the international climate change regime is yet to evolve robust legal safeguards and frameworks which will ensure that carbon projects that violate human rights are not funded by international financial institutions and governments.[89] This book evaluates the importance of integrating rights-based legal safeguards into the international legal regime applicable to carbon finance. It reviews lessons learnt from the implementation of carbon finance instruments such as Kyoto flexibility mechanisms (most especially the CDM), REDD+, carbon funding mechanisms and debt-for-nature swaps to propose legal and institutional frameworks for addressing human rights impacts of carbon projects.

Kyoto flexibility mechanisms As noted earlier, flexibility mechanisms are mechanisms defined under the Kyoto Protocol to achieve emission reductions as part of global efforts to address climate change.[90] The Kyoto Protocol established three trade mechanisms that allow countries to achieve their emission reduction target with much flexibility and suppleness; they are emission trading (ET), joint implementation (JI) and the CDM. These three mechanisms enable countries to pay for emission reductions anywhere on the planet based on the idea that climate change is a global problem and that its reductions are equally good for the climate no matter where they occur.[91]

[89] See B. Richardson, 'Climate Finance and its Governance: Moving to a Low Carbon Economy through Socially Responsible Financing?' *International and Comparative Law Quarterly*, 58(3) (2009), 597–626.

[90] Report of the Conference of the Parties on Its Third Session, held at Kyoto from 1 to 11 December 1997, FCCC/CP/1997/7/ADD.1 (25 March 1998).

[91] Wiener, 'Global Environmental Regulation', (n 12), 677 at 691–92.

While ET, as defined in Article 17 of the Kyoto Protocol, allows countries with commitments under the Kyoto Protocol to buy emission units from other countries with commitments and use them towards meeting a part of their targets, JI and CDM are project-based mechanisms. They allow Annex I countries to meet their GHG emission reduction commitments by investing in emission reduction projects abroad and acquiring emission reduction credits resulting from such projects.[92] JI, as described in Article 6 of the Kyoto Protocol, allows specified countries with emissions targets (Annex B Parties) to earn ERUs towards their targets by participating in emission reduction or removal projects in similarly situated countries (other Annex B Parties), thereby providing a cost-efficient means to fulfil Kyoto Protocol commitments while also facilitating investment and technology transfer.[93] JI is overseen by the JI Supervisory Committee.[94]

The CDM, contained in Article 12 of the Kyoto Protocol, allows countries with emissions targets (Annex B Parties) to earn credit towards their targets by participating in emission reduction or emission removal projects in developing countries. It has two main goals: one is to assist countries without emissions targets (i.e. developing countries) in achieving sustainable development and the other is to help those countries with emission reduction targets under Kyoto

92 The UNFCCC divides countries into three main groups according to differing commitments: *Annex I* Parties include the industrialized countries that were members of the OECD (Organization for Economic Co-operation and Development) in 1992, plus countries with economies in transition (the EIT Parties), including the Russian Federation, the Baltic States, and several Central and Eastern European States. *Annex II* Parties consist of the OECD members of Annex I, but not the EIT parties. They are required to provide financial resources to enable developing countries to undertake emissions reduction activities under the Convention and to help them adapt to adverse effects of climate change. Funding provided by Annex II Parties is channeled mostly through the Convention's financial mechanism. *Non-Annex I* Parties are mostly developing countries. The Convention emphasizes activities that promise to answer the special needs and concerns of these vulnerable countries, such as investment, insurance and technology transfer. The 49 parties classified as *least developed countries* (LDCs) by the United Nations are given special consideration under the Convention on account of their limited capacity to respond to climate change and adapt to its adverse effects. Parties are urged to take full account of the special situation of LDCs when considering funding and technology-transfer activities. The above categorizations are available at UNFCCC, unfccc.int/parties_and_observers/items/2704.php, accessed 23 June 2015.

93 See I. Shishlov, V. Bellassen, and B. Leguet, 'Joint Implementation: A Frontier Mechanism within the Borders of an Emissions Cap', *Climate Report*, 33 (2012); also Art. 6 of the Kyoto Protocol.

94 See the rules of procedure of the Joint Implementation Supervisory Committee, JI-JISC01-A01-PROC.

(i.e. developed countries) in achieving compliance by allowing them to purchase offsets created by CDM projects. The CDM provides an incentive for government and companies in industrialized countries to invest in GHG reduction projects in developing countries and to be credited for GHG reduction achieved. It can act as a basis for developing countries to achieve progress in environmental issues (such as cleaner air and water, reduced deforestation, soil conservation and biodiversity protection) and social development (such as rural development, employment and poverty alleviation). The CDM is therefore an emission 'offset' instrument that allows countries with emission reduction commitments to invest in offset projects in countries without such commitments, that is, most developing countries.

To qualify for accreditation, projects must pass through a rigorous public registration and issuance process designed to ensure that emissions reduced by the project is additional to what would have otherwise occurred without the project. Additionality is calculated using an approved methodology to subtract the estimated emissions of a given project from a hypothetical 'business-as-usual' emissions baseline. GHG emissions after the implementation of the project have to be lower than in the business-as-usual case.

Once registered, projects are then issued Certified Emissions Reductions (CERs), with each CER unit equal to a reduction of one tonne of CO_2 equivalent. These CERs, or offsets, can be bought and used by developed countries to meet their Kyoto commitments. Companies can also purchase CERs to contribute towards their own emission reduction targets under mandatory emissions trading schemes (such as the EU Emissions Trading Scheme, ETS) or voluntary schemes. The CDM is overseen by the CDM EB.

The CDM is arguably the most popular of the three flexibility mechanisms as it is the only mechanism that allows project investments in developing countries. There are currently over 3,000 registered projects delivering an average of 500 million CERs per year.[95] The most significant demands for CERs have so far from the EU

[95] UNFCCC, CDM Statistics, cdm.unfccc.int/Statistics/index.html, accessed 12 May 2015.

Emissions Trading Scheme, the world's largest carbon market.[96] Between 2008 and 2015, European companies acquired over 300 million CERs to meet their emissions reductions targets.[97]

This demand trend is, arguably, because while it requires US$50 to mitigate one tonne of CO_2 eq. in developed countries, the same reduction could be accomplished in developing countries at US$15 per tonne of CO_2 eq.[98] This explains why an investing country may be willing to carry through a project even in the face of gross human rights violations. There have been allegations of collusion by developed countries with despotic regimes in developing countries to violate human rights during the CDM approval process; there have also been allegations of corruption against certain developed countries that have gone as far as bribing accredited CDM project verifiers (called Designated Operational Entities (DOEs)) to verify projects that do not meet project eligibility requirements, with the CDM EB having to suspend four DOEs.[99] What has been described as the 'emission reduction at all cost' motivations

[96] A. Delbosc, N. Stephan, V. Bellassen, A. Cormier, and B. Leguet, 'Assessment of supply-demand balance for Kyoto offsets (CERs and ERUs) up to 2020', CDC Climate Research Working Paper No. 10 (Paris, 2011); also N. Linacre, A. Kossoy, and P. Ambrosi, *State and Trends of the Carbon Market 2011* (Washington, DC: The World Bank, 2011).

[97] See UNFCCC, Statistics, cdm.unfccc.int/Statistics/Public/CDMinsights/index.html, accessed 02 June 2015.

[98] See Ackerman, 'Financing the Climate Mitigation and Adaptation Measures in Developing Countries', (n 9). See also United Nations Framework Convention on Climate Change, 'Investment and Financial Flows to address Climate Change: An Update', (n 9).

[99] In 2009, SGS United Kingdom, arguably the world's largest DOE was suspended from the CDM scheme amidst allegations it had failed to properly vet projects that it had approved for CDM registration and that it received bribes to verify a CDM project. J. Murray 'Carbon Market under Fire as UN suspends British CDM Project Accreditor' (2009), www.businessgreen.com/bg/news/1806434/updatedFcarbonFmarketFunFsuspendsFbritishFcdmFprojectFaccreditor. In 2008, DNV a Norwegian DOE was suspended after it was found to have authorized five projects for registration, despite not having surveyed them. In 2010, the CDM EB also suspended German DOE-TUEV SUED and Korea's Energy Management Company (KEMCO) for various allegations of procedural irregularities. As the WWF notes, 'verifiers are paid by the project operators, so there is a vested interest to approve as many projects as possible. We would like to see the scheme changed so that the CDM Executive Board randomly appoints the verifiers to the projects'. See Reuters, 'UN Panel Suspends Two More Carbon Emissions Auditors' (26 March 2010); see also World Wildlife Fund, 'A Rating of Designated Operational Entities (DOEs) Accredited under the Clean Development Mechanism (CDM): Scope, Methodology and Results' (A Report by the Oko-Institute for the WWF 2009). See also Greenpeace International, 'Carbon Scam, Noel Kempff Climate Action Project and the Push for Sub-national Forest Offsets' (2009), www.greenpeace.org/france/PageFiles/266591/carbonscam.pdf, accessed 12 January 2015.

behind these projects call for legal frameworks beyond the broad spectrum of FDI or developmental projects.[100]

REDD and REDD+ Although the focus of this book is to review how lessons learnt from the implementation of project-based mechanisms under the Kyoto Protocol may help strengthen the international climate change regime to address human rights questions, the book also provides other examples of carbon projects that have attracted significant human rights questions and concerns. One such carbon project mechanism that has already presented difficulties in balancing climate protection and human rights concerns is the REDD+.

Deforestation and forest degradation are the second leading causes of climate change.[101] Deforestation – the clearing or removal of forests in order to make land available for other use – is a major source of GHG emissions, estimated to contribute about 20 per cent of global emissions.[102] According to the United Nations' Food and Agriculture Organization (FAO), an estimated 18 million acres (7.3 million hectares) of forest are lost each year due to deforestation.[103] Deforestation not only lessens the amount of carbon absorbed or stored, it also releases CO_2 into the air. This is because forests generally absorb CO_2 from the atmosphere. Once absorbed, the carbon will remain sequestered in the trees as

[100] See F. Allen *et al.*, 'Niger Delta Oil Flares, Broken Laws, Pollution and Oppression', in *Why the Carbon Trading Gamble and the Clean Development Mechanism Won't Save the Planet from Climate Change and How African Civil Society Is Resisting* (A Report by the University of KwaZulu Ntual Center for Civil Society and Dartmouth College Climate Justice Research Project, April 2012), Ch 4.

[101] G.R. van der Werf *et al.*, 'CO_2 Emissions from forest loss', *Nature Geoscience*, 2(11) (2009), 737–738.

[102] See A. Angelsen *et al.*, *Reducing Emissions from Deforestation and Forest Degradation: An Options Assessment Report* (Meridien Institute, 2009), 3–4, www.redd-oar.org/links/REDD-OAR_en.pdf, accessed 12 December 2015. In Brazil, for example, deforestation in the Amazon is responsible for about 75 per cent of the country's greenhouse gas emissions each year. For an excellent and detailed analysis of the Brazilian situation, see K. Taylor, 'Improving Substantive and Procedural Protections for Indigenous Rights in REDD+ Projects: Possible Lessons from Brazil', *Afe Babalola University Journal of Sustainable Development Law and Policy*, 5(1) (2015), 32–54. See also H.M. Osofsky and L.K. McAllister, *Climate Change Law and Policy* (Philadelphia, PA: Aspen Publishers Inc., 2012), 243. Other Countries with significant deforestation include Indonesia, Thailand, the Democratic Republic of Congo, and Nigeria. See FAO, The 2015 Global Forest Resources Assessment, www.fao.org/forestry/fra/fra2015/en/, accessed 12 May 2015.

[103] See FAO, The 2015 Global Forest Resources Assessment, *ibid.*

long as they are not cut down or destroyed.[104] If a forest is destroyed, the carbon that was once sequestered in the trees is then released into the atmosphere.[105] According to the 2015 Global Forest Resources Assessment, deforestation releases nearly a billion tonnes of carbon into the atmosphere per year.[106]

The United Nations therefore recommends actions aimed at reducing emissions from deforestation and forest degradation, sustainable management of forests and conservation, and enhancement of forest carbon stocks (i.e. REDD+). REDD+ is a financial mechanism that compensates countries for the prevention of deforestation and forest degradation that would otherwise occur. The REDD+ concept is predicated on the need to reduce emissions from the forest sectors, occasioned by deforestation and forest degradation.[107] It allows industrialized countries to gain credits by engaging in forest conservation efforts and efforts that protect trees from being cut down.[108] REDD+ projects link the mitigation potential of forests to financial incentives for forest conservation, reforestation and afforestation.

REDD+ provides mechanisms for investing countries to pay countries with sizeable amounts of forest money to adopt efforts that will conserve their trees and keep their forests from indiscriminate logging.[109] In return, the investing country obtains carbon credits as a way to achieve their national emission goals. REDD frameworks therefore support forest conservation and provide a comparably low-cost mechanism for reducing carbon emissions. REDD+ projects are developed, financed and administered in conjunction with various organizations, including the World Bank

[104] M. Bapna, 'What is REDD? Climate Change and the Challenge of REDD', *World Resources Institute* (3 September 2010), www.wri.org/stories/2010/03/forests-climate-change-and-challenge-redd, accessed 15 September 2014.

[105] See S. Baez, 'The Right REDD Framework: National Laws That Best Protect Indigenous Rights in a Global REDD Regime', *Fordham Law Review*, 80 (2011), 822.

[106] See FAO, The 2015 Global Forest Resources Assessment.

[107] *Ibid.* See also B. Strassburg, A. Kelly, A. Balmford, R. Davies, *et al.* 'Global Congruence of Carbon Storage and Biodiversity in Terrestrial Ecosystems', *Conservation Letters*, 3 (2009).

[108] See 'An Introduction to REDD' (*The REDD Desk*, 2014), theredddesk.org/resources/an-introduction-redd, accessed 22 October 2014.

[109] Baez, 'The Right REDD Framework', 827; also C. Okereke and K Dooley, 'Principles of Justice in Proposals and Policy Approaches to Avoided Deforestation: Towards a Post-Kyoto Climate Agreement', *Global Environmental Change*, 20 (2010), 82, 95.

(Forest Carbon Partnership Facility) and the United Nations (UN-REDD).

The Bali Action Plan (BAP) identified REDD+ as a prominent potential emission mitigation strategy.[110] Estimates suggest that REDD+ projects have a potential of producing emission reductions of up to 7.8 billion tonnes per year.[111] The BAP calls for 'policy approaches and positive incentives on issues relating to reducing emissions from deforestation and forest degradation in developing countries; and the role of conservation, sustainable management of forests and enhancement of forest carbon stocks in developing countries'.[112] Further, UNFCCC decisions have reaffirmed the crucial role of REDD+ and have called for more consideration on the necessary methodologies to enable its expansion beyond a project-based approach to possibly include sectoral approaches.[113] REDD+ was again recognized in the Copenhagen Accord in 2009 as having a 'crucial role' in global mitigation efforts.[114] A more detailed REDD+ decision was agreed in Cancun.[115] Paragraph 2 of Appendix 1 provides that when undertaking REDD+ projects, the following safeguards should be promoted and supported:

> (c) Respect for the knowledge and rights of indigenous peoples and members of local communities, by taking into account relevant international obligations, national circumstances and laws, and noting that the United Nations General Assembly has adopted the United Nations Declaration on the Rights of Indigenous Peoples; (d) The full and effective participation of relevant stakeholders, in particular indigenous peoples

[110] At the United Nations Climate Change Conference held in 2007 in Bali, Indonesia, the Conference of the Parties to the United Nations Framework Convention on Climate Change (COP13) adopted the Bali Road Map that included the Bali Action Plan (BAP). See UNFCCC Decision 1/CP.13.

[111] Angelsen *et al.*, *Reducing Emissions from Deforestation and Forest Degradation*, (n 102).

[112] UNFCCC Decisions 2–4/CP.13, 14 March 2008. Decision 2/CP.13 was dedicated to REDD.

[113] See UNFCCC, Decision 1/CP.16, 15 March 2011. Decision 1/CP.16. The Cancun Agreements: Outcome of the work of the Ad Hoc Working Group on Long-term Cooperative Action under the Convention; which affirmed that the implementation of REDD-plus activities should include the promotion and support of a number of safeguards, including consistency with the objectives of national forest programmes and relevant international conventions and agreements.

[114] See UNFCCC, Decision 2/CP.15. Copenhagen Accord, 18 December 2009. Paragraph 6.

[115] See Decision 1/CP.16, unfccc.int/resource/docs/2010/cop16/eng/07a01.pdf, accessed 12 December 2015. Policy approaches and positive incentives on issues relating to reduction emissions from deforestation and forest degradation in developing countries, and the role of conservation, sustainable management of forests and enhancement of forest carbon stocks in developing countries.

and local communities. . .; (e) That actions are consistent with the conservation of natural forests and biological diversity.[116]

The emergence of REDD+ as a very important part of global climate change mitigation efforts has, however, come with significant concerns with respect to its potential human rights, environmental and social impacts on local communities.[117] This includes forced displacement of indigenous communities from ancestral lands and forests; weak enforcement of land tenure rights; lack of opportunities for participation by citizens in project planning and implementation;[118] lack of governmental accountability on projects; and the absence of judicial and quasi-judicial remedies for victims of the above-mentioned problems.[119] These concerns continue to cast doubts on the integrity of REDD+ and raise questions on its desirability or sustained implementation as part of the international climate change regime.

Funding and investing in REDD+ projects that violate international human rights law is a threat to international law and raises concerns on the need to reconcile incentives for low carbon development and forest carbon sequestration, with the protection of the rights of the numerous communities that rely upon forests for their livelihood, shelter and survival. This book examines how lessons learnt from the design, approval, funding and implementation of REDD+ projects may inform thoughts on the need for rights-based frameworks for carbon finance to address unintended consequences and incongruous outcomes.

Carbon finance instruments The financial resources needed to implement carbon projects are significant and extensive.[120] Carbon projects

[116] Decision 1/CP.16

[117] See Taylor, 'Improving Substantive and Procedural Protections for Indigenous Rights in REDD+ Projects', (n 102) , 35–37. See also A. Long, 'REDD+ and Indigenous Peoples in Brazil', in R.S. Abate and E.A. Kronk (eds.), *Climate Change and Indigenous Peoples: The Search for Legal Remedies* (Northampton, MA; Cheltenham, UK: Edward Elgar Publishing, 2013), 151.

[118] See Seymour, 'Forests, Climate Change, and Human Rights', (n 24), 207; Jodoin, 'From Copenhagen to Cancun', (n 24).

[119] L.A. Miranda, 'Introduction to Indigenous Peoples' Status and Rights Under Int'l Human Rights Law', in R.S. Abate and E.A. Kronk (eds.), *Climate Change and Indigenous Peoples: The Search for Legal Remedies* (Northampton, MA; Cheltenham, UK: Edward Elgar Publishing, 2013), 45; also A. Savaresi, 'REDD+ and Human Rights: Addressing Synergies between International Regimes', *Ecology and Society*, 18(3) (2013), 2–4.

[120] Estimates suggest that USD139–175 billion per year will be needed in 2030 to fund carbon mitigation projects, while USD70–100 billion per year will be needed to fund adaptation. See UNFCCC: Investment and Financial Flows to Address Climate Change:

are often financed through resources from governments, loans, financial instruments from development agencies, centralized funds managed by the UNFCCC, World Bank, international organizations and private sector funding. For example, at the 2010 COP in Cancun, the UNFCCC established the Green Climate Fund (GCF) to redistribute money from developed to developing countries to assist adaptation and mitigation projects. The GCF is expected to provide US$100 billion a year by 2020 in climate finance for projects, programmes, policies and other activities in developing countries.[121]

Furthermore, the UNFCCC maintains two trust funds to support the development of carbon projects: the Special Climate Change Fund (SCCF) and the Least Developed Country Fund (LDCF).[122] These funds are administered by the Global Environment Facility (GEF), an independent financial institution established by a collaboration between national governments, international institutions, civil society organizations (CSOs) and the private sector, to assist countries to defray costs of carbon projects.[123] Since its inception in 1991, the GEF has allocated over 3.84 billion US dollars in support of climate mitigation and adaptation projects. The GEF has financed about 755 projects on mitigation and enabling activities, mostly renewable energy and energy-efficiency projects, with 3.39 billion US dollars in funding to 156 developing countries and economies in transition.[124] The COP provides regular guidance to the GEF on policies, programme priorities and eligibility criteria for funding, and the GEF reports annually to the COP on all GEF-financed activities.[125]

An Update (2008) Document No: FCCC/ TP/2008/7, Nov, unfccc.int/resource/docs /2008/tp/07.pdf., accessed 21 May 2015.

121 See UNFCCC (2011): Cancun Agreements: Outcome of the Work of the Ad Hoc Working Group on Long-term Cooperative Action under the Convention, Decision 1/ CP.16. Document No: FCCC/CP/2010/7/Add.1, March; UNFCCC (2011): Green Climate Fund – Report of the Transitional Committee (advanced unedited version), Nov., unfccc.int/files/meetings/durban_nov_2011/decisions/application/pdf/cop17_gcf.pdf; also The Green Climate Fund, news.gcfund.org, accessed 10 February 2015.

122 See the Global Environment Facility, www.thegef.org, accessed 21 May 2015.

123 United Nations Framework Convention on Climate Change, Subsidiary Body for Implementation, Thirty-First Session, 'The operation of funds under the Global Environment Facility as an operating entity of the financial mechanism, as well as the Special Climate Change Fund and the Least Developed Countries Fund, Submissions from Parties', unfccc.int/resource/docs/2009/sbi/eng/misc10a01.pdf, accessed 21 May 2015.

124 The Global Environment Facility, www.thegef.org, accessed 21 May 2015.

125 UNFCCC: Report of the Global Environmental Facility to the Conference of the Parties. Document No: FCCC/ CP/2011/7, 19 September 2011.

Similarly, the World Bank Carbon Finance Unit (CFU) provides loans and financing for carbon projects, and also purchases carbon credits emanating from such projects on behalf of contributing countries.[126] The CFU manages different carbon funds and facilities that finance carbon projects, for example the Prototype Carbon Fund - the first carbon fund, now with about 24 projects across the world generating emission reduction - Umbrella Carbon Facility T1, Community Development Carbon Fund, Carbon Fund for Europe and BioCarbon Fund Initiative for Sustainable Forest Landscapes, each with a specialized focus and together with a total capital of 2.5 billion US dollars.[127] The World Bank Climate Investment Fund (CIF) also provides developing and middle-income countries with financial resources to fund investments in clean technology, renewable energy, sustainable management of forests and climate-resilient development projects.[128]

Other international climate funds managed by international organizations, multilateral banks and regional development banks that finance carbon projects, include the African Development Fund (ADF) of the African Development Bank; the Cool Earth Partnership of the Government of Japan; the Global Climate Change Alliance funded by the European Commission Development Bank; Climate Investment Funds co-funded by the Asian Development Bank (ADB), African Development Bank, European Bank for Reconstruction and Development, Inter-American Development Bank and the World Bank Group; and the Abu Dhabi Fund for Development. The UNFCCC interface provides a platform to access and screen information on funding options available for carbon projects worldwide.[129]

A number of private sector institutions have also designed voluntary carbon finance instruments to meet their reputational and corporate social responsibility (CSR) needs. For example, some companies finance or purchase carbon credits directly from projects and companies (e.g. direct purchases from Ecosecurities or the World Bank BioCarbon

126 For example, in 2012, the World Bank Group provided a total of USD11.3 billion in lending with mitigation and adaptation co-benefits. See Climate Finance Overview, The World Bank, www.worldbank.org/en/topic/climatefinance/overview, accessed 10 February 2015.

127 See World Bank, Carbon Finance at the World Bank: List of Funds, wbcarbonfinance.org/Router.cfm?Page=Funds&ItemID=24670, accessed 21 May 2015.

128 See Climate Investment Funds, www.climateinvestmentfunds.org/cif/node/11900, accessed 12 June 2015.

129 See UNFCCC, Funding for Adaptation, unfccc.int/adaptation/workstreams/implementing_adaptation/items/4638.php, accessed 21 May 2015.

Fund), in order to lower their carbon footprints, or in some cases to meet local, sectoral or international carbon reduction targets. Furthermore, some private sector entities offer carbon offset programmes that encourage clients to lower their carbon emissions resulting from activities such as driving a car, flying or heating a building (e.g. British Airways offers carbon neutral flights and also has a special fund designed to finance renewable energy projects in the United Kingdom).[130] Private sector involvement and voluntary funding of carbon projects are generally geared towards supporting carbon projects that carry environmental benefits such as climate change mitigation and adaptation.

However, even though these carbon finance instruments were established to fund mitigation and adaptation projects, they have come under scrutiny from civil society leaders and indigenous rights groups for failing to provide sufficient human rights safeguards to uphold land and forest rights of local and indigenous peoples. For example, one of the key human rights issues at stake is that several participants in voluntary and compliance carbon markets fail to adequately respect and protect land rights of indigenous peoples and local communities, when investing in carbon projects.[131] Furthermore, the GEF, which offers financing for REDD+ and sustainable forest management projects, has been increasingly criticized for failing to address environmental, social and human rights impacts of its financed projects.[132] In response to these criticisms,

130 See British Airways, *Customer Carbon Fund*, responsibleflying.ba.com/environment/climate-change/one-destination-carbon-fund/, accessed 21 May 2015. See also Food and Agricultural Organization, 'Carbon Markets – Which types Exist and How They Work', www.fao.org/docrep/012/i1632e/i1632e02.pdf, accessed 21 May 2015.

131 See J. Tyson, 'The World Bank's Carbon Fund: Undermining indigenous rights or saving the planet?', www.devex.com/news/the-world-bank-s-carbon-fund-undermining-indigenous-rights-or-saving-the-planet-85809, accessed 23 May 2015; also Long, 'REDD+ and Indigenous Peoples in Brazil', 151.

132 A. Ballesteros, S. Nakhooda, J. Werksman, and K. Hurlburt, *Power, Responsibility and Accountability: Re-thinking the Legitimacy of Institutions for Climate Finance* (Washington, DC: World Resources Institute, 2010); J. Werksman, 'Consolidating Global Environmental Governance: New Lessons for the GEF?' in N. Kanie and P. Haas, (eds.), *Emerging Forces in Environmental Governance* (Tokyo: United Nations University Press, 2004), 35–50; J. Werksman, 'Consolidating Governance of the Global Commons: Insights from the Global Environment Facility,' in G. Handl (ed.), *Yearbook of International Environmental Law* 27 (Oxford: Oxford University Press, 1995), 6; C. Streck, 'Financing Climate Change: Institutional Aspects of a Post-2012 Framework,' Paper prepared for World Wildlife Fund (WWF, April 2009), para. 3(D), 20; D. Reed, A. Ballesteros, *et al.*, 'The Institutional Architecture for Financing a Global Climate Deal: An Options Paper' (June 2009), www.usclimatenetwork.org/resource-database/Options%20Paper%20Final%20May%2028.pdf; B. Müller, '2009 Bonn Seminar: On Future

the GEF in 2011 released the GEF Policy on Environmental and Social Safeguards (GEF Safeguards Policy), a set of environmental and social safeguard policies designed to prevent or mitigate unintended social and environmental harms in financed carbon projects. This Policy, largely derived from the World Bank's Operational Policy 4.00, requires GEF agencies to establish mechanisms to ensure that projects are designed, financed and implemented in ways that social and environmental well-being of the public are not negatively impacted. In addition to its Safeguards Policy, the GEF also adopted a Policy on Gender Mainstreaming that commits the GEF and its implementing agencies to ensure equal treatment of men and women in financed projects. The GEF Policy on Gender Mainstreaming requires GEF agencies to establish policies that ensure that projects are designed and implemented in such a way that both women and men receive culturally compatible social and economic benefits and do not suffer adverse human rights impacts during the development process.[133]

Although these GEF policies are significant steps in integrating social and environmental considerations in carbon finance, they fail to provide robust human rights safeguards. While gender discrimination, environmental pollution and lack of stakeholder accommodation and participation are some of the key sources of human rights risks in carbon projects, social and environmental safeguards introduced by GEF neglect other salient human rights issues at stake. For example, how do projects place disproportionate burdens on poor and vulnerable communities; what are the roles of responsible business cultures, mandatory human rights reporting, transparency and accountability in identifying, preventing and addressing human rights risks; and how can communities impacted by carbon projects seek and obtain legal redress for their loss? Being a social and environmental safeguard policy, the GEF Safeguards Policy fails to offer robust guidance on these key human rights questions.

Financial Architecture & Governance', European Capacity Building Initiative (ECBI) Policy Brief (July 2009), www.oxfordclimatepolicy.org/publications/documents/ecbiBrief-2009BonnSeminar.pdf, accessed 23 June 2015; L. Gomez-Echeverri and B. Muller, '2009 Bonn Seminar: On Future Financial Architecture & Governance,' Based on the Proceedings of a Meeting at La Redoute, Bad Godesberg (ecbi: 9 September 2009); www.oxfordclimatepolicy.org/publications/mueller.shtml., accessed 23 June 2015.

133 See GEF Council, GEF Policy on Agency Minimum Standards on Environmental and Social Safeguards, (2011) Document No: GEF/C. 41/10/Rev. Nov.; also GEF Council, GEF Policies on Environmental and Social Safeguards and Gender Mainstreaming. Document No GEF/C.40/10/Rev.1.

Furthermore, not all human rights risks in REDD+ projects are manifest at the financing stage. Consequently, while the GEF may finance a project based on it meeting its financing guidelines, there is a need for continuous monitoring and appraisal to ensure that project proponents continue to apply GEF Safeguards Policy over the lifetime of the project. The GEF Safeguards Policy, arguably, fails to establish robust systems through which human rights violations can be effectively monitored over the lifetime of a project. It is one thing to stipulate rules and safeguards; it is another to supervise and follow up on the actual practice of GEF project participants to ensure they have adopted and are continuously applying laid-down environmental and social safeguards to their projects. Without a system for effective monitoring and random spot-checking to ensure that prescribed safeguards and standards are followed, it has been difficult to measure and evaluate the overall impact of the GEF Safeguards Policy. Similarly, the GEF Safeguards Policy fails to integrate mandatory human rights impacts assessment into project financing safeguards. As shown in latter parts of this book, social and environmental impact assessments alone are arguably insufficient to detect human rights vulnerabilities. For example, how will a project affect my right to information, right to life or subsistence rights? These are key questions that human rights impacts assessment process could uncover in greater detail. Without providing mechanisms for assessing likely human rights implications of a project through a comprehensive impact study, it is difficult to see how the GEF policies will 'prevent and mitigate any unintended negative impacts to people and the environment'.

Furthermore, the GEF Safeguards Policy fails to provide robust grievance redress mechanisms or avenues for affected peoples or persons to raise these concerns with the GEF directly in cases where projects implemented with GEF financing fail to meet the objectives outlined in the GEF Safeguards Policy. The GEF Safeguards Policy requires implementing agencies to establish mechanisms for receiving and responding to complaints from local communities about projects and appoint a Conflict Resolution Commissioner to resolve disputes that cannot be addressed at the agency level. A more robust dispute resolution mechanism should provide opportunities for local communities to raise these concerns with the GEF directly.

The loopholes in the GEF policies emphasize the importance of developing and adopting robust and coherent human rights frameworks, as part of the international legal regime on climate change, to provide operational guidelines for finance agencies on how to anticipate and prevent negative human rights impacts of carbon projects. A situation whereby governments and multilateral bodies, such as the World Bank and GEF, finance carbon projects that engender land grabs and forced displacements undermines international law and threatens local rights. The international climate change regime must integrate robust human rights standards with requirements for accessing carbon finance and funding for carbon projects.

Debt for nature swaps Though not directly part of the international legal regime on climate change, debt-for-nature swaps (DfNS) are financial mechanisms through which conservation projects have been financed since 1987. Under this mechanism, debt owed by a country can be waived in exchange for a commitment, by the debtor country, to invest in local environmental conservation projects or measures. Such a debt exchange transaction is based on the simple notion of a reduction in external debt in return for domestic conservation actions.[134]

DfNS, which evolved historically as a debt management instrument, are now used to address a broad range of environmental challenges. Payments generated by DfNS are often directed to finance carbon projects through local climate change mitigation or adaptation efforts, biodiversity and conservation trust funds, which disburse grants to specific projects or fund parks and protected areas. Since the first swap occurred between Conservation International and Bolivia in 1987, over US$1 billion in environmental funding has been generated through DfNS, benefiting conservation in nearly thirty countries.[135] DfNS have been an important source of financing for over thirty conservation trust funds, many of which provide funding for carbon projects that abate carbon-intensive activities and to reinvest in sustainable energy.

Although the relevance and utility of DfNS have significantly waned over the last decade, they continue to play important roles in the financing and funding of carbon projects in developing countries. One criticism of DfNS, however, is the perennial misdirection of funds by national authorities to fund activities that violate human rights. Due to weak

[134] See J.C. Juergensmeyer and J.C. Nicholas, 'Debt for Nature Swaps: A Modest but Meaningful Response to Two International Crises', *Florida International Law Journal*, 5(20) (1990), 193, 198; T.B. Hamlin, 'Debt-for-Nature Swaps: A New Strategy for Protecting Environmental Interests in Developing Nations', *Ecology Law Quarterly*, 16 (1989), 1065, 1067.

[135] P. Sheikh, 'Debt-for-Nature Initiatives and the Tropical Forest Conservation Act: Status and Implementation' (Congressional Research Service Report, 30 March 2010).

democratic and regulatory structures in several emerging democracies, DfNS-funded projects have resulted in social exclusions, forced resettlements and land grabs. For example, DfNS projects have allegedly resulted in local resentments in countries such as Madagascar and Indonesia, due to forced relocations of people living on land set aside for conservation and the exclusion of local communities from planning processes.[136]

While DfNS arrangements are ingenious methods of financing carbon projects, it is important to develop holistic legal frameworks that would prevent social and human rights impacts associated with misdirected or improperly implemented swap arrangements to fund carbon projects under the international climate change regime. Most importantly, it is necessary to integrate stringent eligibility preconditions for debtor nations to participate in such agreements, including respect for human rights, stakeholder engagement and accountability.[137]

The above-discussed carbon finance mechanisms are not meant to be exhaustive; they, however, provide general examples of how carbon projects may result in human rights risks, if not effectively designed, implemented and monitored.[138]

1.2.2.4 Sustainable development

The twin objectives of the Kyoto Protocol are to achieve emission reductions and promote sustainable development.[139] As such, the concept of sustainable development lies at the heart of this book. The most commonly cited definition of sustainable development is the definition provided by the Brundtland Commission: 'development that meets the needs of the present without compromising the ability of future generations to meet their own needs'.[140] Article 3 (1) (a) of the 2002 Convention for Cooperation in the Protection and Sustainable Development of the

[136] See N. Hassoun, 'The Problem of Debt-for-Nature Swaps from a Human Rights Perspective', *Journal of Applied Philosophy*, 29(4) (2012), 359–377.

[137] This pre-condition approach was adopted by the US Government in the *1990 Enterprise for the Americas Initiative Act*. Under Section 703 (4) of the Act, only countries 'whose government (including its military or other security forces) does not engage in a consistent pattern of gross violations of internationally recognized human rights' are eligible to participate in US-funded DfNS.

[138] Other examples of land-based carbon market offset projects are available at www.forestcarbonportal.com/project/, accessed 21 May 2015.

[139] See Arts. 2 and 12 of the Kyoto Protocol.

[140] The United Nations World Commission on Environment and Development Report (the Brundtland Report) *Our Common Future* (Oxford: Oxford University Press, 1987).

Marine and Coastal Environment of the Northeast Pacific also defines sustainable development as:

> the process of progressive change in the quality of life of human beings, which places them as the centre and primary subjects of development, by means of economic growth with social equity and transformation of production methods and consumption patterns, sustained by the ecological balance and life support systems of the region. This process implies respect for regional, national and local ethnic and cultural diversity, and full public participation, peaceful coexistence in harmony with nature, without prejudice to and ensuring the quality of life of future generations.[141]

The above definition places humans as the subject of efforts to protect the environment while pursuing economic growth and development. As the ICJ noted in the *Gabcikovo-Nagymaros Case*, the need to reconcile economic development with the protection of the environment is aptly expressed in the concept of sustainable development.[142] Furthermore, the Rio+20 United Nations World Summit on Sustainable Development held in 2012 specifically identifies the three mutually reinforcing pillars of sustainable development as economic development, social development and environmental development.[143] The UNFCCC in a recent evaluation report takes a more expansive view of sustainable development by classifying respect and consideration of the rights of local/indigenous people and community or local/regional involvement in decision-making as part of the social indicators of sustainable development.[144] This interpretation embeds respect for human rights as part of social development.[145]

141 United Nations Environment Program (UNEP), 2002 Convention for Cooperation in the Protection and Sustainable Development of the Marine and Coastal Environment of the Northeast Pacific (Antigua Convention) adopted 18 February 2002, www.unep.ch/regionalseas/main/nep/nepconve.html, accessed 03 January 2015.

142 *Case Concerning the Gabcikovo-Nagymaros Dam* (25 September 1997) *(Hungary v. Slovakia)* I.C.J. Rep., 37 ILM (1998) 162.

143 2012A/CONF.216/L.1, United Nations, Rio de Janeiro, Brazil 20–22 June. See also the 2009 United General Assembly Resolution A/RES/64/236 which also describes economic development, social development and environmental protection as interdependent and mutually reinforcing pillars of sustainable development, daccess-dds-ny.un.org/doc/UNDOC/GEN/N09/475/99/IMG/N0947599.pdf?OpenElement, accessed 12 January 2015.

144 UNFCCC, 'Benefits of the Clean Development Mechanism 2012', (n 12).

145 See the 2009 United General Assembly Resolution (A/RES/64/236) which also describes economic development, social development and environmental protection as interdependent and mutually reinforcing pillars of sustainable development, daccess-dds-ny.un.org/doc/UNDOC/GEN/N09/475/99/IMG/N0947599.pdf?OpenElement, accessed 12 January 2015.

This book adopts this expansive understanding of sustainable development that considers respect for human rights as part of social development. Sustainable development is discussed as development that promotes economic development, environmental protection and social development (which includes the fulfilment of fundamental human rights) in a balanced, mutually supportive and integrated manner. Arguably, the protection and fulfilment of human rights is a fundamental prerequisite and pre-condition for achieving sustainable development.[146] Studies point to ongoing trade-offs in carbon projects wherein projects that generate huge emission reductions have not been the most sustainable.[147] To avoid such trade-offs, there is a need to balance emission reduction goals with the goal of respecting and protecting fundamental human rights.

This leads to the second question: what is the legal status of the concept of sustainable development under international law? Understanding the legal status of the concept is critical to understanding the normative basis and *raison d'être* of a coherent and integrated management of environmental, social and economic goals in all aspects of development. This question, whether sustainable development falls into any of the formal categories of international law identified by Article 38(1) of the ICJ, has not been without debate.[148] While some scholars argue that the persistent recognition of the

[146] Ever since the 1986 *United Nations Declaration on Development* described the lack of attention for the implementation, promotion and protection of civil, political, economic, social and cultural rights as one of the obstacles to development; and that the promotion of, respect for and enjoyment of certain human rights and fundamental freedoms cannot justify the denial of other human rights and fundamental freedoms, many scholars have argued that respect, protection and fulfilment of human rights is crucial to achieving sustainable development. This book aligns with this argument. This book is premised on the argument that without explicit safeguards, policies and projects intended to advance environmental or development goals can have serious negative impacts on fundamental human rights. It shows that emission reduction projects that are designed to promote environmental development have resulted in massive land grabs, forced eviction and relocations of communities to allow for projects, social exclusion for the poor and vulnerable groups, killings, maiming and oppressions. This book argues that real sustainable development can only be achieved if human rights are protected in the international climate change regime.

[147] Pottinger, *The CDM's Hydro Hall of Shame, Bad Deal for the Planet*, (n 26); McInerney-Lankford, 'Climate Change and Human Rights', (n 26), 431–437; Oxfam, Climate Wrongs and Human Rights: Putting People at the Heart of Climate-Change Policy (2008), Oxfam Briefing Paper on Climate Change and Human Rights, www.oxfam.org.uk/resources/policy/climate_change/downloads/bp117_climatewrongs.pdf, accessed 12 July 2014; see also Aminzadeh, 'A Moral Imperative', (n 26); Limon, 'Human Rights and Climate Change', (n 26).

[148] 1945 Statute of the International Court of Justice (Adopted on 26 June 1945) 3 Bevans 1179; 59 Stat. 1031; T.S. 993; 39 AJIL Supp. 215 (1945).

principle of sustainable development in national, regional and international laws and treaties, case law and state practice point to the fact that it has become a general principle of international law and is maturing into a custom of international law, others opine that the reluctance to formalize its distinctive legal status means it is too early to describe it as a customary rule or principle of international law.[149] It has also been argued that while sustainable development may not be a customary principle of international law, one of the principles of international law related to sustainable development includes the 'principle of integration'.[150]

This book will not delve into an exhaustive discussion of these debates. It instead aligns with the views that even though it might be too early to speak of sustainable development as part of customary international law, the concept of sustainable development can be described as a general principle of international environmental law recognized by civilized nations.[151] The concept of sustainable

149 See V. Lowe, 'Sustainable Development and Unsustainable Arguments' in A. Boyle and D. Freestone (eds.), *International Law and Sustainable Development: Past Achievements and Future Challenges* (Oxford: Oxford University Press, 1999), 36. See also A. Boyle and D. Freestone, *International Law and Sustainable Development: Past Achievements and Future Challenges* (Oxford: Oxford University Press, 1999), 16–18.

150 See 2002 *International Law Association New Delhi Declaration on Principles of International Law Related to Sustainable Development* (ILA, 2002); see also 'ILA New Delhi Declaration of Principles of International Law Relating to Sustainable Development' in *International Environmental Agreements: Politics, Law and Economics* (Kluwer Academic Publishers, 2002), 209–216, www.kluweronline.com/issn/1567-9764/current. See also M.C. Cordonier Segger and A. Khalfan, *Sustainable Development Law: Principles, Practices and Prospects* (Oxford: Oxford University Press, 2004) 45–50; D. French, *International Law and Policy of Sustainable Development* (Manchester: Manchester University Press, 2005), 51.

151 This book aligns with the views expressed by Alston and Simma that general principles derive legitimacy from their recognition by states, and that what is required for the establishment of general principles in international law is an evidence of general recognition by states, even if the norm is widely violated. Evidence of state practice is not a pre-requisite for establishing the existence of general principles in international law. This narrows the query to whether sustainable development has been sufficiently recognized by states? Arguably, in the light of the evidence of recognition adduced in this section, this will receive an affirmative answer. See generally, P. Alston and B. Simma, 'The Sources of Human Rights Law: Custom, Jus Cogens and General Principles', *Australian Yearbook of International Law*, 12 (1991), 102. This point was emphasized in *South West African Cases*, ICJ Report 1966, where Tanaka J held that 'the behaviour and consent of States are thus not a necessary prerequisite to the recognition of general principles'. See P. Hulsroj, 'Three Sources – No River: A Hard Look at the Sources of Public International Law with Particular Emphasis on Custom and General Principles of Law', *Austrian Journal of Public and International Law*, 54 (1999), 245.

development has become one of the recognized general principles of international environmental law, and there is evidence to suggest that it is maturing into a custom of international law.

The strongest support for the view that sustainable development is arguably a general principle of international law is the fact that it has been vastly recognized as a yardstick for measuring development in many countries and internationally.[152] The concept of sustainable development is an integral object and purpose of international environmental law and many environmental treaties. Currently, there are over 300 international legal instruments in addition to other soft law declarations, guidelines and General Assembly Resolutions that have identified and recognized the principle of sustainable development.[153] For example, the idea of sustainable development lies at the centre of efforts to mitigate climate change. The UNFCCC is a direct product of the 1992 United Nations Rio Conference on Environment and Development which laid down sets of principles (Rio Declaration) and conventions intended to deliver sustainable development.[154] Principle 3 of the Rio Declaration states that the right to development must be fulfilled so as to equitably meet developmental and environmental needs of present and future generations.[155] Principle 27 of the Rio Declaration states that states shall cooperate in good faith in the further development of international law in the field of sustainable development. Similarly, even though the aim of the UNFCCC is to achieve stabilization of GHGs at a level that would prevent dangerous interference with climate systems, the UNFCCC calls on parties to ensure that responses to climate change are coordinated with social and economic development in an integrated manner, taking into full account the legitimate priority needs of developing countries for the

[152] Environmental Sustainability is recognized as part of the Millennium Development Goals. See the United Nations, *Summit Outcome of the 65th Session of the UN General Assembly*, 17 September 2010, UN/A/ 65/L.1; See also United Nations, *Claiming the Millennium Development Goals: A Human Rights Approach* UN/HR/PUB/08/3.

[153] For a comprehensive list of these instruments, see the United Nations Sustainable Development Knowledge Platform, 'Major Agreements and Conventions', sustainable development.un.org/index.php?menu=122, accessed 12 January 2015.

[154] The UNFCCC is one of the three 'Rio Conventions', adopted at the Rio Earth Summit. The other two are the UN Convention on Biological Diversity and the Convention to Combat Desertification.

[155] United Nations' Conference on Environment and Development (Rio Declaration) UN Doc A/CONF.48/PC.13(1971), www.unep.org/documents.multilingual/default.asp?documentid=78&articleid=1163, accessed 21 January 2015.

achievement of sustained economic growth and the eradication of poverty.[156]

The concept of sustainable development has also frequently gained recognition in national and international case law[157] and in national laws and policies.[158] Similarly, a cursory look at how many countries are active parties to environmental treaties and conventions which are developed based on the idea of sustainable development shows that the concept has gained sufficient recognition from countries. For example, with respect to climate change, 195 countries have signed and are parties to the UNFCCC, while 191 countries have signed and ratified the Kyoto Protocol.

These examples explain how environmental law and sustainable development are intertwined. As the Permanent Court of Arbitration rightly

156 See the Preamble to the United Nations Framework Convention on Climate Change (Adopted 9 May 1992; entered into force on 21 March 1994).

157 See the *US – Shrimp Case*, where the World Trade Organization (WTO) Appellate Body stated that:

> . . .[t]his concept has been generally accepted as integrating economic and social development and environmental protection. . .and the concept is elaborated. . . so as to put in place development that is sustainable. . . that 'meets the needs of the present generation without compromising the ability of future generations to meet their own needs'.

See also *United States – Import Prohibition of Certain Shrimp and Shrimp Products* (6 November 1998), WTO Doc WT/DS58/AB/R at note 115. See also *Case Concerning the Gabcíkovo-Nagymaros Project (Hungary/Slovakia)* (1997), I.C.J. Rep., 37 ILM (1998) 162 at 140; *United States- Import Prohibition of Certain Shrimp and Shrimp Products* (20 September 1999), WTO Doc WT/DS58/AB/R (Appellate Body Report) and *Permanent Court of Arbitration: In the Arbitration Regarding the Iron Rhine ('Ijzeren Rijn') Railway, between the Kingdom of Belgium and the Kingdom of the Netherlands* (24 May 2005) at 59, 114, Award of the Arbitral Tribunal; and national cases such as *Vellore Citizens Welfare Forum v. Union of India* (1996, Supreme Court of India), *Bulankulame v. Secretary, Ministry of Industrial Development and Others* (the Eppawela Case) (2000, Supreme Court of Sri Lanka), and *Rajendra Parajuli and Others v. Shree Distillery Pvt. Ltd. & Others* (Writ No. 3259, 1996, Supreme Court of Nepal).

158 There is a plenitude of examples. See the Australian NSW Protection of the Environment Administration Act 1991, The Nigerian National Environmental Standards and Regulations Enforcement Act (NESREA) Act; The 2005 UK Sustainable Development Strategy, Securing the Future; Canada`s 1985 Fisheries Act R.S.C. 1985 c. F-14, ss. 34–42; 1999 CEPA (The Canadian Environmental Protection Act) S.C. 1999 c. 33, ss. 39–103; The United States`1980 CERCLA (Comprehensive Environmental Response, Compensation, and Liability Act of 1980) 42 USC ss. 9601–9675, US 1990 OPA (Oil Pollution Act of 1990) 33 USC ss. 2701 – 2761; German Environmental Liability Act 1990; Venezuela`s 1976 Environmental 'Frame' Act.

noted in *Arbitration Regarding the Iron Rhine (Ijzeren Rijn) Railway (Belgium v. Netherlands)*:

> environmental law and the law on development stand not as alternatives but as mutually reinforcing, integral concepts, which require that where development may cause significant harm to the environment, there is a duty to prevent, or at least mitigate such harm. . . This duty, in the opinion of the Tribunal, has now become a principle of general international law. This principle applies not only in autonomous activities but also in activities undertaken in implementation of specific treaties between the Parties.[159]

Agreeably, the increased reflection of the concept of sustainable development in national laws, international treaties and case law, all provide evidence that it has gained recognition as a principle relevant to international law on environment and natural resources.

1.2.2.5 UN resolutions and soft law

This book discusses several UN resolutions, declarations, general comments and normative recommendations of treaty bodies which have no legally binding force under international law;[160] for example, the United Nations HRBA, the Ruggie Guiding Principles for Business and Human Rights, Decisions and Guidelines by the COP to the Convention and General Comments of UN treaty bodies. This raises questions on the legal and normative value and importance of non-binding instruments and documents in international law. This has remained a highly debated issue.[161] These debates can be traced to questions on the legal status and role of soft law under international law.

[159] *Arbitration Regarding the Iron Rhine ('Ijzeren Rijn') Railway (Belgium v. Netherlands)* (24 May 2005).

[160] On the status and normative importance of UN resolutions, declarations, general comments and normative recommendations, see D. Forsythe, *Human Rights in International Relations* (Cambridge: Cambridge University Press, 2006), 12. See also A. Cassese, *International Law* (Oxford: Oxford University Press, 2001).

[161] For example, UN General Assembly resolutions have been described as attempts to impose norms of international law on dissenting minorities and to radically change the way in which international law is made. See O. Schachter, *International Law in Theory and Practice* (The Hague: Kluwer Academic Publishers, 1991). See also A. Peters and I. Pagotto, 'Soft Law as a New Mode of Governance: A Legal Perspective' (28 February 2006) NEWGOV: New Modes of Governance, ssrn.com/abstract=1668531, accessed 23 June 2015; F. Snyder, 'Soft Law and Institutional Practice in the European Community', in S. Martin (ed.), *The Construction of Europe: Essays in Honour of Emile Noël* (The Hague: Kluwer Academic Publishing, 1994), 198.

Soft law comprises those legal instruments that are not directly enforceable in courts and tribunals but nonetheless have an impact on international relations and, ultimately, international law.[162] Broadly put, soft law instruments include resolutions and declarations of the United Nations and its organs, normative recommendations, declarations of principles, codes of conduct, codes of practice, Programmes of Action (PA) and other non-treaty obligations which do not create binding obligations but often propose guidelines and rules of behaviour which are important and useful for responding to emerging challenges. Soft law is often contrasted with hard law.[163] While some scholars have argued that soft law is not law, some question the importance of soft law generally under international law since they do not create binding obligations.[164]

This book will not delve extensively into these debates. Instead, it aligns with the views that even though resolutions adopted by the United Nations General Assembly or General Comments by UN treaty bodies are not international treaties and therefore do not have legally binding force, they have important legal functions and normative effects to the extent that they often elaborate and interpret international law norms.[165] They provide practical interpretations and guidelines on how a country should apply a treaty in order to fulfil its international obligations. Arguably, soft law instruments play significant roles in the development and evolution of international law. As Higgins noted, "the passing of binding decisions by an international body is not the only way in which the development of the law occurs. Legal consequences can also flow from acts which are not, in the formal sense, 'binding'."[166] Not only do soft law instruments provide flexible guidelines on how to tackle emerging concerns, they also provide templates on how international policies can be implemented. Furthermore, they serve as forbearers to binding hard law instruments in the future.[167]

[162] See P. Dupuy, 'Soft Law and the International Law of the Environment', *Michigan Journal of International Law*, 12 (1991), 420, 422–425, 428–431.

[163] M. Reisman, 'Hard Look at Soft Law', *American Society of International Law Proceedings*, 82 (1988), 371–377.

[164] A. Guzman and T. Meyer, 'International Soft Law', *Journal of Legal Analysis*, 2 (2010), 171.

[165] For an excellent discussion of the normative effects and importance of UN Resolutions and Declarations see Forsythe, *Human Rights in International Relations* (3rd edition, Cambridge University Press 2012), (n 160).

[166] R. Higgins, *Problems and Process: International Law and How We Use It* (Oxford: Oxford University Press, 1994), 25.

[167] Guzman and Meyer, 'International Soft Law', (n 164), 171; A. Boyle, 'Some Reflections on the Relationship of Treaties and Soft Law', *International and Comparative Law*

For example, since the HRBA became popularized by the United Nations through the *1997 Programme for Reform* launched by the UN Secretary General,[168] it has prompted a flurry of efforts to mainstream human rights principles within the UN systems and internationally.[169] For example, in 2007, the World Health Organization released sector-specific guidelines for adopting the HRBA in relation to health issues. The Food Agency Organization (FAO) also released its voluntary guidelines to provide 'human rights-based practical tool addressed to all states' to help them develop, design and implement the right to food.[170] The guidelines cover the full range of actions that need to be taken at the national level to build an enabling environment for the provision of food and to ensure food security. Currently, the HRBA has been adopted and applied by a number of other agencies with appreciable success, especially in the areas of development, poverty reduction, social protection and security.[171] In 2011, the UNDG-Human Rights Mainstreaming (HRM) Mechanism released a comprehensive action plan on how

Quarterly, 48 (4) 1999), 901–913. R. Andorno, 'The Invaluable Role of Soft Law in the Development of Universal Norms in Bioethics' (2007) Paper delivered at a workshop jointly organized by the German Ministry of Foreign Affairs and the German UNESCO Commission, Berlin, 15 February 2007, www.unesco.de/1507.html, accessed 12 May 2014. D. Tillinghast, R. David and H. Higenberg, 'A Fresh Look at Soft Law', *European Journal of International Law*, 3 (1999), 502.

168 According to the Secretary General, 'A major task for the United Nations, therefore, is to enhance its human rights programme and fully integrate it into the broad range of the Organization's activities'. The Secretary General called on all *agencies* of the United Nations to commit themselves to the formulation, promotion and implementation of human rights'. See UN *Secretary-General's Programme for Reform,* UN Doc A/51/950 14 July 1997 (paragraphs 78–79).

169 In 1988 there was The Oslo Symposium on 'Human Development and Human Rights', which brought together UN organizations, NGOs and academic experts. The aim was to develop a background on how to apply the human rights-based approach and how to operationalize an integrated human rights-based approach to human development. The Symposium laid the groundwork for the 2000 Human Development Report.

170 Voluntary Guidelines to Support the Progressive Realization or the Right to Adequate Food in the Context of National Food Security, adopted by the 127th Session of the FAO Council, November 2004.

171 For an excellent and detailed discussion of the adoption of the HRBA by the International Labour Organization; United Nations Development Programme; United Nations Children's Fund; United Nations Educational, Scientific and Cultural Organization; United Nations High Commissioner for Refugees; United Nations Human Settlements Programme; World Health Organization; Food and Agricultural Organization; World Bank and International Monetary Fund and the World Trade Organization, see G. Oberleitner, Global Human Rights Institutions (Netherlands: Wiley, 2007) particularly Chapter 5 on 'Mainstreaming Human Rights'; see also See D. Rozga, 'Applying a Human Rights Based Approach to Programming: Experiences of

human rights mainstreaming could be achieved within UN systems. The plan, among other things, called for the adoption of a human rights-based approach within UN systems and teams as a policy approach for mainstreaming human rights within their areas of activities and mandate.[172] The UNDG also provided normative recommendations and guidelines on how to ensure that an HRBA will form the guiding principle and basis for activities and programmes at the national level.

The HRBA has also been recognized in a number of countries as a framework for mainstreaming human rights issues into decision-making.[173] Countries, including Australia,[174] Denmark,[175] Ireland,[176] New Zealand,[177] Mali,[178] Peru[179], India[180] and Scotland,[181] have developed robust national methodologies and guidelines which direct agencies and national authorities to mainstream human rights norms into programming through the HRBA framework.[182] In 2000, the Government of Sweden identified the HRBA as a better and revolutionary strategy for focusing on the root causes of human rights violations and for ensuring

UNICEF' (2001), www.handicap-international.fr/bibliographie-handicap/3ApprocheDroit/OutilProgMEO/ApHRight.pdf, accessed 12 October 2014.

172 See UNDG, www.undg.org/index.cfm?P=2, accessed 20 January 2015.

173 Swedish International Development Agency, Human Rights Based Approach (2012), www.sida.se/English/Partners/Sida-Partnership-Forum/Sida-Partnership-Forum1, accessed 12 December 2014.

174 Australian Human Rights Commission, Human Rights Based Approach, www.hreoc.gov.au/about/media/speeches/speeches_president/2011/20110413_rights_women.html, accessed 12 December 2014.

175 See J. Boesen and T. Martin, *Applying a Rights Based Approach in Denmark: An Inspirational Guide for Civil Society* (Copenhagen: Danish Institute for Human Rights, 2007).

176 See International Human Rights Network, *Our Rights, Our Future Human Rights Based Approaches in Ireland: Principles, Policies and Practice* (Dublin, 2005).

177 UNICEF, *Human Rights Based Programming: The Mali Experience* (UNICEF, 2003).

178 UNICEF, *Peru Case Study: A Human Rights Approach* (UNICEF, 2002).

179 See UNICEF, *Human Rights Based Programming: India Case Study* (UNICEF, 2003).

180 J. Theis, 'Promoting Rights-based Approaches: Experiences and Ideas from Asia and the Pacific' (Child Rights Information Network Working Documents, 2004), www.crin.org/docs/resources/publications/hrbap/promoting.pdf, accessed 15 July 2014.

181 The Scottish Human Rights Commission (SHRC) advocates for the PANEL principles as a way of breaking down what a human rights based approach means in practice. PANEL stands for Participation, Accountability, Non-Discrimination, Empowerment and Legality. See Scottish Human Rights Commission (SHRC), *Human Rights in a Health Care Setting: Making it Work for Everyone, An Evaluation of a Human Rights-based Approach at The State Hospital* (SHRC, 2009), www.scottishhumanrights.com/application/resources/documents/HRHCSFINALVERSION.pdf, accessed December 13, 2015.

182 See OHCHR, *Handbook on National Human Rights Plans of Action*, 29 August 2002, UN Professional Training Series No. 10.

the accountability of agents of the state.[183] It described the HRBA as a more efficient analytical tool based on internationally accepted standards, which provides a legal standard and a framework that clarifies state obligations to protect and realize human rights.

The HRBA is a good example of how soft law instruments could aid the development of international and domestic law. Soft law instruments provide ground rules for the conduct of development policies. Though not binding, their consistent recognitions in national legislation and policies play a vital role in the formation of broadly accepted general principles of international law and international customary law. Consequently, this book discusses both hard law instruments such as treaties and conventions and soft law instruments such as UN declarations, resolutions and norms that have been recognized as part of the international climate change regime.

Consequently, references to 'international climate change regime' in this book mean the web of extant or emerging conventions, protocols, modalities, decisions, including governance structures and institutions established by the international community to limit and reduce the emission of GHGs and to cope with whatever impacts are inevitable.[184] The book specifically identifies the Kyoto Protocol and its amendments,[185] the Framework Convention on Climate Change (FCCC), the Marrakech Accords on the Modalities and Procedures on the CDM and other relevant COP decisions as sources of the rules and frameworks that govern the approval and execution of CDM projects discussed in this book. Similarly, the UNFCCC Secretariat, the COP, the Compliance

[183] A. Frankovits and P. Earle (eds.), *Working Together, The Human Rights Based Approach to Development Cooperation in Sweden* Parts 1 & 2 (Stockholm Swedish International Development Cooperation Agency (SIDA) and Human Rights Council of Australia (HRCA), 2001).

[184] This definition draws from Krasner's classic definition of international regimes as 'a set of implicit or explicit principles, norms, rules, and decision-making procedures around which actors' expectations converge in a given area of international relations'; S. Krasner, 'Structural Causes and Regime Consequences: Regimes as Intervening Variables', in S.D. Krasner (ed.), *International Regimes* (Ithaca, NY: Cornell University Press, 1983), 2.

[185] At the COP meeting held in Doha, Qatar, in December 2012, parties to the Kyoto Protocol adopted the Doha Amendments to the Kyoto Protocol, which amongst other things confirmed the second commitment period to be 2013–2020. The Doha Amendment will enter into force pursuant to Article 20 (4) and (5) of the Kyoto Protocol, which provides that such amendments shall enter into force on the ninetieth day after the date on which parties deposit with the Depositary an instrument of acceptance of the said amendment. See UNFCCC, Doha Amendments to the Kyoto Protocol, unfccc.int/files/kyoto_protocol/application/pdf/kp_doha_amendment_english.pdf, accessed 12 January 2015.

Committee and the CDM Executive Board are the principal governance structures under the international climate change regime discussed.[186]

1.2.3 *Structure of the book*

The book is organized into four parts consisting of ten chapters to reflect a transition from conceptual and theoretical issues to practical and substantive issues. Part I comprises two chapters that provide introductions on the key intersections between climate change and human rights. This introduction is the first chapter.

Using existing projects in Honduras, Nigeria and Panama as examples, Chapter 2 examines how carbon projects, specifically CDM and REDD+ projects, produce human rights concerns in developing countries. These human rights issues include mass displacements of people from their homes to allow for climate change mitigation projects, lack of participation by citizens in project planning and implementation, siting and concentration of projects in poor and vulnerable communities, lack of governmental accountability on projects and the absence of judicial and quasi-judicial remedies for victims of these problems. Chapter 2 also identifies some of the loopholes in the international climate change regime applicable to carbon projects. It reviews how the lack of transparency and accountability in verification and registration of projects, lack of safeguards on the disclosure of relevant project information, lack of opportunities for broad-based stakeholder participation and the absence of a complaints mechanism or procedure entrenched a culture of 'approvals', in which carbon projects are approved and registered irrespective of their human rights consequences. This chapter essentially lays a foundation on the need for mainstreaming human rights standards into the international climate change regime.

A theoretical thread that runs through this entire book is the concept of human rights mainstreaming, that is, the idea of integrating human rights norms into climate change responses and actions. Given the pivotal nature of this concept, it is pertinent from the onset to analyse and discuss the meaning, nature, theoretical basis and value of human

186 The Conference of the Parties is the supreme decision-making body of the Convention. The COP adopts rules and takes decisions necessary to promote the effective implementation of the Convention. The COP serves as the Meeting of the Parties to the Kyoto Protocol (CMP). All states that are parties to the Kyoto Protocol are represented at the CMP, while states that are not parties participate as observers. See UNFCCC, 'Bodies', unfccc.int/bodies/items/6241.php, accessed 21 January 2015.

rights mainstreaming. This is the focus in Part II. Part II comprises two chapters that provide theoretical foundations for the human rights-based approach to carbon finance. It will review, explore and contribute to emerging debates on the human rights mainstream paradigm under international law. First, Chapter 3 reviews, evaluates and contributes to emerging debates on the human rights mainstream paradigm under international law. This chapter discusses theoretical underpinnings, values and problems associated with the human rights mainstream paradigm. Apart from definitional questions, certain theoretical questions that are more specific to the legal basis of the human rights mainstream paradigm as a whole are raised and discussed. Theoretical questions such as the novelty or duplication question, neo-liberalism critique, determinacy and the conflict of rights debates are identified and discussed. This chapter also examines the legal and theoretical foundations and bases for human rights mainstreaming. It will analyse how we could invoke the principle of systemic integration of international law, the inter-doctrinal legal transplant/cross-fertilization theory, participatory development theory and the full belly thesis as legal and theoretical anchor for the human rights mainstream paradigm.

One of the most knotty debates in the human rights mainstream literature is the question of approach. Put simply, if we were to agree that human rights mainstreaming is important and necessary, how should it be done? What is the best policy mix or approach for achieving the goal of human rights mainstreaming? This question is the focus in Chapter 4. It unpacks legal debates on the question of approaches for mainstreaming human rights into the international climate change regime. The chapter examines two main strategies that have been proposed: the substantive rights approach and the procedural rights approach (human rights-based approach). It is argued here that the HRBA provides a procedural framework which focuses on the institutionalization of human rights into the existing international climate change regime. Through procedural rights, the design of the international climate change regime could be strengthened to anticipate, avoid and prevent some of the human rights issues; this is unlike traditional approaches that are reactive in nature. By ensuring that the execution and approval of projects are dependent on the consent and cooperation of stakeholders, individuals and communities, human rights infringements due to climate change projects may be avoided in the first place. This book discusses the HRBA as a preventive, transformative and forward-looking approach that avoids the human rights versus environment arguments, prompts changes in institutional practices and also provides a preventive/long-term process through which human rights are systematically integrated into environmental systems, structures and practices.

Developing theoretical formulations on human rights mainstreaming through the HRBA is only part of the task. Other essential questions relate to understanding the normative content and scope of the HRBA in practical and applied terms, identifying the mechanisms that are to be put in place for effective implementation, how decision-making on project approval and finance are to be functionally operationalized, how progress will be monitored and evaluated, and what further action should be taken in case of unsatisfactory progress. Part III examines the question of 'how', that is what are the legal frameworks required to mainstream human rights norms into the international climate change regime under a human rights-based approach? Part III transits from theoretical issues to consider the practical aspects of implementing a human rights-based approach to carbon finance. This part includes three chapters devoted to exploring the legal and institutional frameworks that would assist in moving the idea of human rights mainstreaming from theory to practice.

Chapter 5 examines the normative contents and elements of a human rights-based approach, specifically what the HRBA entails in practice. There are five inter-connected human rights norms and principles that have been recognized in relevant literatures as forming the core of the HRBA; they are access to information (ATI), participation, accountability, equality and non-discrimination, and access to justice.[187] Chapter 5 examines why these elements form the core of the HRBA. It also evaluates the legal status of each of these elements under international human rights law and answer the following question: 'are these elements human rights?' Chapter 5 will analyse how each of these procedural human rights have been recognized in human rights treaties and conventions and also discuss their normative contents.

Chapter 6 examines legal frameworks required to mainstream human rights safeguards into the design, approval, finance and implementation of carbon projects. It develops a functional rights-based

[187] See United Nations, 'The Human Rights Based Approach to Development Cooperation: Towards a Common Understanding among UN Agencies' (2003), www.undg.org/archive_docs/6959-The_Human_Rights-Based_Approach_to_Development_Cooperation_Towards_a _Common_Understanding_among_UN.pdf, accessed 12 October 2014.

legal framework that could assist in moving the idea of human rights mainstreaming from theory to practice. It proposes a six-part legal threshold that could serve as the minimum core content of a reformed rights-based international climate change regime. They are project eligibility screen, sustainable development screen, human rights impacts assessment, stakeholder identification and consultation, information disclosure system and project review mechanism. This threshold emphasises the importance of an integrated assessment to identify the human rights, social and environmental impacts and risks of projects; the importance of effective community engagement and consultation with local communities on matters that directly affect them; and the need for project proponents to manage and disclose the human rights, social and environmental performance of a project throughout the lifecycle of the project. Adopting this minimum legal threshold as the basis for project approval and registration at the international level could arguably go a long way in addressing some of the human rights issues that have been identified in many carbon projects; they could also provide guidance to policymakers at national levels on how to design balanced and rights-based policy responses and projects for climate change mitigation.

Chapter 7 discusses institutional reforms necessary to spearhead a rights-based legal reform process. Institutions would go a long way to determine how the identified legal threshold would be operationalized and enforced, how progress will be monitored and evaluated, and what further action could be taken in case of unsatisfactory progress. It discusses four complementary institutional reforms that are required to move the idea of human rights mainstreaming within the context of the international climate change regime from theory to practice. They are the creation of a specific human rights and climate change unit (HRCU) within the UNFCCC secretariat to be responsible for the overall mainstreaming strategy; the expansion of the terms of reference and powers of project supervisory boards or institutions; empowering the compliance committee to ensure that after registration, implementation of projects are in line with approved standards and do not create unanticipated problems; and setting up of a public complaint branch of the compliance committee that could hear appeals by stakeholders who may be adversely affected. This institutional framework mirrors the shift away from top-down planning towards more participatory forms of development and recognition for stakeholders in the design and implementation of carbon projects. It provides a window of opportunity and a platform for stakeholders to bring their complaints about a project before an international supervisory body at every stage of project planning and execution, that is prior to registration and post-registration. It is argued here that mobilizing and empowering institutions would enable individuals, communities and NGOs to have a voice in processes of approving carbon projects, in a balanced and transparent way.

Laying down a theoretical framework is one thing; understanding its practical implications, possible real-life limitations and the requisite steps for moving the concepts from the realm of theory to practice is another. Part IV, which is the final part, comprises three chapters that examine the legal and practical strategies and potentials for, and the difficulties that may arise in, actualizing a rights-based reform of the international climate regime applicable to carbon projects. Chapter 8 discusses practical steps and legal guidelines on how human rights-based reforms of carbon markets may be practically realized. It evaluates legal pathways, strategies and prerequisites for achieving a rights-based reform of extant and emerging carbon markets, under the consensus-based international climate regime. As will be discussed in this chapter, to achieve the adoption and recognition of a human rights-based framework on carbon finance, there is a need to adopt a step-by-step approach. This approach would include three complementary steps: the reform or expansion of the scope of the international climate change regime to reflect human rights standards; the adoption of rights-based modalities and rules for project approval; and the implementation of a carbon finance certification scheme that will ensure that only projects that have met prescribed human rights thresholds will be financed. These practical steps will provide a legal basis and frameworks for mainstreaming human rights language and frameworks into the design, approval, finance and implementation of carbon projects and for screening out carbon projects that may violate human rights.

Theorizing practical frameworks is only one essential aspect of the task of promoting a human rights-based approach to carbon finance. It is also important to review and examine practical barriers and limitations that such a proposal may face along the way. Chapter 9 provides real-life checks on practical challenges that may limit the success of efforts aimed at invoking a rights-based reform of the international climate change regime. It will discuss how the lack of political will, political alignments and negotiation blocks, and regime uncertainties may act as practical barriers to a rights-based reform of the international climate change regime applicable to carbon projects.

Chapter 10 provides an overview and summary of key points for policymakers. It offers recommendations and proposals that, in the view of the author, could assist policy makers to navigate practical and logistical barriers that a push for a rights-based carbon market may face.

1.2.4 Conclusion

The increased currency and adoption of human rights language in climate change instruments captures the growing consensus by human rights groups, advocates and pundits, that without a human rights perspective, climate projects and actions risk exacerbating the vulnerabilities of communities and countries where such projects are located. It also reflects, very vividly, the evident need for a functional rights-based risk management framework, through which private actors and business enterprises can be mandated to protect, respect and fulfil human rights in the execution of climate actions and projects. Even though human risks in the extant implementation and delivery of carbon projects have increasingly been rehashed in various international fora, legal and institutional frameworks for addressing these risks have not been exhaustively explored and communicated. To design effective legal and institutional response frameworks, our understanding of the intricacies of how carbon projects affect human rights must, as a starting point, be robust and detailed. The next chapter therefore provides detailed examination of the linkages between carbon projects and human rights and offers detailed analysis of how the design, approval, finance and implementation of emission reduction schemes have fostered human rights violations and concerns.

2

Carbon projects and human rights struggles

2.1 Introduction

The focus of this chapter is to examine the linkages between carbon projects and human rights. Using CDM and REDD+ projects as main reference points, this chapter examines how the lack of adequate human rights protection and enforcement in design, approval, finance and implementation of carbon projects has resulted in gross human rights violations and impacts in a number of registered projects.

This chapter discusses loopholes in the approval processes of CDM projects that have fostered human rights violations in project implementation and delivery, especially how the lack of transparency in the processes for verification and registration of CDM projects; non-disclosure of relevant project information by project proponents; limited opportunities for stakeholder participation; absence of public complaint procedures under the CDM governance framework; and, most importantly, the absence of human rights safeguards in the United Nations Framework Convention on Climate Change (UNFCCC) and its Kyoto Protocol have entrenched a culture of 'approvals' under which projects that negatively impact the rights of affected people and communities have been registered and approved for carbon credits.

To provide some background and context, this chapter begins with a brief description of the processes and procedures involved in designing and developing a carbon project, using the approval processes and procedures under the CDM as a primary reference point.

2.1.1 Design and development of carbon projects: example of the CDM project cycle

Designing and developing a carbon project is a complex process that: involves a diverse set of activities with a wide range of actors; takes a long time; requires in-depth technical expertise; and involves considerable financial resources. There are various technical activities and steps required to complete the design of a carbon project. This includes project scoping; preparing and submitting project design documents for verification; handling project negotiations and agreements; designing efficient project monitoring protocols; setting up carbon monitoring systems and preparing for validation; and verifying, marketing and finalizing emission reduction and carbon sales agreements (ERPA). Using the CDM project cycle as an example, this section reviews key steps involved in implementing a carbon project from conception to registration.

As discussed earlier, the CDM was designed essentially to serve two key purposes: to assist developing countries in achieving sustainable development; and to assist industrialized countries to achieve compliance with their quantified emission limitation and reduction commitments under the Kyoto Protocol.[1] Procedurally, a developed country identifies an investment that would lead to a reduction in the emission of greenhouse gases (GHGs) in a developing country and approaches the government of the country where the investment is to be located for its approval. The host government then decides whether such a project meets its sustainable development criteria and then approves it as a CDM project.

This is best illustrated with a practical example. Let us consider two countries: United Kingdom and Nigeria. Under the Kyoto Protocol, the United Kingdom agreed to ensure its emissions of GHGs were at least 12.5 per cent lower than base year levels, on average, over the 2008–2012 period (1st commitment period). This equates to annual emissions of 682.4 million tonnes of carbon dioxide equivalent ($MtCO_2e$) on average over the period.[2] For the second commitment period, parties have so far pledged to reduce GHG emissions by at least 18 per cent below 1990 levels in the eight-year period from 2013 to 2020.[3] If the United Kingdom

[1] See Art 12 of the Conference of the Parties to the Framework Convention on Climate Change, *Kyoto Protocol*, 10 December 1997, UN Doc FCCC/CP/1997/L.7/add.1, 37 ILM 22 (1998) [*Kyoto Protocol*].

[2] See UNFCCC, 'The United Kingdom's Kyoto Commitments', www.defra.gov.uk/environment/climate/, accessed 10 January 2015.

[3] For the second period the European Union (EU) has taken on emission reduction commitments in line with its domestic target of cutting emissions by 20 per cent of 1990 levels by 2020. The EU also registered a conditional offer to move to a 30 per cent reduction by 2020 compared to 1990 levels, provided that other developed countries commit themselves to comparable emission reductions and developing countries contribute adequately according to their responsibilities and respective capabilities. The EU and its member states, which includes the United Kingdom, will fulfil the reduction commitment

is to achieve its targets through domestic action alone, it would require legislation and policies that might compel local businesses and industries to embark on a drastic cut down on emissions of GHGs. This might put a lot of pressure on industries that may have to cut down their levels of production. This might ultimately strain the economy of the United Kingdom. Nigeria, on the other hand, is a developing country with no specific emission reduction commitments under the Kyoto Protocol. Nigeria's major priorities include boosting its economy, achieving sustainable development and proffering solutions to its numerous environmental problems such as gas flaring, oil spillage and air pollution, to mention just a few.[4] Thus, for the United Kingdom to meet its emission reduction target under Kyoto Protocol, it could invest in projects that reduce gas flaring in Nigeria or that lead to solutions to any of its environmental problems. GHG reduction achieved by such a project will then be credited to the United Kingdom, while Nigeria could benefit from cleaner air, better environment, job opportunities and more income through such projects. The CDM thus allows both parties involved to benefit from the mechanism: while it promotes sustainable development in the host country, it allows the investing country to attain their emission reduction targets in a flexible manner.[5]

There are generally six key stages in the typical life cycle of a carbon project. They are: project identification and scoping; host government approval; project design; project verification and validation; project

jointly. The targets of all countries participating in the second commitment period will be revisited by 2014 with a view to considering raising ambition. According to the UK Energy and Climate Change Secretary, the United Kingdom, as part of the EU, will be working to meet these commitments. See COP18: Statement from UK Energy and Climate Change Secretary (9 December 2012), www.gov.uk/government/speeches/cop18-statement-from-uk-energy-and-climate-change-secretary, accessed 19 January 2015; See also, Doha Amendment to the Kyoto Protocol to the United Nations Framework Convention on Climate Change, C.N.718.2012.TREATIES-XXVII.7.c (Adopted 8 December 2012, not yet in force). Pursuant to Article 20, paragraph 4, and Article 21, paragraph 7 of the Kyoto Protocol, the Amendment shall enter into force for those parties having accepted it, on the ninetieth day after the date of receipt by the Depositary of an instrument of acceptance by at least three fourths of the parties to the Kyoto Protocol.

[4] See the Nigerian National Planning Commission, *Nigeria's NEEDS*, siteresources.worldbank.org/INTPRS1/Resources/Nigeria_PRSP(Dec2005).pdf, accessed 12 June 2015.

[5] See I. Rowlands, 'The Kyoto Protocol's Clean Development Mechanism: A Sustainability Assessment', *Third World Quarterly*, 22 (2001), 795; see also D. Olawuyi, 'Beautifying Africa for the Clean Development Mechanism: Legal and Institutional Issues Considered', in B. Richardson, Y. Bouthillier, H. Mcleod-Kilmurray and S. Wood (eds.), *Climate Law and Developing Countries: Legal and Policy Challenges for the World Economy* (Cheltenham, UK: Edward Elgar Publishing, 2009).

registration; project implementation, monitoring, verification; and credit issuance.

2.1.1.1 Project identification and scoping

A project proponent must, as a first step, determine the type of carbon project to be developed. There are several types of carbon projects ranging from wind, hydro, biomass energy, methane, solar, landfill gas, fossil fuel switch, afforestation, reforestation, geothermal, transport, improved farming techniques (soil carbon sequestration), renewable energy and avoided deforestation projects to mention a few.[6] For example, projects that have been developed and registered under the CDM include industrial plants that destroy hydrofluorocarbons and nitric acids in facilities; renewable electricity projects such as hydro, wind, solar and biomass power; projects that destroy the GHG methane, for example, from landfills or in agriculture; energy efficiency (EE) projects that, for example, make coal power plants more efficient or provide compact fluorescent light bulbs for buildings; projects that support switching from a fuel with high GHG emissions (e.g. coal) to one with fewer emissions, and storing CO_2 in trees and soils through afforestation and reforestation.[7]

One part of project scoping is determining the location of the project. Which country will be selected as the host country for the project? Project proponents tend to select countries and locations that will generate the greatest returns for projects in terms of emission reduction potentials, investment climate and adequate legal and institutional framework for the investment.[8] Furthermore, the project area and boundary (geographically) has to be established. For REDD+ projects, the project area needs to be large enough to generate significant emission reductions.[9] The project identification stage involves an initial selection of a country and region as the site for the project. It also includes assessing host country risks (legal and political risks); evaluating counterparty risks (experience, capacity, reputation, financial standing, contract management and

[6] See United Nations Center on Energy, Climate and Sustainable Development, 'CDM Projects by Type', www.cdmpipeline.org/cdm-projects-type.htm#3, accessed 23 May 2015.

[7] See United Nations Center on Energy, Climate and Sustainable Development, CDM/JI Pipeline Analysis and Database, www.cdmpipeline.org/, accessed 23 May 2015.

[8] See Olawuyi, 'Beautifying Africa for the Clean Development Mechanism', (n 5).

[9] Generally, the minimum project size area is around 30–40,000 ha for REDD+ projects and 10,000 ha for an afforestation (A/R) project. Small-scale A/R CDM projects must result in GHG removals of less than 16,000 tCO_2 per year.

ownership structure); testing project risks (feasibility, implementation, permits and legal structure); estimating carbon credit yields (registration, additionality, methodology, carbon credit estimates and carbon credit ownership); and assessing carbon credit generation and delivery risks.[10]

Determining a project start date is also important, to create a clear project plan. Setting a target start date enables a project proponent to plan for the project in more concrete and definitive terms. Furthermore, at this stage it is important for a project proponent to develop a business and economic plan, which identifies the cost of the project and potential sources of funding. At the project identification stage, a project proponent must consider and analyse why it would be attractive to engage in undertaking a carbon project, what are the potential volumes of emission credits to be generated and what are the potential practical problems. Furthermore, a project proponent must identify and select project collaborating partners and funding agencies; develop a collaborative work plan with such partners; estimate financial projections; and develop emission credit/benefit-sharing mechanisms with the participating parties.

It is good practice for a project proponent to, at this stage, enter into carbon agreements such as emission allowances trade agreements, emissions reduction purchase agreement, offset agreements, licensing and intellectual property agreements that document the nature and scope of the project and the allocation of resulting carbon credits among participating entities. In addition to carbon agreements, other agreements may also be important to secure land tenure recognition and utilization under local laws.

A common practice at the scoping stage is for project proponents to put together a project identification note (PIN) or project checklist to summarize key information on the potential location, date, budget and potential funding sources of a carbon project.[11] A project developer

[10] See Norton Rose Fulbright, 'Regulations and Investigations: Dealing with Legal and Regulatory Risk', www.nortonrosefulbright.com/files/regulation-and-investigations-112016.pdf, accessed 12 December 2015.

[11] A PIN provides indicative information on the type and size of the project; its location; the anticipated total amount of Greenhouse Gas (GHG) reduction compared to the 'business-as-usual' scenario (which will be elaborated in the baseline later on at Project Design Document [PDD] level); suggested crediting life time; suggested Certified Emission Reductions (CER), Emission Reduction Unit (ERU), Removal Unit (RMU) or Verified Emission Reduction (VER) price in US$/ton CO_2eq reduced from a project financial structuring (indicating which parties are expected to provide the project's financing); and the project's other socio-economic or environmental effects/benefits. See BioCarbon Fund, *Project Idea Note (PIN) for*

may also be selected to assist with the formulation of the project. Project developers are typically entities that have significant experience and expertise in project identification, scoping and funding. They offer services that help a project proponent to design a carbon project and move it from conception to registration.[12]

2.1.1.2 Host government approval

After selecting a potential location for the carbon project, a project proponent must move to the formal process of obtaining formal written approval from the host country for the proposed project.[13] For example, under CDM Rules, the Designated National Authority (DNA) of the host country must endorse the project as meeting its national sustainable development goals and must affirm that the project will assist the host country to achieve sustainable development. A written approval constitutes authorization and confirmation by the DNA that the country is a willing participant in a project activity. The DNA issues a statement confirming that the host country has ratified the Kyoto Protocol; that its participation in the proposed project activity is voluntary; and that the proposed project activity contributes to sustainable development of the

Land Use, Land-Use Change and Forestry (LULUCF) Project https://wbcarbonfinance.org/docs/New_PIN_Template_for_LULUCF_Projects_10-2007.doc, accessed 12 December 2015.

[12] Examples include firms such as Ecosecurities, Ecopositive, EC Carbon, Terra Global Capital, Carbon Neutral Company, and the World Bank Carbon Finance Unit.

[13] The CDM Rules require every developing country to establish a Designated National Authority (DNA), which will serve as its CDM monitoring body responsible for approving or rejecting a prospective CDM investment. The DNA serves as the focal point between investors and the host country government. Its functions include providing potential projects for investors; processing framework agreements with investors; ensuring that an environmental impact assessment is carried out before approving projects and considering the assessment reports carefully; providing legal advice for project investors; coordinating with other relevant official entities and authorities within the host country; drawing up standardized baselines; monitoring ongoing CDM projects; granting export of emissions rights (Certified Emission Reductions: CERs); conducting public relations and providing information on CDM implementation in the host country through advertisements in the media and through the web. It must also design and establish an evaluation procedure that adopts international eligibility criteria to assess the contribution of the prospective CDM projects to sustainable development in the host country. See P. Manso, *Establishing a National Authority for the Clean Development Mechanism (CDM): The Costa Rican Experience* (Paris: Organization for Economic Cooperation and Development, 2003), 5–6.

host country. A written approval from a party may cover more than one project provided that all projects are clearly listed in the letter.

CDM rules allow every developing country to develop its sustainable development priorities and criteria for granting 'host country approval'. As such, the responsibility for determining whether a CDM project contributes to national sustainable development, as defined by the host country, rests with its DNA. Once the DNA endorses a project as meeting the national sustainable development criteria, it grants its host country approval.[14]

2.1.1.3 Project design

The next stage is for the project proponent (i.e. the investing party) to develop a project execution template or outline that details key features of the project, including the project's GHG emission reduction or removal activity(ies), adopted methodology, resources and actions required to implement project activities. Under the CDM, this information is outlined in a Project Design Document (PDD). The PDD contains general descriptions and details of the project activity, duration of the project activity, crediting period, estimation of potential GHG emissions by sources, proposed monitoring methodology and baseline, project participants, environmental impacts, stakeholder comments and the method by which the participants will communicate with the CDM Executive Board (CDM EB).[15] In preparing the PDD, the project design team must collect and document a range of information on the socio-economic, biodiversity and carbon aspects of the project.

The PDD must demonstrate that the project utilizes approved methodologies to calculate emission reductions and other socio-economic baselines that can be used to monitor changes resulting from the project. A methodology is an application of an approach to an individual project activity, reflecting aspects such as sector and region. Generally, no methodology is excluded a priori so that project participants have the opportunity to propose any methodology. The PDD must state clearly the methodological approach taken by the project and how that

[14] See paragraphs 28–30, Annex to Decision of the COP-7 (Clean Development Mechanism Modalities and Procedures).

[15] See CDM – Executive Board, *Guidelines for Completing the Project Design Document (CDM-PDD), The Proposed New Methodology: Baseline (CDM-NMB) and The Proposed New Methodology: Monitoring (CDM-NMM)*, cdm.unfccc.int/Reference/Docu ments/Guidel_Pdd/English/Guidelines_CDMPDD_NMB_NMM.pdf, accessed 12 December 2014.

methodology is the most appropriate approach for the project activity. The PDD is then sent to a Designated Operational Entity (DOE) for verification.[16]

2.1.1.4 Project verification and validation

The next stage is the project verification and validation stage. The focus at this stage is to review, verify and validate the project design against applicable rules and requirements. Supervisory bodies, or their accredited verifiers, will at this stage review the PDD, the planned approaches and methodologies to determine whether they comply with relevant modalities and procedures under the international climate change regime. For example, under the CDM, project verification and validation is undertaken by DOEs duly accredited by the CDM EB to validate the PDD or verify whether projects have been designed in accordance with applicable rules and procedures.[17] DOEs are responsible for verifying the accuracy of the PDD; ensuring that proposed CDM project activities meet all requirements established by the Conference of the Parties (COP)/Meeting of the Parties (MOP) and the CDM EB; and submitting the project proposals to the CDM EB for registration.[18]

The validation process includes a global stakeholder consultation process during which the DOE collects public comments on the project via the CDM website. As part of the validation, the DOE reviews project documentation, conducts site visits and produces draft and final validation reports. A DOE can also request clarifications (CLRs) and corrective actions (CARs) to the project documentation in order to collect sufficient information to assess the project's compliance with the CDM requirements and the applied methodology.

After the completion of validation procedures, the DOE presents a final validation report, which is then submitted to the UNFCCC. If the DOE concludes that the project design is compliant with all CDM

[16] For an excellent and detailed description of the CDM Process, see United Nations Development Programme, *Going through the CDM Process*, content.undp.org/go/cms-service/stream/asset/?asset _id=2222587, accessed 12 January 2015; see also UNEP, *Legal Issues Guide Book to the CDM*, cd4cdm.org/Publications/ImplementingCDM_GuidebookHost CountryLegalIssues.pdf, accessed 12 December 2014.

[17] See UNFCCC, 'Designated Operational Entities (DOE)', cdm.unfccc.int/DOE /index.html, accessed 12 June 2014. See Para 20 of the CDM Modalities & Procedures, which provides that the Executive Board shall maintain a publicly available list of designated operational entities.

[18] See Para. E, CDM Modalities & Procedures, 3/CMP.1 (2005), cdm.unfccc.int /Reference/ COPMOP/08a01.pdf, accessed 12 July 2014.

requirements, the project is submitted to the CDM EB for registration as a CDM project activity.

2.1.1.5 Project registration

For emission reductions achieved under a carbon project to be certified or credited for the purpose of meeting emission reduction obligations or targets, it must be registered with the international climate change regime. Registration of a project can be requested when the project has completed project validation. For example, under the CDM, at the registration stage, successfully validated projects are submitted to UNFCCC to request registration. The CDM EB was established by the COP to serve as the supervisory authority that approves and registers CDM projects.[19] Once the CDM EB receives the final validation report, it undertakes a review process to determine whether to register the project or not.

CDM Rules contain project eligibility requirements that must be met by all projects before registration.[20] The rules stipulate that any project proposed as a CDM project must promote sustainable development in the country where it is to be located. This requirement is simply met by showing that it has been endorsed by the developing country as meeting its sustainable development criteria. Other requirements are that the project must be real, measurable, long term and additional to reductions that would have occurred without the project.[21] Similarly, funding for

[19] The main tasks of the CDM EB include making detailed rules for modalities and procedures for the CDM according to the decision and recommendation of the COP, suggesting simplified rules for small-scale CDM projects, developing and approving new methodologies for CDM, making recommendations to the COP for the designation of operational entities, developing procedures for the CDM, accrediting DOEs, registering CDM projects, issuing CERs, making publicly available information on proposed CDM project activities in need of funding and investors seeking opportunities, and maintaining a public database of CDM project activities containing information on registered PDDs. See United Nations Environmental Programme, *CDM Information and Guide Book* (3rd Edition, UNEP 2004), 40–43.

[20] Decision 17/COP.7 of the Seventh Session of the Conference of Parties to the UNFCC (COP-7) in Marrakech.

[21] The requirements also include that the project has to generate CERs compared to the baseline emission; the methodology applied by the project has to be approved; and if the project has a great environmental impact, a solution has to be developed to minimize that impact. The project baseline should also be established on a project-specific basis, in a transparent and conservative way, taking into account the relevant national or sectoral policies; and a reasonable system boundary has to be set and the leakage issue needs to be fully considered. These requirements have been criticized as very ambiguous, most especially the additionality requirement. See A. Michaelowa, 'Determination of Baselines and Additionality for the CDM: A Crucial Element of Credibility of the

CDM projects must not divert funding from existing official development assistance (ODA).[22] The CDM EB registers a project once it meets these requirements. According to the modalities and procedures for CDM projects, a successfully completed registration process should take a maximum of eight weeks. A validated project can be delayed at registration by the CDM EB if a party involved in the project or at least three members of the CDM EB request a technical review of the project to address concerns. Upon registration, emission reductions generated from the carbon project will be issued and credited to the project proponent.

2.1.1.6 Project implementation, monitoring, verification and credit issuance

The project implementation stage is when the carbon project is actually executed. It includes executing the project and generating emission reductions in accordance with approved approaches and methodologies. Generally, a project proponent will be credited for emission reductions achieved by the project after such reductions have been verified and certified as meeting approved guidelines and methodologies.

Under the CDM, the crediting period of a project starts with the project implementation. The crediting period may only start after the date of registration of the proposed activity and it may not extend beyond the operational lifetime of the project activity. Projects are often monitored by DOEs after registration to ensure that the project is operated in a manner that reduces, abates or sequesters GHGs. The monitoring plan, often included in the PDD, is the basis for implementing the monitoring procedures. It lists the variables that need to be monitored and measured as per applied methodology at specified intervals during project implementation. It also specifies the procedures that developers must undertake to assure the quality of both measurements and data storage. The project developer compiles the monitoring results in a

Climate Regime', in F. Yamin (ed.), *Climate Change and Carbon Markets* (New York: Routledge, 2005), chp. 7.

22 ODA refers to a concessional financial assistance from official government agencies to developing countries and multilateral institutions, which aims primarily to promote the economic development and welfare of developing countries. Developed countries are not permitted under CDM Rules to use the CDM as a substitute for existing forms of ODA. This requirement seeks to underscore the differences between the CDM and other forms of ODAs, www.oecd.org/department/0,2688,en_2649_33721_1_1_1_1_1,00.html, accessed 21 January 2015.

monitoring report, which is published and subject to verification and periodic independent review by the DOE. The DOE is responsible for undertaking a periodic review and verification of the achieved emission reductions of the registered project since its start. As per the CDM modalities and procedures, only one verification is expected to take place per commitment period of the Kyoto Protocol. Project developers can decide on the date of the first verification; subsequent verifications have to be carried out at five-year intervals. The DOE assesses the monitoring report and checks compliance with the registered project design, monitoring plan and the applied methodology. It then certifies to the CDM EB that the project's emission reductions have been verified and submits a request to the CDM EB to issue certified emission reductions (CERs) for the volume of GHG abatement that occurred during the verification period. The culmination of the process is that the CDM EB issues CERs as evidence of the emission reductions achieved by the project.

2.1.2 Evaluation of the project approval process

The foregoing procedures arguably contain several loopholes that have brought about complex human rights risks and concerns. Notably, CDM Rules do not include the requirement that projects must not infringe on the human rights of the host communities, neither do the rules establish a detailed human rights threshold to be met by a project before registration and approval. The rules also fail to establish procedures that allow individuals and members of the community to play active roles in blocking or reporting projects that infringe on human rights after registration.

Furthermore, the CDM EB confers excessive discretion on host countries to approve projects that meet their respective sustainable development criteria. The CDM as a market mechanism involves many developing countries competing to host projects. Its market outlook is underscored by the fact that developed countries have wide discretion in deciding on which developing country to invest in or in which country to pursue emission reduction investments, based on the level of derivable profit.[23] Thus developed countries, like prudent investors, are often in

[23] Profits are measured in terms of which developing country generates more CERs at the least cost. See CDM Modalities & Procedures, 3/CMP.1 (2005), cdm.unfccc.int /Reference/COPMOP/08a01.pdf, accessed 12 July 2014; for a discussion of these see Olawuyi, 'Beautifying Africa for the Clean Development Mechanism', (n 5), 262.

search of investment locations that guarantee maximum profits and offer a safe investment climate. Such countries are perceived as attractive investment locations. In a situation where countries are under pressure to attract foreign CDM investments or to deliver cheap projects, states often tend to lower sustainability standards in order to attract projects, approve carbon-intensive technologies and often force through project approvals irrespective of legitimate opposition from stakeholders and citizens.[24] These have resulted in a long list of human rights concerns as demonstrated by some approved CDM projects.

Similarly, the CDM EB has over the years outsourced some of its most important oversight roles to DOEs, which are mostly for-profit audit companies, to determine and verify the accuracy of the PDD and to ensure that proposed CDM project activities meet all requirements established in the CDM Rules. While this might appear a decent arrangement on its face, practical implementations of CDM projects have been linked with allegations of bribery, corruption and lack of transparency by DOEs.[25] In a number of cases, projects that have been resisted for violating human rights have been allegedly approved by DOEs.[26] This arrangement also raises questions as to whether auditors and verifiers

[24] See M. Jung, 'Host Country Attractiveness for CDM Non-sink Projects', *Energy Policy* (2006), 2174; See A. Silayan, 'Equitable Distribution of CDM Projects Among Developing Countries', *HWWA-Institut für Wirtschaftsforschung-Hamburg (HWWA) Report*, 255 (2006), 1.

[25] See T. Dyck, 'Enforcing Environmental Integrity: Emissions Auditing and the Extended Arm of the Clean Development Mechanism',(n 5), *Columbia Journal of Environmental Law*, 36(2) (2011), 259–358; also J. Green, 'Delegation and Accountability in the Clean Development Mechanism: The New Authority of Non-State Actors', *Journal of International Law and International Relations*, 4(2) (2008), 21.

[26] In 2009, SGS United Kingdom, one of the world's largest DOEs, was suspended from the CDM scheme amidst allegations that it had failed to properly vet projects and that it received bribes to verify a CDM project. See J. Murray, 'Carbon Market under Fire as UN suspends British CDM Project Accreditor' (2009), www.businessgreen.com/bg/news/1806434/updatedFcarbonFmarketFun FsuspendsFbritishFcdmFprojectFaccreditor. In 2008, DNV, a Norwegian DOE, was also suspended after it was found to have authorized five projects for entry into the CDM, despite not having surveyed them. As the World Wildlife Fund (WWF) notes, 'verifiers are paid by the project operators, so there is a vested interest to approve as many projects as possible. We would like to see the scheme changed so that the CDM Executive Board randomly appoints the verifiers to the projects.' See 'A Rating of Designated Operational Entities (DOEs) Accredited under the Clean Development Mechanism (CDM): Scope, Methodology and Results' (Freiburg: A Report by the Oko-Institute for the WWF 2009). See also Greenpeace International, 'Carbon Scam, Noel Kempff Climate Action Project and the Push for Sub-national Forest Offsets' (2009), www.greenpeace .org/france/PageFiles/266591/carbonscam.pdf, accessed 12 January 2015.

would have the capacity and training to identify and assess human rights risks in a proposed project before granting it an 'all clear'.[27]

These and many other loopholes in the CDM Rules have entrenched a culture of approvals in which questionable projects have been approved without a thorough examination of their human rights impacts on host communities.

2.2 How carbon projects violate human rights: lessons from approved CDM projects

It is impossible to discuss all carbon projects that have attracted human rights litigation and protests. Therefore, this book draws examples from approved CDM projects across key sectors to examine how human rights concerns may derail the sustainable development promises of carbon projects. This book draws examples from a gas flaring reduction project in Nigeria, a biogas power generation project in Honduras, a hydroelectric power project in Panama and REDD+ projects.[28] It also makes references to some non-CDM projects to show that the tendency to achieve emission reduction credits at all cost, with less attention to the human rights impacts of projects, is arguably the foundation for human rights violations in emission reduction schemes. These examples emphasize the importance of mainstreaming human rights safeguards into international legal frameworks and governance structures on climate change to make them more accountable, transparent and rights-based so as to better protect human rights of the public in climate change actions and projects.

[27] For example, SGS United Kingdom is run by accountants, economists, environmental scientists and business administrators. This raises questions on the capacity of the firm to identify human rights concerns and implement human rights standards. See Board and Executive Management, www.sgs.com/en/Our-Company/About-SGS/Board-and-Executive-Management/Operations-Council.aspx, accessed 12 January 2015.

[28] The choice of case studies for this book is influenced by an attempt to provide examples that cut across different types of GHG emission reduction projects in different developing countries. Each of these case studies illustrates the nature of projects that constitute emission reduction projects. They also highlight how the absence of a solid human rights culture in these jurisdictions and of a protective framework in the international climate change regime has reduced the integrity of emission reduction projects and the efficacy of the international climate change regime as a whole. For other examples and studies on the impacts of climate change projects on human rights see the Xioaxi Hydropower Project and the Guangxi Xiafu Hydropower Project in China. See H. Jiang, 'Carbon-Trading Scheme Undermined by Chinese Projects', *Epoch Times* (Washington, 9 December 2009).

2.2.1 The Aguan Biogas project in Honduras

The Aguan biogas project is, perhaps, one of the most controversial CDM projects ever registered. This project, sponsored by the UK government, is located in the Bajo Aguan region in Honduras. It was designed to reduce GHG emissions by collecting biogas from methane emissions, and replacing fossil fuels utilized for heat generation in the mill of a palm oil plantation of Grupo Dinant's subsidiary Exportadora del Atlantico. Its emission reduction projections are significant; estimates suggest that the project would reduce about 23,000 tonnes of CO_2 annually, generating about US$2.8 million between February 2010 and January 2017.[29] However, this project has been heavily criticized internationally for its gross human rights violations.[30] According to a petition submitted to the Inter-American Commission on Human Rights, the local project developer, Grupo Dinant, is alleged to have been at the centre of violent conflicts with local people, mass displacements of people from their ancestral lands, violent repressions of protesters and the killing of about twenty-three local farmers and human rights defenders.[31] A coalition of over seventy international human rights groups called on the United Kingdom to withdraw sponsorship for the project and for the CDM EB not to approve or register the project.[32] The International

[29] CDM Watch, 'United Nations Under Pressure to Denounce Human Rights Abuses in Carbon Offsetting Scheme' (2011), www.cdm-watch.org/wordpress/wpcontent/uploads/2011/04/180411_PR_UN_under_Pressure_to_denounce_human_rights_abuses_in_carbon_offsetting_scheme.pdf, accessed 12 December 2014.

[30] See BIOMASS Hub, 'Human Rights Violations Linked to CDM Biogas Project in Honduras' (July 2011), biomasshub.com/human-rights-violations-linked-cdm-biogas-honduras/, accessed 12 July 2014.

[31] For comprehensive details of human rights violations by this project, see Inter-American Commission on Human Rights (IACHR), 'Human Rights Situation in the Bajo Aguán Honduras' (24 October 2011); IACHR, 2011 Annual Report, ch. IV, 297–302; IACHR, 2010, Country Report, Honduras, 118–112; see also CDM Watch, *Petition to the CDM Executive Board on Aguan Gas project* (February 2011), www.cdm-watch.org/wordpress/wp-content/uploads/2011/02/unsolicited_letter_cdmproject_application_3197_honduras.pdf., accessed 02 August 2015; See also FIAN, Human Rights Violation in Bajo Arguan (July 2011), www.fian.org/resources/documents/others/honduras-human-rights-violations-in-bajo-aguan/pdf, accessed 02 August 2014.

[32] CDM Watch, 'Open Letter: UK Government must withdraw authorization for Aguan and Lean CDM projects linked to assassinations and other human rights abuses in Honduras' (2011), www.cdm-watch.org/?p=1648, accessed 02 August 2015. In response to protests by international human rights groups German public development bank DEG (Deutsche Entwicklungsgesellschaft) declared that it will not pay out an already approved loan of US $20 million for the project. Similarly, EDF Trading, a wholly owned subsidiary of Electricité de France SA's and one of the biggest CDM investors, pulled out from a

Finance Corporation of the World Bank also launched a full-scale investigation into the activities of the project developer.[33]

Despite petitions and protests, this project was approved by the government of Honduras and subsequently registered by the CDM EB on 18 July 2011.[34] The CDM Board's official response was that it has no mandate to investigate human rights abuses and that any matter related to the sustainable development of the project or human rights is determined by the government that hosts the project.[35]

The approval of this project by the CDM EB in the face of human rights risks such as repressive land grabs without compensation, loss of lives and the environmental pollution brought associated with the project (to mention but a few) raises fundamental questions on the need for a governance regime through which supervisory authorities for carbon projects at international and national levels are empowered to consider the human rights consequences of a project while granting or withholding project approval and to screen out projects which violate human rights. Approving such projects diminishes the gains that have been made under international human rights law; it also erodes the value and integrity of carbon projects.

2.2.2 The Kwale-Okpai CDM project in Nigeria

Nigeria's Kwale-Okpai project provides another illustration of how carbon projects could result in human rights violations.[36] The Kwale CDM

contract to buy carbon credits from the project. See CDM Watch, 'German Bank Won't Lend to Honduran CO_2 Project' (April 2011), www.bloomberg.com/news/2011-04-18/german-bank-won-t-lend-to-honduran-co2-project-cdm-watch-says.html, accessed 1 August 2015. For the official response of the UK Government, see Letter by Rt. Hon Chris Huhne M.P, (April 2011), www.cdm-watch.org/wordpress/wp-content/uploads/2011/02/UK_Gov_reponse_on_aguan_130411.pdf, accessed 31 July 2015.

[33] Office of the Compliance Advisor/Ombudsman, 'Audit of IFC's Environmental and Social Performance in relation to Dinant, Honduras' (26 February 2013), www.caoombudsman.org/documents/CAOcommuniqueDinant_February2013.pdf, accessed 20 March 2015.

[34] See UNFCCC, Lists of Registered CDM projects (2011), cdm.unfccc.int/Projects/DB/TUEV-SUED1260202521.42/view, accessed 02 August 2014.

[35] See 'The Response of the CDM Executive Board to the Petition', www.cdm-watch.org/wordpress/wp-content/uploads/2011/02/EB59-14_CDM-Watch_Concerns-on-CDM-project-in-Honduras-Ref.-3197_Response.pdf, accessed 02 August 2015.

[36] Nigeria is a party to the Kyoto Protocol having ratified the instrument on 10 December 2004. Nigeria is thus eligible to host and execute projects under the Clean Development Mechanism. A number of CDM projects are currently registered and ongoing in Nigeria. These include the Kwale Project, Ovade Ogharefe project, the Lafarge Cement Project;

project in Ndokwa land, Delta State, was jointly developed by Royal Dutch Shell, the Netherlands; National Agip Oil Company, Nigeria; Phillip Oil, Nigeria; and the Nigerian National Petroleum Corporation as a CDM project aimed at reducing GHG emissions caused by gas flaring.[37] It was granted approval by the Nigerian government in 2006.

The main objective of the project is to address perennial problems of gas flaring in Nigeria through the capture and recovery of associated gas that would otherwise flare up at the Kwale Oil-Gas Processing Plant (OGPP).[38] It involves the capture and transportation of a substantial volume of the flared associated gas via a pipeline to the Okpai combined cycle gas turbine power plant, to generate electricity.[39] The recovery and utilization of flared gas through this project started in 2006. As of 2010, the total emissions reduction generated by the project was estimated to be above 791,325 tonnes of CO_2 eq.[40]

The Kwale project promises significant emission reductions and, most importantly, good prospects of resolving the gas-flaring conundrum that

and recently the Asuokpu/Umutu Gas Recovery and Marketing Facility. Reducing Emissions from Deforestation and Degradation (REDD) projects are also currently being executed in Nigeria by Shell. Some of these projects have been subjects of protests and condemnation by stakeholders and host communities. See F. Allen et al, ' Oil Flares, Broken Laws, Pollution and Oppression',(n 36), in *Why the Carbon Trading Gamble and the Clean Development Mechanism Won't Save the Planet from Climate Change and How African Civil Society is Resisting* (A Report by the University of KwaZulu Ntual Center for Civil Society and Dartmouth College Climate Justice Research Project, April 2012), chp.4. See also N. Bassey, 'Foreword', in *Mired in a Fossil Trap, the Nigerian CDM Report* (Environmental Rights Action/Friends of the Earth, May 2011); Carbon Trade Watch, 'Groups slam Nigeria's Submission of Gas Flare Reductions for Carbon Credits' (Amsterdam, 29 March 2006).

37 One of the main environmental issues in Nigeria is the problem of gas flaring. Nigeria and Russia are reputed to be the two largest gas-flaring countries on earth. Estimates suggest that of the 3.5 billion cubic feet of associated gas produced annually in Nigeria, 2.5 billion cubic feet, or about 70 per cent is wasted via flaring. This equals about 25 per cent of the UK's total natural gas consumption, and is the equivalent to 40 per cent of the entire African continent's gas consumption. See 'Flames of Hell: Gas Flaring in the Niger Delta', www.saction.org/home/saction_image/flames_of_hell.pdf, accessed 12 February, 2015; See also ERA, Facts Sheet: *Harmful Gas Flaring in Nigeria*; Vanguard, 'Nigeria loses $150 billion to gas flare in 36 yrs', *Vanguard*, (Lagos, 12 July 2008); See also Environmental Rights Action/Friends of the Earth Nigeria, 'Gas flaring in Nigeria: A Human Rights, Environmental and Economic Monstrosity' (June 2005), www.climatelaw.org/cases/case-documents/nigeria/gas-flaring-in-nigeria.pdf, accessed 12 February 2015.

38 See UNFCCC, 'Kwale CDM Project Monitoring Report: Recovery of Associated Gas that would otherwise be flared at Kwale Oil-Gas Processing Plant, Nigeria' (2007) UNFCCC/ Document ID: CDM0553- MR01.

39 *Ibid.* 2. 40 *Ibid.* 3–4.

Nigeria has struggled unsuccessfully to combat for over a decade.[41] Despite its 'clean' potential and promise, it was the lack of transparency in the approval process, inadequate opportunities for stakeholder consultation and participation and the lack of a transparent environmental impact assessment (EIA) on the short- and long-term human rights impacts of the project that gained the most attention and publicity.[42] These human rights concerns about the project have resulted in widespread street protests, blockades, lawsuits, project obstruction and formal petitions by local communities and civil groups.[43]

First is the absence of a transparent environmental impact assessment on the short- and long-term impact of the project. Environmental groups in Nigeria have raised concerns about the prospects of a leakage problem in the long term. Considering the weak and aged pipeline transportation infrastructure available in Nigeria, there are concerns about the high possibility of leakage due to low storage capacity formations or low-density seals that could ultimately result in the escape of captured gas. Leakage of gas could pose health problems for the public. Large concentrations of CO_2 in the soil could also lead to acidification, thus harming soil fertility and killing plants. This could cause more problems than good in a country where 70 per cent of the population depend on agriculture and farming for food.[44]

There were therefore questions whether appropriate infrastructure had been put in place to deal with the possibilities of pipeline rupture and leakage of the captured gas. Furthermore, community leaders sought to know whether project proponents would utilize associated gas, as was

41 Gas flaring in Nigeria has remained a major contributor to global climate change and a significant source of Africa's net contribution to greenhouse gas emission. See ERA, Facts Sheet: *Harmful Gas Flaring in Nigeria*, (n 36); see also Vanguard, 'Nigeria loses $150 billion to gas flare in 36 yrs', (n 36); Environmental Rights Action/Friends of the Earth Nigeria, 'Gas Flaring in Nigeria', (n 36).

42 See F. Onojiribholo, 'Kwale Chief Laments Plight of Communities' (*Daily Independent*, 16 August 2011); Allen et al., 'Niger Delta Oil Flares, Broken Laws, Pollution and Oppression',(n 36), chp.4; see also Bassey, 'Foreword', (n 36); Carbon Trade Watch, 'Groups Slam Nigeria's Submission of Gas Flare Reductions for Carbon Credits', (n 36).

43 See Environmental Justice Atlas, *Kwale-Okpai CDM Project, Nigeria*, ejatlas.org/conflict/kwale-okpai-cdm-project-nigeria, accessed 21 February 2015.

44 International Fund for Agricultural Development (IFAD), 'Enabling Poor Rural People to Overcome Poverty in Nigeria', www.ifad.org/operations/projects/regions/pa/factsheets/ng.pdf, accessed 5 February 2015; see also Food and Agricultural Organization of the United Nations, Statistics on Production and Trade, faostat3.fao.org/home, accessed 12 January 2015.

laid out in the PDD, or the cheaper non-associated gas in the gas recovery and utilization phase of the project. Additionally, there were claims that in its previous independent power plant projects the company mostly used easier-to-process and less expensive non-associated gas instead of the agreed-upon associated gas. There were also concerns that the project could foster already existing displacement from traditional lands resulting in the loss of traditional knowledge, practices and cultures.[45]

These questions were not adequately addressed by project proponents or by the Nigerian government, which consequently fuelled public mistrust about CDM implementation in Nigeria in general.[46] Project proponents failed to address key environmental concerns associated with the project by not conducting a transparent EIA that could provide comprehensive details on how short- and long-term impacts of the project have been mitigated or addressed. Allegedly, the EIA for this project was conducted as a 'smokescreen' after the project had already been approved by the Nigerian designated national authority for CDM and registered by the CDM EB.[47] Furthermore, the consultation process was allegedly deceptive as community leaders were told by the Nigerian government that the project was an electrical power project aimed at bringing electricity to the community.[48] The Nigerian government was also accused of corruption and colluding with project proponents and increasing police presence to quell public resistance and force through the project.

Most importantly, the 'additionality' of this project was not sufficiently demonstrated to the local communities; therefore, it was perceived as an opportunity for oil companies operating in the Niger Delta to receive carbon credits for reducing gas flaring - an act (i.e gas flaring) expressly prohibited by Nigerian law in the first place.[49] Project proponents failed to demonstrate how the

[45] See Onojiribholo, 'Kwale Chief Laments Plight of Communities', (n 42); Allen et al., 'Niger Delta Oil Flares, Broken Laws, Pollution and Oppression', (n 36), chp. 4.

[46] Onojiribholo, 'Kwale Chief Laments Plight of Communities', *Ibid.* Allen et al., *Ibid.*; also Carbon Trade Watch, 'Groups Slam Nigeria's Submission of Gas Flare Reductions for Carbon Credits', (n 36).

[47] K. Adeyemo, 'Nigerians Oppose Climate Development Projects' *The Tribune* (Ibadan, 12 September, 2010) 3.

[48] Allen et al., *ibid.*

[49] The *Associated Gas Re-Injection Act Number 99 of 1979* (with an amendment in 1985) expressly prohibits gas flaring. Section 3 (2) of the Act required oil corporations operating

Kwale CDM project - which had existed since 1987 as a gas recovery and utilization project and which was simply repackaged and labelled as a CDM project in 2005 - would bring additional emission reduction and sustainability benefits to the country when compared with its pre-CDM scenario.[50] This lack of detailed information fuelled public perception that the project was an indirect effort by project proponents to gain cheap carbon reduction credits for an already illegal activity under Nigerian anti-gas flaring law, and for a problem the proponents were meant to address without the CDM.[51] These are questions that arguably could have been addressed with more transparent stakeholder participation and information disclosure system.[52]

Despite campaigns and protests by Environmental Rights Action (ERA) and other environmental groups in Nigeria, the Nigerian government approved the Kwale project as a sustainable CDM project. The CDM EB subsequently registered the project.[53] This project demonstrates the dangers of entrusting national authorities with the final authority to determine the sustainability of CDM projects and how the absence of an international review process for approved projects has undermined the sustainable development potentials of CDM projects. It also underscores the need for a more transparent, accountable and rights-based process for approving and registering CDM projects.

2.2.3 *The Barro Blanco hydroelectric power plant project in Panama*

Debates on the human costs of large hydroelectric power projects, and whether they are truly cleaner sources of energy have been ongoing for several decades. For example, environmental and human rights impacts

in Nigeria to guarantee zero flares by 1 January, 1984 and to submit detailed plans for gas utilization. The Kwale Project was first developed in 1987 as part of efforts to comply with the requirement of this legislation.

50 D. Idowu, 'ERA condemns Nigeria's Kwale CDM Project' *Vanguard* (Lagos, 02 September 2009).

51 Environmental Rights Action, 'Nigeria: Learning Clean Development Mechanism (CDM) Lessons', (September 2012), www.eraction.org/publications/learningcdmlessons.pdf, accessed 10 June 2015.

52 Environmental Rights Action/Friends of the Earth Nigeria, 'Gas flaring in Nigeria', (n 36).

53 See UNFCCC; 'Statistics of CDM projects', cdm.unfccc.int/Statistics/index.html, accessed 18 January 2015.

of the Changuinola (Chan 75) hydroelectric dam project in Panama have been extensively analysed and debated.[54] Even though it is not a CDM project, the Chan-75 project demonstrates how the execution of a project designed to reduce GHG emissions, cut down the high cost of energy in Panama, generate cleaner energy and to foster a reduced dependence on fossil fuels could produce negative human rights impacts.[55] Perhaps the most significant interference of hydroelectric projects with human rights is the massive resettlement of people from ancestral lands and homes.

Efforts to register hydroelectric power plant projects as CDM projects have only intensified these debates.[56] A good example is the Barro Blanco hydroelectric power plant project in Panama that was registered as a CDM project on 26th January 2011.[57] The Barro Blanco project has been

[54] B. Mayer, 'Judicial Review of Human Rights Impacts of Hydroelectric Projects' (Center for International Sustainable Development Law, 2012); K. Hoshour and J. Kalafut, *A Growing Global Crisis: Development-Induced Displacement and Resettlement* (International Accountability Project Issue Paper, 2010). For example, the United Nations Special Rapporteur on human rights and indigenous people has criticized the project as creating 'a significant impact on the indigenous communities in the surrounding area' and concluded that none of the communities were adequately consulted nor had the opportunity to give their consent in relation to relocation. See *Chan-75 dam*, 2009 *Report of the Special Rapporteur on the Situation of Human Rights and fundamental freedoms of indigenous peoples* (A/HRC/12/34/Add.5). See also UN Criticizes displacements of Panama Dam Tribe, *Terra Daily* (2008), www.terradaily.com/reports/UN_criticises_displacement_of_Panama_dam_tribe_999, accessed 12 December 2014.

[55] See *Chan-75 dam*, 2009 *Report of the Special Rapporteur on the Situation of Human Rights and fundamental freedoms of indigenous peoples*. See also UN Criticizes displacements of Panama Dam Tribe; 'Petition, Human Rights violations by the Government of Panama against the Ngöbe indigenous communities and individuals in the Changuinola River Valley, Bocas del Toro, Panama', (28 March 2008) 5.

[56] See the Michaelowa and Lazarus Debate, in A. Michaelowa and R. Lazarus, 'Arguing the Point: Should Large-Scale Power Projects Have a Future under the CDM?' *Carbon Mechanisms Review* (2013), 14–15 (Michaelowa insists 'Do not throw out the child with the bathwater!' while Lazarus argues that 'maybe this kid has become too big for the bathtub!').

[57] International Rivers, 'Letter to the CDM Executive Board Regarding the Barro Blanco Hydroelectric Project' (9 February 2011), www.internationalrivers.org/resources/letter-to-the-cdm-executive-board-regarding-the-barro-blanco-hydroelectric-project-3078, accessed 1 April 2015; Carbon Market Watch, 'Barro Blanco Hydropower Project in Panama Violates CDM Rules' (2011), carbonmarketwatch.org/barro-blanco-hydro-power-project-in-panama-violates-cdm-rules-newsletter-12/, accessed 12 March 2015; see also Agence France-Presse, 'Protests Over Planned Dam Turn Violent in Panama' (9 March 2013); Intercontinental Cry, Nagare Barro Blanco, 17 March 2013; Intercontinental Cry, 'Panama: Police Brutality Signals Impending Storm over Barro Blanco hydroelectric project' (21 March 2013), intercontinentalcry.org/panama-police-

criticized as violating several human rights of the Ngöbe indigenous people of Panama.[58] The Barro Blanco project is a 28.84 MW hydroelectric power project in the district of Tole, within the Chiriqui province in Panama. The project was designed to reduce the high cost of energy in Panama, to generate cleaner energy and to foster a reduced dependence on fossil fuels in power generation in Panama.[59] The project is developed by Generadora del Istmo S. A. (GENISA) and financed by the German Investment Corporation (DEG), the Netherlands Development Finance Company (FMO), and the Central American Bank for Economic Integration (CBIE). Total funding from the three banks amounts to US$78.3 million.[60]

Despite its clean energy prospects, this project has been criticized internationally and by the people of Panama.[61] According to a submission filed on 25 March 2013 with the CDM EB, some of the human rights violations experienced by the indigenous people include arbitrary displacements from homes and farms, loss of

brutality-signals-impending-storm-over-barro-blanco-hydroelectric-project/, accessed 1 April 2015; D. Cullenward and D. Victor, 'An Editorial Comment: The Dam Debate and Its Discontents', *Climatic Change*, 75 (2006), 81; I.B.T. Lima, F. Ramos, L. Bambace and R. Rosa, 'Methane Emissions from Large Dams as Renewable Energy Resources: A Developing Nation Perspective', *Mitigation and Adaptation Strategies for Global Change*, 13(2) (2006), 193–206.

58 The Ngöbe, numbering about 170,000 people, are the largest indigenous group in Panama with a vast majority still living in their traditional lands in Western Panama. See International Rivers, 'Letter to the CDM Executive Board Regarding the Barro Blanco Hydroelectric Project', *ibid.*; see also 'Petition, Human Rights violations by the Government of Panama against the Ngöbe Indigenous Communities and Individuals in the Changuinola River Valley, Bocas del Toro, Panama', (n 55), 32–33.

59 The project is one of the concession projects designed by the Government of Panama to increase the country's energy capacity by 15 per cent and energy production by 18 per cent. The Barro Blanco project aims to add to the country's installed energy generation capacity by approximately 2.35 per cent; increase the presence of energy generated with autochthonous natural resources, as compared to the generation of energy with hydrocarbons; protect the national energy market from the rising cost of hydrocarbons in the international markets; reduce or mitigate the CO_2 emissions derived from the generation of electricity with hydrocarbons in the medium and long-term, thus contributing to the sustainable development of the region and the country. See 'The CDM Project Design Document: Barro Blanco Hydroelectric Power Plant Project' (27 April 2010), cdm.unfccc.int/filestorage/B/O/L/BOLWGNA7FS4YKM23JCHPQ695ZIU8ET/Barro%20Blanco%20PDD%20v3.08?t=Rmt8bWtsYzZnfDD39eq-NE9PUI8ll0bveLu5, accessed 1 April 2014.

60 See 'The CDM Project Design Form: Barro Blanco Hydroelectric Power Plant Project', *ibid.*

61 International Rivers, 'Letter to the CDM Executive Board Regarding the Barro Blanco Hydroelectric Project', (n 58).

housing and destruction of crops, as well as the excessive use of force and detention of opponents of the project. It is projected that the project will displace residents of four Ngöbe villages that are home to approximately 1,005 people. Another 4,000 Ngöbe living in neighbouring villages will also be negatively affected because the dam will destroy their transportation routes, harm their agricultural plots, cut off their access to their farmlands or open up their territories to non-Ngöbe settlers.[62]

The environmental side effects of the project have also been highlighted. Already, the project has reportedly led to excessive flooding, erosion and destruction of the vegetation of one of the world's most notable diversity hotspots.[63] Over 25,000 people have allegedly been forced to abandon their farmlands, the majority of them being indigenous peoples.[64] Flooding is also expected to kill and decrease wildlife in the affected areas.[65]

Scientific studies have also determined that there is a high risk of losing the endangered *Tabasará* rain frog species and the *diadromous* – a rare migratory river fish species in Panama – because the dam will destroy their migration routes. Some of these species, such as the *Joturos Pitchardi*, locally called 'bocachica', are a central source of protein in the Ngöbe diet.[66] The project is expected to prevent these endangered fish species from moving up the river to their spawning grounds, thereby killing them. High emission of methane by the project could also pose a great threat to the fish, thus aiding their extinction.[67]

[62] See 'Submission Regarding Human Rights to the CDM' (25 March 2013), www.internationalrivers.org/resources/submission-regarding-human-rights-to-the-cdm-7899, accessed 28 March 2015; see also Petition on 'Human Rights violations by the Government of Panama against the Ngöbe indigenous Communities and Individuals', (n 55), 5.

[63] See 'Submission Regarding Human Rights to the CDM', *ibid.*

[64] Amnesty International, 'Panama Dam Flooding threatens Indigenous Families' (26 May 2011), www.unhcr.org/refworld/docid/4ddf3cd32.html, accessed 12 December 2014.

[65] J. Stein, 'Resistance to Dam Nation: An Analysis of the Stance and Strategies of the Opposition Movement to the Chan-75 Hydroelectric Project in Bocas del Toro, Panama' (Wesleyan University, 2008).

[66] Petition on the 'Human Rights Violations by the Government of Panama directed against Ngöbe Indigenous Communities and Individuals', (n 55).

[67] Estimates show that hydroelectric power plants such as the Chan-75 and the Barro Blanco plant emit 104 million metric tonnes of methane annually from reservoir surfaces, turbines, spillways and rivers downstream. The direct effect of vegetation, plants and soil flooded by reservoirs and hydroelectric dams is that when such plants rot they produce methane. When methane is released back into the river, it increases water

There have also been widespread human rights repressions, assaults, arbitrary detentions, public humiliations, threats and illegal destruction of crops and homes at the hands of the Panama police and the project executors - Generadora del Istmo, S.A. (GENISA). According to the petition, about fifty-four protesters have been arrested, including thirteen minors (two of whom were infants.) The police allegedly killed several farmers, broke the nose of a nine-year-old and injured the arm of a twelve-year-old during an anti-Barro Blanco protest.[68] The government of Panama was also accused of increasing military presence on the project site, in order to suppress protests, by farmers and indigenous groups, against the continued implementation of the project.

Human rights groups have also highlighted the lack of a participatory process or the prior informed consent of the traditional people by the government or the project proponents. According to the different human rights submissions and petitions on the project, project proponents forcefully acquired family lands without consulting with the representatives of the community.[69] PDDs that were submitted for the project also failed to document the impacts of the Barro Blanco reservoir on the Bakama area of the Ngöbe-Bugle community.

Despite these concerns, the accredited DOE for this project - the Asociación Española de Normalización y Certificación (AENOR) - allegedly issued a validation report that recommended the approval and registration of the project. The project was subsequently registered by the CDM EB. The absence of a participatory process through which some of the issues of displacement, relocation and compensation could have been discussed and agreed upon further highlights the lack of transparency and accountability in the planning and execution of this project. Not only does it diminish the emission reduction gains of the project, but it also casts doubts on the credibility of the Panama government and on the sincerity of the project proponents.[70] Addressing these issues within

toxicity and threatens the survival of fish and other aquatic animals and plants. Building materials used during dam construction are also an important cause of methane emissions into dams. See Lima et al., 'Methane Emissions from Large Dams as Renewable Energy Resources', (n 57).

68 See 'Panama: Police Brutality Signals Impending Storm over Barro Blanco Hydroelectric Project', (n 57) intercontinentalcry.org/panama-police-brutality-signals-impending-storm-over-barro-blanco-hydroelectric-project/. accessed December 12, 2015.

69 See 'Submission Regarding Human Rights to the CDM', (n 62).

70 'Role of The United Nations Questioned - Barro Blanco Hydro Electric Project' (12 May 2012), www.panama-guide.com/article.php/20120531154806624, accessed 12 March 2015.

the framework of the fundamental human rights of the Ngöbe people could have, perhaps, avoided some of the criticisms, protests and petitions that have trailed the project. The manner of approval and execution for this project and the attendant negative publicity it has generated further underscore the importance of integrating human rights considerations into the planning and implementation of climate change projects.[71]

2.2.4 REDD+ projects

REDD+ is an international mechanism designed to combat deforestation and climate change through socially sustainable use and conservation of forest resources.[72] REDD+ frameworks aim to achieve this goal by paying countries, with sizeable amounts of forest, money to adopt efforts that will conserve their trees and forests.[73] In return, the investing countries receive resulting emission reduction credits to achieve their national emission goals.[74] If implemented with robust human rights and environmental safeguards, REDD+ activities can increase investments in forest conservation, improve forest governance and help to protect important ecosystems that are critical to the livelihoods of many of the country's poorest communities.[75] Currently, REDD+ funding and initiatives are predominantly spearheaded by multilateral initiatives, such as the United Nations REDD Programme (UN-REDD) and the Forest Carbon Partnership Facility (FCPF), that finance and support nationally led REDD+ projects. Over US$195.7 million have been provided in funding to developing countries to promote REDD+ initiatives.[76]

[71] See CDM Policy Dialogue, 'Climate Change, Carbon Markets and the CDM: A Call to Action, Report of the High-Level Panel on the CDM Policy Dialogue', (September 2012), 56.

[72] 'An Introduction to REDD' *The REDD Desk*, 2014, theredddesk.org/resources/an-introduction-redd, accessed 22 October 2014.

[73] S. Baez, 'The Right REDD Framework: National Laws That Best Protect Indigenous Rights in a Global REDD Regime', *Fordham Law Review*, 80 (2011), 827.

[74] *Ibid.*

[75] I. Mustalahti, A. Bolin, E. Boyd and J. Paavola, 'Can REDD+ Reconcile Local Priorities and Needs with Global Mitigation Benefits? Lessons from Angai Forest, Tanzania', *Ecology and Society*, 17(1) (2012), 16; see also N. Roht-Arriaza, 'Human Rights in the Climate Change Regime', *Journal of Human Rights and the Environment*, 1(2) (2010), 215–219.

[76] See UN Collaborative Program on Reducing Emissions from Deforestation and Forest Degradation in Developing Countries, UN-REDD Program Strategy: 2011–2015 (UN-

Although REDD+ is one the leading proposals to address climate change, its practical implementation has raised a number of significant questions on its potentially harmful effects on indigenous communities that traditionally rely on ancestral lands and forests for their livelihoods. In Brazil and Peru, for example, there have been concerns regarding forceful land grabs and lack of proper consultation and engagement by governments.[77] For example, in 2008 and 2009, Amazonian indigenous peoples in Peru organized uprisings and blockades over highways, rivers and airports, as well as some energy installations, to demand greater accountability, access to information and participation in projects and activities that affect indigenous lands and forests. On 5 June 2009, local police authorities unsuccessfully attempted to clear the protests and blockade, leading to the killing of thirty-four people and serious injuries to several others.[78] The events in Peru and Brazil and the repressive response of national authorities underscore the need for international action to ensure that national authorities protect, respect and fulfil human rights when designing and implementing REDD+ activities.

These concerns are even more serious in countries with ambiguous land tenure and native title laws, Nigeria is one good example.[79] In Nigeria, REDD+ has had a very rough start. Nigeria's first REDD project is financed by Shell Canada and Gazprom.[80] The mere mention of Shell as a sponsor of REDD+ projects in Nigeria triggered

REDD, 2011) 1, also UN-REDD, www.un-redd.org/AboutUN-REDDProgramme, accessed 13 December 2014.

77 See K. Taylor, 'Improving Substantive and Procedural Protections For Indigenous Rights in REDD+ Projects: Possible Lessons From Brazil', *Afe Babalola University Journal of Sustainable Development Law and Policy*, 5(1) (2015), 32–54; A. Long, 'REDD+ and Indigenous Peoples in Brazil', in R. S. Abate and E. A. Kronk (eds.), *Climate Change and Indigenous Peoples: The Search for Legal Remedies*, (Massachusetts, USA; Cheltenham, UK: Edward Elgar Publishing, 2013), 151; Baez, 'The Right REDD Framework', 827; A. Fincke, 'Indigenous Peoples and Climate Change/REDD: An Overview of Current Discussions and Main Issues', *Int'l Union for Conservation of Nature* (March 2010), www.iucn.org/about/work/programmes/social_policy/sp_themes_ip/?5709/indigenousredd-plus, accessed 15 September 2014; C. Van Dam, 'Indigenous Territories and REDD in Latin America: Opportunity or Threat', *Forests*, 2 (2011), 396, www.mdpi.com/1999–4907/2/1/394/pdf, accessed 15 September 2014.

78 Center for World Indigenous Studies, 'Police Violently Attack Peaceful Indigenous Blockade in Peru', intercontinentalcry.org/police-violently-attack-peaceful-indigenous-blockade-in-peru/, accessed 15 September 2014.

79 S. Ogidan, 'Nigeria to Earn N 34.44 Billion yearly in Tree Planting', *Nigerian Compass* (Lagos, 14 February 2011).

80 D. Forgarty, 'Shell Bankrolls REDD Project in Nigeria', climatevoices.wordpress.com/2010/09/07/shell-bankrolls-redd/, accessed 13 January 2014.

widespread unrest and demonstrations across Nigeria. This is due to the perception of Shell as a major contributor to the many environmental problems of Nigeria's Niger Delta. There is a general perception of current REDD+ efforts by Shell as potentially an attempt at the 'largest land grab of all time' and a wholesale attempt to legalize violations of human rights, threats to cultural survival, militarization, oppression and servitude.[81]

Arguably, the process that led to the approval of Shell for REDD+ projects in Nigeria was not sufficiently participatory. Not only did project proponents fail to carry along relevant stakeholders, but approval of the project was the decision of the Nigerian authorities alone, concluded without the consent and approval of the indigenous communities who own the lands and trees in question. The people of Cross River State, whose ancestral lands are targeted for the REDD+ project, allegedly woke up to hear of it in the news.[82] What followed was immediate protest.[83]

For projects that could alter the lifestyles and existence of inhabitants of these areas who depend on these forests for farming and for food, lack of information and participation have led to a misunderstanding of the mitigation prospects of REDD+. Instead, as noted above, farmers and community members, most of whom are uneducated, understandably construe it as an attempt to forcefully grab their lands, subject them to servitude and destroy their means of existence. A transparent, informative and participatory approach could have arguably addressed these concerns.

Equally problematic is the absence of a framework for compensation or providing alternatives for the indigenous people of these areas who depend on the forests for their lives. While REDD+ projects could contribute to climate change mitigation, acquiring a

[81] C. Lang, 'Shell Project Slammed in Nigeria', (September 2010), www.redd-monitor.org/2010/09/08/indigenous-environmental-network-and-friends-of-the-earth-nigeria-denounce-shell-redd-project/, accessed 19 January 2015; see also 'REDD under fire in Nigeria', (September 2010), uk.oneworld.net/article/view/165950/1/246, accessed 13 January 2015.

[82] See 'REDD under fire in Nigeria', *ibid.*

[83] As the official spokesperson of the indigenous community noted: 'Forests in Cross Rivers State – some of the few remaining tracks of mangrove and rainforest reserve in the world targeted for REDD are in grave danger due to the scheme, hence the need for participants to uncover any cover-ups by government which is detrimental to community forests and the environment. To us, carbon trading/market mechanism promoted by the REDD are false solutions to climate change; REDD promotes deforestation, more plantation and corporate land grabs'. See Lang, 'Shell Project Slammed in Nigeria', (n 81).

large portion of land from indigenous communities by brute force without providing compensation or settlement options is a major human rights concern.[84] Another concern is the absence of an EIA on the impacts of REDD+ projects on the inhabitants of these areas.[85]

The absence of a robust EIA, lack of participation and inclusion and the lack of remedies for victims all raise genuine concerns on the potentials of REDD+ to contribute to sustainable development in developing countries. If dedicated funding mechanisms for REDD+ are to be included as part of the international climate finance architecture, there is a need to establish a robust participatory framework that would allow citizens to take part in and influence decision-making on REDD+ projects.[86] There is a need to further define and integrate human rights safeguards which are to be applied when financing and undertaking REDD+ activities under the international climate change regime.[87]

[84] A. Sampson, 'Don't Sell Forests: Group urge Nigerian Government', www.scoop.co.nz/stories/WO1008/S00467/dont-sell-forests-groups-urge-nigerian-govts.htm, accessed 18 January 2014.

[85] *Ibid.*

[86] See L. Crippa and G. Gordon, 'International Law Principles for REDD+: The Rights of Indigenous Peoples and the Legal Obligations Of REDD+ Actors' (Indian Law Resource Centre, May, 2012); E. Fenton (ed.), 'Realising Rights, Protecting Forests: An Alternative Vision for Reducing Deforestation Case studies from the Accra Caucus', (June 2010); R. Lyster, 'REDD+, Transparency, Participation and Resource Rights: The Role of Law', *Environmental Science and Policy*, 14 (2010).

[87] The COP at the Cancun conference adopted a set of general guidelines and safeguards to be applied when financing and under-taking REDD+ activities. Despite their very general nature, they provide broad foundations and normative frameworks for safeguarding human rights of indigenous peoples in REDD+ projects implemented within and outside the international climate change regime. The legal framework discussed in this book builds on and advances the Cancun decisions by answering questions on how such normative goals could be applied and implemented in practice. See Appendix I of the Cancun Agreements: Outcome of the Work of the Ad Hoc Working Group on Long-term Cooperative Action under the Convention, Decision 1/CP.16. Document No: FCCC/CP/2010/7/Add.1, March, which provides that the following safeguards, among others, should be promoted and supported by developing country parties: (c) Respect for the knowledge and rights of indigenous peoples and members of local communities (...); (d) The full and effective participation of relevant stakeholders, in particular, indigenous peoples and local communities in REDD+ activities.

2.3 Human rights impacts of carbon finance schemes: an evaluation

The above examples demonstrate how carbon projects are producing human rights impacts and risks in developing countries. The approval processes for carbon projects have arguably focused more on the emission reduction prospects and economic benefits while neglecting the human rights consequences on host communities. This threatens the respect and protection of several human rights under international law. It is important that emission reduction goals of carbon projects are harmonized and balanced with the respect and protection of fundamental human rights, most especially the right to development (RTD).[88] This section evaluates some of the core human rights violations and risks resulting from improperly implemented carbon projects.

2.3.1 *Forced displacement and the right to property*

One of the most repressive aspects of some of the above-discussed carbon projects is the arbitrary displacement of people from their ancestral lands and homes without compensation. Apart from displacement and land grabs, the right to property is also often affected in hydroelectric power projects such as the Chan-75 and the Barro Blanco projects in Panama where water allegedly escaped from reservoirs, flooding the homes of

[88] Even though scholars have questioned the status of the RTD under international law, it remains an important reference point on the need to frame development within the context of respect and fulfilment of human rights. The right to development emphasizes the need to balance development with the respect and protection of human rights. The pursuit of development goals cannot justify the violation of human rights. In this book, we will not go into the debates on the status of the RTD, instead we align with the view that the RTD, as promoted in many UN declarations and resolutions, is crucial for mainstreaming human rights principles into the international climate change regime. The UN has promoted the need for all its agencies to balance development goals with the respect and protection of fundamental human rights. The extent to which the UNFCCC as a part of the UN system has identified and recognized human rights principles within the context of climate change actions and projects, however, remains obscured. The *UN Declaration on the Right to Development* emphasizes that failure to observe human rights in development processes constitutes an obstacle to development. An aspect of the RTD is the right of individuals to participate in decision-making processes. The RTD also encourages states to treat development as a comprehensive process; the attainment of one goal should not be at the expense of the other. Within the African context for example, the right to development is recognized in Article 22 of the 1981 African Charter on Human and Peoples' Rights as a definitive individual and collective human right. Article 22(1) provides that all peoples shall have the right to their economic, social and cultural development with due regard to their freedom and identity and in the equal enjoyment of the common heritage of mankind.

citizens in the project areas. This affects the rights of individuals to exclusively own and enjoy their properties.

The right to exclusively own and possess property is guaranteed under international law. It is one of those rights recognized in John Locke's human rights tripod of 'life, liberty and possessions'.[89] Though not recognized in the International Covenant on Civil and Political Rights (ICCPR) and the International Covenant on Economic, Social and Cultural Rights (ICESCR), it is recognized in a number of other core international human rights instruments. For example, Article 5 of the International Convention on the Elimination of All Forms of Racial Discrimination (ICERD) states that everyone has the 'right to own property alone as well as in association with others' and 'the right to inherit'.[90] Article 16 of the Convention on the Elimination of All Forms of Discrimination against Women (CEDAW) also establishes rights to ownership, acquisition, management, administration, enjoyment and disposition of property for both men and women. Similarly, Article 17 of the Universal Declaration of Human Rights (UDHR), recognizes the right of individuals to the exclusive possession and use of property.[91] Though not legally binding, the United Nations Declaration on the Rights of Indigenous Peoples (UNDRIP) also stipulates that states shall, *inter alia*, provide effective mechanisms for the prevention of and redress for any action which has the aim or effect of dispossessing indigenous peoples of their lands, territories or resources.[92] The right to property is also recognized in regional human rights instruments such as the American Convention on Human Rights,[93] the

[89] J. Locke, *Two Treatises of Government* (1690), www.gutenberg.org/files/7370/7370-h/7370-h.htm, accessed 21 January 2015.

[90] See International Convention on the Elimination of All Forms of Racial Discrimination, 7 March 1966, 660 UNTS 195.

[91] Though not binding, the UDHR is a foundation of many human rights under international law. It provides that 'Everyone has the right to own property alone as well as in association with others'. Under Article 17 (2) 'No one shall be arbitrarily deprived of his property'. See UDHR, adopted 10 December 1948, GA Res.217A9 (III), 3 UN GAOR, UN Doc A/810 at 71 (1948).

[92] United Nations Declaration on the Rights of Indigenous Peoples A/RES/61/295, adopted 2 October 2007. See also J. Ruggie, 'United Nations Guiding Principles on Business and Human Rights: Implementing the United Nations "Protect, Respect and Remedy" Framework' (2011) UN Document A/HRC/17/31 (21 March 2011), paras. 8–12.

[93] Article 21 provides that 'Everyone has the right to the use and enjoyment of his property. The law may subordinate such use and enjoyment to the interest of society'. While Art. 21(2) provides that 'No one shall be deprived of his property except upon payment of just

African Charter on Human and Peoples' Rights (AfCHPR)[94] and in Protocol No. 1 to the European Convention on Human Rights.[95]

The right to property is not absolute; states have a wide degree of discretion and a 'margin of appreciation' to restrict the right of individuals to the use of their property for public interest. Abuses of such discretionary powers by autocratic governments are commonplace in less democratic countries.[96] For example, in Nigeria, the Land Use Act vests the absolute ownership, management and control of land in each state of the Federation in the governor, who is to hold the land in trust and administer it for the benefit of all Nigerians.[97] Section 28 of the Act further provides that land may be appropriated for 'overriding public interests'. This includes 'the requirement of the land for mining purposes or oil pipelines or for any purpose connected therewith'.[98] This provision has been at the heart of many arbitrary land deprivation cases in Nigeria, for example in the Kwale case and in REDD+ projects in Cross River state, where lands were allegedly forcefully acquired from citizens without compensation. There have also been instances where land is taken for reasons that do not apparently fall under the overriding public interest category.[99]

In the absence of a defined right to property in the ICCPR, the United Nations Human Rights Committee has only admitted or commented on

compensation, for reasons of public utility or social interest. . .' See American Convention on Human Rights (pact of San Jose) signed 22 November 1969, entered into force 18 July 1978, 1144 UNTS 123, OASTS 36, OASTS 36.

94 Article 14 provides that 'The right to property shall be guaranteed. It may only be encroached upon in the interest of public need or in the general interest of the community and in accordance with the provisions of appropriate laws'. Charter on Human and Peoples Right, adopted 27 June 1981, entered into force 21 October 1986, OAU Doc CAB/LEG/67/3 Rev. 5.

95 Article 1 provides that 'every natural or legal person is entitled to the peaceful enjoyment of his possessions. No one shall be deprived of his possessions except in the public interest and subject to the conditions provided for by law and by the general principles of international law'.

96 See K. Roth, 'Defending Economic, Social and Cultural Rights: Practical Issues Faced by an International Human Rights Organization', *Human Rights Quarterly*, 26(1) (2004), 70, where the author discusses some of the practical challenges facing human rights in developing countries, and cited examples of a government that arbitrarily refuses to apply available resources for the benefit of its population's health.

97 *Land Use Act of 1979*, Laws of the Federation of Nigeria 2004, c L5, s 1.

98 For an excellent and detailed evaluation of this section and the Land Use Act in general, see R. Ako, 'Nigeria's Land Use Act: An Anti-Thesis to Environmental Justice', *Journal of African Law*, 53(2) (2009), 289–304.

99 *Ibid.*

petitions on the violation of the right to property where it relates to the violation of any of the rights recognized in other core international human rights instruments, for example non-discrimination under CEDAW.[100] However, the European Court has held in a number of cases that even though states have a wide degree of discretion and a 'margin of appreciation' to restrict the rights of individuals, they are to provide just compensation whenever deprivation of property occurs.[101] The African Commission on Human Rights also provided the same ruling in the case of *SERAC v. Nigeria*, where it held that the forceful acquisition of the land belonging to the Ogoni people of Nigeria without compensation was in violation of Article 14 of the AfCHPR.[102] The African Commission provided the same ruling in the *Malawi African Association case*, where the Commission found that the confiscation and looting of the property of black Mauritanians and the expropriation or destruction of their land and houses constituted a violation of the right to property.[103]

In *Mayagna (Sumo) Awas Tingni Community v. Nicaragua*, the Inter-American Court recognized the principle of communal land ownership, as a way of addressing some of the difficulties faced by individuals who have never been given formal title to their lands and are therefore, unable to enforce their property rights. The court found that the right to property imposes positive obligations on states to delimit, demarcate and provide formal titles to lands and that failure to register ancestral land cannot justify a denial of the right to property. The court held that the government of Nicaragua had violated the rights of the Awas Tigni

[100] See, for example, *Adam, Blazek and Marik v. The Czech Republic; Oló Bahamonde v. Equatorial Guinea; Ackla v. Togo and Diergaardt of Rehoboth Baster Community et al. v. Namibia.*

[101] See *Sporrong and Lonnrith v. Sweden* Series A52 (1982); See also *Marckx v. Belgium* Case No 00006833/74, (13 June 1979); *Jacobson v. Sweden* Series A 163; *Hentrich v. France*, A 296-A (1994); *Holy Monasteries v. Greece*, A 301-A (1994); *Pressos Compania Naviera SA v. Belgium* A332 (1995); *Aka v. Turkey* 1998-VI (1998); *Papachelas v. Greece* (25 March 1999); *Brumarescu v. Romania* (28 October 1999); *Immobiliare Saffi v. Italy* (28 October 1999); *Spacek v. the Czech Republic* (9 November 1999); *Beyeler v. Italy* (5 January 2000); *Chassagnou v. France* (29 April 2000); *Carbonara and Ventura v. Italy* (30 May 2000); *Former King of Greece and Others v. Greece* (23 November 2000).

[102] (2001) AHRLR 60 (ACHPR 2001). Communication 155/96.

[103] *Malawi African Association Amnesty International*, Comm. Nos. 54/91, 61/91, 98/93, 164/97 à 196/97 and 210/98 (2000); *Ms. Sarr Diop, Union Interafricaine des Droits de l'Homme and RADDHO, Collectif des Veuves et Ayants-droit and Association Mauritanienne des Droits de l'Homme v. Mauritania, Communications 54/91, 61/91, 98/93, 164/97–196/97 and 210/98.*

people to their property by granting concessions to a foreign company to log on traditional lands without consulting them and without getting their consent.[104] It found that the protection of right to property extends to the rights of members of the indigenous communities within the framework of communal property. This principle is further illustrated in *Yakye Axa Indigenous Community v. Paraguay*.[105] These cases suggest that justifications for arbitrary deprivation of property rights, particularly non-payment of compensation, are very unlikely to be accepted as parts of the margins of appreciation before an international forum.[106]

Despite these rulings, however, the extent to which property rights are protected in host countries, especially the less democratic countries, remains doubtful. Even though many carbon projects will arguably fall under the category of public interest projects, whenever such projects could displace individuals from homes, governments in many developing countries, for example in Nigeria, have not done much to provide minimum procedural guarantees to ensure the active participation of individuals in such decisions and to ensure just compensation is provided for them. This weak enforcement of the right to property at national level raises the need for increased protection and enforcement at the international level, particularly prior to the approval and registration of such projects by the CDM EB. For example, the question whether CDM project proponents in Panama could use CERs to compensate victims of land grabs as proposed by the project proponents in the Barro Blanco project has been put before the Inter-American Human Rights Commission. Even if the Commission answers this in the affirmative, it does not prevent or address the causes and sources of these land grabs and the other human rights questions linked to the project.[107]

This underscores the need for the international climate change regime to lay down clear criteria that will mandate project proponents to prevent such arbitrary displacements, and to address resulting loss, by providing adequate compensation, where a project inevitably infringes on property

[104] (2001) 79 (c) Int. Am. Ct. H.R.

[105] Judgment of 17 June 2005. (Merits, Reparations and Costs).

[106] See also Art 21(2) of the American Convention on Human Rights 'No one shall be deprived of his property except upon payment of just compensation, for reasons of public utility or social interest, and in the cases and according to the forms established by law'. See American Convention on Human Rights (pact of San Jose) signed 22 November 1969, entered into force 18 July, 1978, 1144 UNTS 123, OASTS 36, OASTS 3.

[107] CDM Watch, 'Massive Protests against Barro Blanco Hydropower Project in Panama', (2011), carbonmarketwatch.org/wp-content/uploads/2011/04/cdm_watch_newsletter_2_april_2011.pdf, accessed 12 March 2015.

rights. This also emphasizes a mandatory requirement that host countries would have to demonstrate that project sites and lands have not been arbitrarily acquired and that compensation has been paid to individuals and communities whose lands are affected as part of a project plan whenever such is inevitable. This will go beyond the mere ticking of a box in the Project Design Document, instead mechanisms must be put in place to verify such claims before projects are approved and registered. This will emphasize the need for states to take human rights seriously in the design and execution of climate change mitigation projects and activities.

2.3.2 Loss of benefits of culture and tradition

Displacement of people from ancestral lands, forests and homes is intricately intertwined with the loss of some important aspects of their traditional lifestyles and culture. For example, the people of Cross River State in Nigeria are known to be farmers that depend on subsistence farming for food and as part of their cultural life. Their massive displacements due to REDD+ projects have resulted in the loss of these elements of their culture.[108] Similarly, studies show that wind turbine projects in Panama could kill off migratory birds, thereby threatening the rights of indigenous people who depend on them for food and cultural use.[109] The right to religion and culture would also be affected in Panama by hydroelectric projects, which flood the ceremonial sites of the indigenous peoples.[110] Land grabs due to REDD+ projects reduce access to the resources through which indigenous people could enjoy benefits of their culture, thereby threatening their cultural rights.

Cultural rights have been recognized in a number of core international human rights instruments. For example, Article 27 of the ICESCR provides that persons belonging to minorities have the right to enjoy their own culture.[111] Article 27 of the ICCPR also provides that 'persons belonging to minorities shall not be denied the right in community with other members of the group to enjoy their own cultures'.[112]

[108] See Bassey, 'Foreword', (n 36), 4.

[109] See World Bank, *Environmental Matters* (World Bank Group, 2009), 19–20.

[110] *Ibid.*

[111] See the International Covenant on Economic, Social and Cultural Rights, 16 December 1966, 993 UNTS 3 (hereinafter ICESCR).

[112] International Covenant on Civil and Political Rights, 19 December 1966, 999 UNTS 171 (hereinafter ICCPR).

Though not one of the core UN human rights treaties, Convention 169 on Indigenous and Tribal Peoples, a legally binding convention that was adopted by the International Labour Organization in 1989, also establishes the right of indigenous peoples to 'exercise control, to the extent possible, over their own economic, social and cultural development'.[113] The Convention requires parties to respect and protect the cultural rights of indigenous peoples that relate to their ways of making a living, language, customs and traditional lifestyles. Article 4 of the Convention calls for special measures to be adopted to safeguard the persons, institutions, property, labour, cultures and environment of indigenous peoples. Article 14 also requires states to identify lands traditionally occupied by indigenous peoples and guarantee their ownership and protection.

In 2007, the UN General Assembly recognized the importance of lands to indigenous peoples when it adopted the Declaration on the Rights of Indigenous Peoples. While not legally binding, the Declaration recognizes that indigenous peoples have the right to the lands, territories and resources, which they have traditionally owned, occupied or otherwise used or acquired.[114] It also recognizes the rights of indigenous peoples to own and develop resources on their land and a right to redress for the lands, territories and resources which they have traditionally owned or otherwise occupied or used, and which have been confiscated, taken, occupied, used or damaged. Regionally, the AfCHPR and American Declaration of the Rights and Duties of Man also guarantee the right of individuals to take part in the cultural life of their community.[115]

There is currently a dearth of cases in which violation of cultural rights have been commented upon under the complaint mechanism of the ICESCR.[116] Decisions by the Inter-American Commission on Human

[113] The Convention has been ratified by 20 countries, including Honduras. The Convention encourages ratifying countries to 'align legislation, policies and programmes to the Convention'. See, www.ilo.org/indigenous/Conventions/no169/lang–en/index.htm, accessed 12 March 2015.

[114] See Art. 26(2), 26(3) & 28. United Nations Declaration on the Rights of Indigenous Peoples.

[115] Art. 17 of the African Charter and Article 13 of the American Declaration of the Rights and Duties of Man.

[116] This Optional Protocol was adopted by the UN General Assembly on 10 December 2008 and it opened for signature on 24 September 2009. It entered into force on 05 May 2013. According to the UN High Commissioner for Human Rights, Navi Pillay, 'The Protocol will provide an important platform to expose abuses linked to poverty, discrimination and neglect, which up until now victims have had to endure without any possible recourse at the international level. It will provide a way for individuals, who may otherwise be isolated and powerless, to make the international community aware of

Rights are, however, instructive. The right has been interpreted to include the right to maintain close ties with cultural lands. The Inter-American Human Rights Court held in the *Case of the Mayan Indigenous Communities from the Toledo District, Belize*, that certain indigenous peoples maintain special ties with their traditional lands and a close dependence; as such, the use and enjoyment of the land and its resources are integral components of the survival of the indigenous communities.[117] Similarly, in *Saramaka People v. Suriname* the court held that the use of traditional lands was critical to the cultural structure of the Saramaka Tribe, as such, the government of Suriname violated fundamental human rights law when it denied them these benefits through the forceful acquisition of their land.[118] In the *Case of the Indigenous Community Yakye Axa v Paraguay*, the court came to the same conclusions.[119]

Based on the understanding that all rights are interdependent and important, there is a need to protect cultural rights in climate change mitigation efforts at the international level, particularly prior to the approval and registration of carbon projects. The impacts of emission reduction schemes, such as REDD+, on the cultural rights of the people of the Cross River community in Nigeria, particularly the massive dislocations of citizens from their communal and ancestral lands, have to be examined and addressed by the international climate change regime. Through robust human rights impact assessments, project proponents can discover how a project can impact the culture of the people and how best to minimize or avoid such impacts. There is a need for a mandatory requirement that compels host countries to demonstrate that a comprehensive human rights impact assessment has been completed and that cultural lands would not be affected before projects are approved and registered. This will emphasize the need for states to take human rights seriously in the planning and design of carbon projects and activities.

their situation'. See 'Pillay Welcomes Major Breakthrough Enabling Individual Complaints on Economic, Social and Cultural Rights' OHCHR, www.ohchr.org/en/NewsEvents/Pages/DisplayNews.aspx?NewsID=12968&LangID=E, accessed 10 February 2015.

117 Report No. 40/04, Case 12.053, Merits, Mayan Indigenous Communities from the Toledo District, Belize, 12 October 2004.

118 *Saramaka People v. Suriname*, Judgment of 12 August, 2008 *(Interpretation of the Judgment on Preliminary objections, Merits, Reparations, and Costs)*.

119 Merits, Reparations and Costs, Judgment of 17 June 2005 Series C No. 125, 96.

2.3.3 *Loss of subsistence rights*

Cultural rights aside, carbon projects that do not safeguard local rights could deny members of the public their rights to earnings from farming, especially in rural communities where most of the local food supplies and income are generated from farming. When citizens are ejected from ancestral lands, they lose the ability and right to make their living from subsistence farming and hunting. For example, farmers in the Kayapó territory in Brazil that are known for subsistence farming and whose only source of income is from selling agricultural produce are facing increased pressure due to massive dislocations from REDD+ projects.[120] The Ngöbe community in Panama face similar threats from hydroelectric power plant projects that affect hunting and farming activities of local farmers that traditionally depend on subsistence farming for food. REDD+ projects in Cross Rivers State, Nigeria may also lead to a lowering of the water table resulting in water shortage and loss of crop yields.[121] These concerns threaten the ability of farmers in these communities to retain their sources of livelihood and subsistence.

Access to the resources through which indigenous people could enjoy dietary elements of their culture is a major part of subsistence rights. The right to food is expressly recognized in Article 11 of the ICESCR as part of the right to an adequate standard of living. Pursuant to the 2009 Optional Protocol to the International Covenant on Economic, Social and Cultural Rights, the right to food can now be enforced at the international level before the Committee on Economic, Social and Cultural Rights. Furthermore, though they do not enjoy any binding legal status under international law, several UN general comments, declarations and resolutions provide significant and influential normative interpretations of

[120] The Kayapó territory is located in the southern Pará and northern Mato Grasso states of Brazil. See D.A. Posey and K. Plenderleith, *Kayapó Ethnoecology and Culture* (London: New York: Routledge, 2002), 33; see Taylor, 'Improving Substantive and Procedural Protections For Indigenous Rights in REDD+ Projects', (n 77).

[121] This concern has also been raised about the impacts of biofuel projects. Firstly, by contributing significantly to the increase in food prices recent studies show that biofuel production has already contributed to the steep increase in food prices. Secondly, by causing land concentration for plantation-type production, due to considerations of economy of scale, biofuel projects would result in evictions of indigenous groups from their farmlands. See A. Eide, 'The Right to Food and the Impact of Liquid Biofuels', (Food and Agriculture Organization, 2008), 18–20; R. Doornbusch and R. Steenblik, 'Biofuels: Is the Cure Worse than the Disease?' Background Paper for OECD meeting in Paris 11–12 September 2007. SG/SD/RT (2007), 3.

subsistence rights to include access to traditional foods.[122] For example, Article 25 of the 1948 UDHR recognizes the right to food as part of the right to an adequate standard of living.[123]

Summarily, when a community's food-growing capacity is affected by a carbon project or when their abilities to secure foods that are traditionally part of their diet is undercut, elements of their culture may also be at risk. A good example is the Barro Blanco project in Panama, which has allegedly affected the rights of the Ngöbe communities to eat *Joturos Pitchardi*, locally called 'bocachica', a rare and central source of protein in the Ngöbe diet.[124] The high level of illiteracy and low level of human rights awareness in many traditional communities often hinder the capabilities of individuals to demand their human rights; this exacerbates and deepens human rights oppressions. A number of such communities

[122] See for example General Comment No. 23: The Rights of Minorities (Art. 27): 04/08/1994. CCPR/C/21/Rev.1/Add.5 which recommends that culture manifests itself in many forms, including a particular way of life associated with the use of land resources, especially in the case of indigenous peoples. That right may include traditional activities such as fishing, or hunting or the right to live in reserves protected by law. The enjoyment of those rights mat require positive legal measures of protection and measures to ensure participation of members of minority communities in decisions which affect them. See also General Comment 12 which notes that the food a community or people grow and how they grow it, their choice of food, the way they prepare and eat it – all of these are imbued with cultural values. See also Similarly, Art. 1 of the UN Declaration on the Right to Development which provides that the right to development is an inalienable human right by virtue of which every human person and all peoples are entitled to participate in, contribute to, and enjoy economic, social, cultural and political development, in which all human rights and fundamental freedoms can be fully realized. See also Art. 1 of the UNESCO Principles on International Cultural Co-operation, which provides that each culture has a dignity and value, which must be respected and preserved. Every people have the right and the duty to develop their culture. In their rich variety and diversity, and in the reciprocal influences they exert on one another, all cultures form part of the common heritage belonging to all mankind.

[123] The UN Special Report on the Right to Food has provided influential and persuasive explanations on the content of the right to food under international law. Though this report does not have a binding legal status under international law, it interprets the right to adequate food is a human right, inherent in all people to have regular, permanent and unrestricted access, either directly or by means of financial purchases, to quantitatively and qualitatively adequate and sufficient food corresponding to the cultural traditions of people to which the consumer belongs, and which ensures a physical and mental, individual and collective fulfilling and dignified life free of fear. See J. Ziegler, *The Right to Food* (Report by the Special Rapporteur on the right to food to the Commission on Human Rights 57th session, 2001) UN Doc E/CN.4/2001/53 at 2' daccessdds.un.org/doc/UNDOC/GEN/G01/110/35/PDF/G0111035.pdf?OpenElement, accessed 23 April 2015.

[124] See Petition on 'Human Rights violations by the Government of Panama against the Ngöbe indigenous Communities and Individuals', (n 55).

have with the cooperation of national governments, been targeted as sites for carbon projects. When such national governments (e.g. the government of Panama) are then entrusted with having a final say on project eligibility, it, arguably, leaves ordinary individuals with little or no opportunities to block the approvals of such projects or to demand its rethink. There is a need to provide international supervisory mechanisms and platforms through which such human rights issues could be addressed prior to the approval and registration of projects such as emission reduction projects.

Loss of farmland also infringes on the rights of members of the community to work. As the Food and Agricultural Agency (FAO) has aptly noted, 'rural landlessness is often the best predictor of poverty and hunger'.[125] Farmland constitutes the main asset from which the rural poor are able to derive a livelihood. Arbitrary land grabs therefore affects the rights of individuals to work. Several international human rights instruments and other non-binding interpretive documents therefore recognize the need to protect access of peasants to farmlands as part of the right to work.[126]

Article 6 of the ICESCR recognizes the right to work, which it defines as the opportunity of everyone to gain their living by freely chosen or accepted work. Parties to the ICESCR are required to take appropriate steps to safeguard this right by adopting policies that enable citizens to engage in their freely chosen occupation. States are enjoined to pursue reforms that would ensure that citizens and particularly the rural communities continue to have access to arable lands.[127] Article 23 of the UDHR also provides that everyone has the right to work, to free choice of employment, to just and favourable conditions of work and to protection against unemployment.

The jurisprudence evolved by the Inter-American Commission on Human Rights on the right to work is instructive on the content of this right. In the *Toledo Maya* case, the Commission noted that 'development activities must be accompanied by appropriate and effective measures to ensure that they do not proceed at the expense of the fundamental rights

125 Food and Agricultural Agency (FAO), *Voluntary Guidelines on Responsible Governance of Tenure of Land and other Natural Resources* (Rome: FAO, 2012), i.

126 See The Vancouver Declaration on Human Settlements, UN Conference on Human Settlements, Adopted June 11 1976, General Principles: Land; Voluntary Guidelines of the Food and Agriculture Organization of the United Nations.

127 See Art. 11(2) and Art 14(2) (g) of the ICESCR, which deal with agricultural reform to ensure access to arable land.

of persons who may be particularly and negatively affected, including indigenous communities and the environment upon which they depend for their physical, cultural and spiritual well-being'.[128] The Commission ruled that the state's contraventions had a negative impact on the natural environment upon which they depend for subsistence.[129] Similarly, in *Yanomami Case* the Inter-American Commission recognized the right of indigenous people in Brazil to use their land for farming, hunting and subsistence. The Commission therefore called on states to take the measures aimed at restoring, protecting, and preserving the rights of indigenous peoples to their ancestral territories, which they depend upon for subsistence, and continuity of life.[130] These cases suggest that access to farmland is crucial to the protection of the right to work of traditional farmers particularly in rural areas.

Carbon projects that deny farmers the opportunity of farming and hunting pose major threats to the enjoyment of their subsistence rights.[131] It is not enough for the COP to declare that parties should in all climate actions respect human rights. The COP must go a step further by providing practical frameworks and guidelines that would prevent the approval and registration of mitigation projects that violate human rights. Mainstreaming human rights safeguards into the approval and registration processes for carbon projects would revitalize the international climate governance architecture to become more responsive, accountable and rights-based.

[128] *Toledo Maya Case;* see also Inter-Am. C.H.R., *Report on the Situation of Human Rights in Brazil*, OEA/Ser.L/V/II.97 Doc. 29 rev.1, ch. VI (29 September, 1997).

[129] The petitioners claimed that international instruments, including the UN and Inter-American draft declarations on indigenous peoples and the Rio Declaration, acknowledge the need for states to protect the natural environments on which indigenous peoples depend and that such an obligation is 'implicit in the provisions of the American Declaration in the context of indigenous land claim issues'. The Commission did not expressly address this point, but considered that environmental damage had occurred and had a negative impact on indigenous property rights. The Commission found that the damage resulted in part from the fact that the state failed to put into place adequate safeguards and mechanisms, to supervise, monitor and ensure that it had sufficient staff to oversee that the execution of the logging concessions would not cause further environmental damage to Maya lands and communities. *Ibid.* 53 and 147.

[130] *See, e.g., Yanomami Case*, Rep. No. 12/85, Case 7615 (Brazil), (1985).

[131] Greenpeace International, 'Human Rights and the Climate Crisis: Acting Today to Prevent Tragedy Tomorrow' A Briefing Paper by Greenpeace (2008), 4, www2.ohchr.org/english/issues/climatechange/docs/submissions/Greenpeace_HR_ClimateCrisis.pdf, accessed 12 March 2015.

2.3.4 Loss of life

The fact that carbon projects could result in direct and indirect loss of life was brought to the fore by the Aguan biogas project in Honduras. It is estimated that over twenty-three local farmers and human rights defenders in the Bajo Aguan region where the project is located were killed by national authorities for refusing to vacate their lands for the project to commence and for leading protests against the project.[132] Despite evidence of loss of life, the CDM EB approved and registered the project. National authorities, especially in many developing countries, have often demonstrated the propensity to execute economically and socially rewarding projects at all cost, leading to human rights repressions. Such repressions, mutilations and detentions were also alleged in the Panama case.[133]

The right to life is a fundamental human right under international law. Article 6(1) of the ICCPR prohibits the arbitrary deprivation of life.[134] It provides that every human being has the inherent right to life and that this right shall be protected by law. The right to life has also been identified as one of the rights for which no state shall derogate from, even in times of emergency threatening the life of the nation.[135] Article 3 of the UDHR also upholds the right to life, liberty and security of the person.[136] Regionally, Article 4 of the AfCHPR recognizes the right to life.[137]

[132] See CDM Watch, *Petition to the CDM Executive Board on Aguan Gas project* (2011), www.cdm-watch.org/wordpress/wp-content/uploads/2011/02/unsolicited_letter_cdm-project_application_3197_honduras.pdf, accessed 23 March 2015.

[133] See Agence France-Presse, 'Protests Over Planned Dam Turn Violent in Panama', (n 57); see also Intercontinental Cry, 'Panama: Police Brutality signals impending storm over Barro Blanco hydroelectric project', (n 57).

[134] ICCPR, (n 112).

[135] See The Siracusa Principles on the Limitation and Derogation of Provisions in the International Covenant on Civil and Political Rights (1984). The Siracusa Principles developed by the UN Sub-Commission on the Prevention of Discrimination and Protection of Minorities in 1984 states that no state party shall, even in time of emergency threatening the life of the nation, derogate from a number of key guarantees which include the right to life.

[136] Universal Declaration on Human and Peoples Rights, (Adopted on 10 December 1948) GA Res. 217A (III), UN Doc A/810 at 71 (1948) (UDHR).

[137] African Charter Human and Peoples Rights (Adopted 27 June 1981, entered into force 21 October 1981) OAU Doc CAB/LEG/67/3 Rev. 5, reprinted in 21 ILM 58 (1982); See also Art. 2 of the European Convention on Human Rights (Convention for the Protection of Human Rights and Fundamental Freedoms 1949) and American Convention on Human Rights (Adopted 22 November 1969, entered into force 18 July 1978) OAS Res. 447 (IX-0/79).

The UN Human Rights Committee has held in numerous cases that state repressions, summary executions by police and army are violations of the right to life.[138] Though not legally binding, the UN Guiding Principles on Business and Human Rights also condemns the taking of life and the deprivation of the right to life during business operations by corporations and other business enterprises.[139] The protection of the right to life has been interpreted to include the obligation of governments to take all possible measures to reduce mortality and to increase life expectancy.[140] It also includes the duties of government to prevent projects that threaten human life and survival.

A number of cases establish a direct link between environmental quality and the right to life.[141] For example, the Inter-American Commission in the *Yanomami's case* held that failure of the Brazilian government to preserve environmental quality, by approving a construction project which led to the spread of disease and loss of life, was a direct violation of the fundamental right to life guaranteed by Article 1 of the American Declaration. The court held *inter alia* that the right to have one's life respected is not limited to the protection against arbitrary killing, and that the realization of the right to life is in many ways dependent upon one's physical environment.[142] In the Nigerian case of *Gbemre v. Shell*, the court, in a radical stance, held that Shell's action in continuing to flare gas in the course of their oil exploration and production activities in the applicant's community is a violation of their fundamental right to life (including healthy environment) and dignity of human persons guaranteed by the Nigerian Constitution and the AfCHPR.[143] This case sets precedence in Nigeria as the first and only case to declare gas flaring as a breach of the fundamental human right to life.

The Stockholm Declaration is a major influence for the expansionist view that environmental protection is a prerequisite for enjoying all forms of rights, including the right to life. It links environmental

[138] See Annual Reports of the UN Special Rapporteur on Summary and Arbitrary Executions UN Doc E/CN.4/1995/61.

[139] See Ruggie, 'United Nations, Guiding Principles on Business and Human Rights', (n 92).

[140] See United Nations General Comment 6/16 of 27 July 1982, para. 5.

[141] Indian courts have provided a number of landmark decisions that link environmental protection to the right to life. See for example, *Subhash Kumar v. State of Bihar* AIR 1991 SC 420.

[142] *Yanomami Case*, Rep. No. 12/85, Case 7615 (Brazil), (1985).

[143] Federal High Court, Benin 14 November 2005, Unreported Suit No FHC/B/CS/53/05. The officially certified full-text of the judgment in this case can be found at www.climatelaw.org/cases.

protection to the ability of human beings to enjoy and realize human rights. This Declaration has led to a flurry of declarations, resolutions and reports, which also establish that environmental protection is a *sine qua non* for the enjoyment of the right to life.[144] Agreeably, without adequate protection for the environment, the human right to life would be substantially threatened.[145] Carbon projects that pollute the environment pose direct threats to human survival and to the enjoyment of the right to life. As such, there is a need for an international approach that mainstreams the protection of the right to life and other human rights into project approval processes specifically in the context of climate change mitigation projects. Similarly, there is a need to prevent direct threats to life, for example killing citizens that oppose projects, repressions and executions, in the rush to attract emission projects, or in a bid to push through the registration of as many CDM projects as possible. This diminishes the integrity of the international climate change regime and reduces support for emission reduction schemes as a whole.[146]

2.3.5 Health concerns

The construction of projects without detailed environmental impact assessments significantly raises the probability of health complications for local residents from such projects. There is also the issue of siting projects in residential areas, especially in poor neighbourhoods. These practices have resulted in air and water pollution leading to health related concerns. For example, in Nigeria, it has been reported that the gas flaring reduction project in Kwale has caused health concerns to residents of the area. It has also been estimated that many residents of that

[144] United Nations' Conference on the Human Environment (Stockholm Declaration) UN Doc. A/CONF.48/PC.13(1971), www.unep.org/Documents.Multilingual/Default.asp?DocumentID=97&ArticleID=1503, accessed 10 March 2015.

[145] See also, The Preamble to the Aarhus Convention, which links human rights and environment. It notes that adequate protection of the environment is essential to the enjoyment of basic human rights. UNECE, Aarhus Convention on Access to Information, Public Participation in Decision-Making and Access to Justice in Environmental Matters (Adopted 25 June 1998, entered into force 30 October 2001) No. 37770, treaties.un.org/doc/Publication/MTDSG/Volume%20II/Chapter%20XXVII/XXVII-13.en.pdf, accessed 19 March 2015 (Aarhus Convention).

[146] See S. Kravchenko, 'Right to Carbon or Right to Life: Human Rights Approaches to Climate Change', *Vermont Journal of Environmental Law*, 9 (2008), 513; see also CDM Policy Dialogue, 'Climate Change, Carbon Markets and the CDM', (n 71).

community would be diagnosed with cancer in their lifetimes due to effluents from the project.[147]

Studies have also shown that biogas and other biofuel projects could produce adverse long-term effects on human health.[148] Pesticides and other chemicals connected with biofuel production are reported to contaminate water resources and give rise to health problems.[149] Wind power projects have also been reported to produce unprecedented levels of noise to residents leading to sleep deprivation and insomnia thereby affecting their health. These concerns violate the rights of the public to the highest attainable standard of health.

The human right to health has been recognized in Article 12 of the ICESCR, which explicitly provides for 'the right of everyone to the enjoyment of the highest attainable standard of physical and mental health'.[150] The ICESCR also defines steps that states should take to realize progressively the 'highest attainable standard of health', including the improvement of all aspects of environmental and industrial hygiene. Though not legally binding, Article 25 of the UDHR provides that everyone has the right to a standard of living adequate for the health and well-being of themselves and their family.[151]

At the regional level, Article 16(1) of the AfCHPR provides that every individual shall have the right to enjoy the best attainable state of physical and mental health. Article 12 of the European Social Charter recognizes the 'right of everyone to the enjoyment of the

[147] Carbon Trade Watch, 'Groups Slam Nigeria's Submission of Gas Flare Reductions for Carbon Credits', (n 36); Bank Information Center, 'Local Groups Say Project Will Not End Gas Flaring, Could Exacerbate Conflicts in the Niger Delta', www.bicusa.org/en/Project.39 .aspx, accessed 14 January 2014.

[148] See Lucas and Patzek, 'The Disastrous Local and Global Impacts of Tropical Biofuel Production', *Energy Tribune* (March 2007), 19. Some of the highlighted challenges of biofuels include the environmental impact of expanded crop production in the manufacturing of biofuels, land use conflicts, soil erosion issues, loss of biodiversity and the impact on water availability. Sugarcane, which is among the most efficient feedstock for biofuel, is heavily dependent on extensive use of water. Competition between biofuel production and requirements for water for other needs, including household use is thus likely to occur. See D. Mitchell, *Biofuels in Africa: Opportunities, Prospects and Challenges* (The World Bank, 2010).

[149] C. Furgal and J. Seguin, 'Climate Change, Health, and Vulnerability in Canadian Northern Aboriginal Communities', *Environmental Health Perspectives*, 114 (2006), 1964–1968.

[150] ICESCR, see also B. Toebes, *The Right to Health as a Right in International Law* (Amsterdam: Hart/Intersentia 1999).

[151] See UDHR, (n 136).

highest attainable standard of physical and mental health',[152] while Article 10 of the First Protocol to the American Convention on Human Rights provides that 'everyone shall have the right to health, understood to mean the enjoyment of the highest level of physical, mental and social wellbeing'.

In order to foster the protection of the right to health, states are to take public health measures to prevent epidemic, endemic, occupational or environmental diseases.[153] This has been interpreted to include removing as far as possible all causes of ill health.[154] One way to reduce the short- and long-term impacts of carbon projects is to ensure that transparent environmental impact assessments (EIA) are conducted before project approval, to show clearly their likely environmental and health impacts.[155] In the 2006 case of *Sawhoyamaxa Indigenous Community v. Paraguay*, the Inter-American Court of Human Rights unanimously found Paraguay to be in violation of rights to health, life, and property of the Sawhoyamaxa indigenous community for failing to remove projects that caused pollution, ill health and diseases in these indigenous communities.[156] The Court called on the government of Paraguay to demarcate the indigenous lands and provide a development fund, among other remedies.[157]

To take human rights seriously in the design, approval, finance and implementation of carbon projects, it is critical for national authorities to ensure that projects that threaten human health are not granted host country approval; and for supervisory institutions, such as the CDM EB, to ensure that projects approved without an EIA are not registered. The same applies to projects such as the Kwale project that was allegedly approved by the host country after it allegedly connived with project proponents to stage-manage an EIA. Without good health, the ability to enjoy other human rights would be significantly threatened.[158]

[152] See also European Convention for the Protection of Human Rights and Fundamental Freedoms (Signed 4 November 1950, entered in to force 3 September 1953, 213 UNTS 221, ETS 5).

[153] See Art. 12(2), ICESCR, (n 111).

[154] See A. Chapman, 'Core Obligations Related to the Right to Health', in A. Chapman and S. Russell (eds.), *Core Obligations: Building a Framework for Economic, Social and Cultural Rights* (Antwerp: Intersentia 2002), 85–216.

[155] See UNECE, The Espoo Convention on Environmental Impact Assessment in a Transboundary Context adopted 25 February 1991 EmuT 991:15.

[156] *Case of the Sawhoyamaxa Indigenous Community v. Paraguay*, 2006 Inter-Am. Ct. H.R. (ser. C) No. 146, 248(1)–(3) (29 March 2006).

[157] *Ibid.*, 239–241.

[158] As Herophilus, the physician to Alexander the Great, wrote in his often quoted poem: 'When health is absent Wisdom cannot reveal itself, Art cannot become manifest,

2.3.6 *Inequalities and discrimination against minority communities and vulnerable groups*

Discriminatory trends in climate actions and projects take two main dimensions: the targeting of poor communities as locations for projects; and the exclusion of women and vulnerable minority groups from stakeholder meetings and consultations.[159] In Nigeria and Panama for example, large-scale carbon projects have been sited in poor and disadvantaged communities where stakeholders do not have the means and resources to demand accountability or to resist such projects.[160] Such trends build on well-entrenched power imbalance, low level of awareness and other discriminatory practices such as limited opportunities for education, stereotypes and bias against women, poverty, inequitable resource distribution, and traditional caste systems amongst others, to place the burden of climate change mitigation actions and projects on poor and marginalized groups.[161]

Studies have shown that women and vulnerable minority groups that could be affected by carbon projects are often excluded or marginalized from project decisions and stakeholder meetings. According to a 2011 Report of the United Nations, skewed power relations coupled

Strength cannot fight, Wealth becomes useless and intelligence cannot be applied' *Encyclopedia of World Biography Supplement* (Vol. 25, Thomson Gale, 2005).

159 See N. Eddy and G. Wiser, 'Public Participation in the Clean Development Mechanism of the Kyoto Protocol', in C. Bruch (ed.), *New Public: The Globalization of Public Participation* (Environmental Law Institute, 2002), 203; G. Terry, 'No Climate Justice Without Gender Justice: An Overview of the Issues', *Gender and Development*, 17(1) (2009), 5–18; see also G. Terry, *Climate Change and Gender Justice* (Oxford: Oxfam, 2009), 3; Roht-Arriaza, 'Human Rights in the Climate Change Regime', (n 75), 215–219; S. Jodoin, 'Human Rights and Climate Change Governance' (Center for International Sustainable Development Law, Legal Working Paper, December 2010).

160 See Allen et al., 'Niger Delta Oil Flares, Broken Laws, Pollution and Oppression', (n 36), chp. 4.

161 A good example of discrimination in environmental planning was the case of environmental racism, which was the subject of intense debates and litigation in the United States in the late 1990s. It included the intentional siting of hazardous waste sites, landfills, incinerators, and polluting industries in communities inhabited mainly by African-American, Hispanics, Native Americans, Asians, migrant farm workers, and the working poor. For example the South Side of Chicago, which is predominantly African-American and Hispanic, has the greatest concentration of hazardous waste sites in the United States. See C. Beasley 'Of Pollution and Poverty', *The Environmental Journal*, 2(4) (1990), 38–45; R. Bullard, *Confronting Environmental Racism: Voices from the Grassroots* (Cambridge, MA: South End Press, 1993); K. Colquette, K. Michele, A. Elizabeth, and H. Robertson 'Environmental Racism: The Causes, Consequences, and Commendations', *Tulane Environmental Law Journal*, 5(1) (1991), 153–207.

with inequitable, cultural and social norms, often mean that the perspectives of women are neglected in the planning of mitigation projects with only few programmes designed for them to carry them along in the process and to ensure their active participation in climate change mitigation.[162]

These discriminatory trends in climate change mitigation are violations of the fundamental human rights against discrimination, a right recognized in a number of core international human rights instruments.[163] Article 26 of the ICCPR provides that all persons are equal before the law, and are entitled without any discrimination to equal protection of the law. It requires states to prohibit discrimination and guarantee to all persons equal and effective protection against discrimination on any ground such as race, colour, sex, language, religion, political or other opinion, national or social origin, property, birth or other status.[164] Article 2(3) and 3 of the ICESCR also contains similar provisions on non-discrimination.[165]

The UN Sub-Commission on the Prevention of Discrimination and Protection of Human Rights, described prevention of discrimination as the prevention of any action which denies to individuals or groups of people the equality of treatment which they may

[162] According to the report, several societal dynamics contribute to this, they include: lack of access to formal education, poverty, discrimination in job distribution, food insecurity, limited access to resources, exclusion from policy and decision-making institutions and processes and other forms of social marginalization. These dynamics put women at a distinct disadvantage. See UNEP, *Women at the Frontline of Climate Change: Gender Risks and Hopes: A Rapid Response Assessment* (United Nations Environment Programme, 2011), 6. See also M. Hemmati, 'Gender Perspectives on Climate Change', Emerging Issues Panel, Commission on the Status of Women Fifty-second session New York, 25 February/7 March 2008. See also O. Oyebode, 'Nigeria Fails Commonwealth's Mark for Women Representation' (*Saturday Punch*, 30 June 2007), A1; Nigerian CEDAW NGO Coalition, *The Situation of Nigerian Women Rights in Nigeria* (A Report Submitted to the 41st session of the United Nations Committee on the Elimination of all Forms of Discrimination Against Women, New York, 2008).

[163] Non-discrimination provisions are contained in several core international human rights instruments and in specialized international instruments such as the United Nations Charter of 1945 (arts. 1 and 55), UDHR (art. 2) and the ICCPR (art. 2); ILO Convention concerning Discrimination in Respect of Employment and Occupation No. 111 of 1958 (art. 1); ICERD (art. 1); UNESCO Convention against Discrimination in Education of 1960 (art. 1); UNESCO Declaration on Race and Racial Prejudice of 1978 (arts. 1, 2 and 3); Declaration on the Elimination of All Forms of Intolerance and of Discrimination based on Religion or Belief of 1981 (art. 2); and the Convention on the Rights of the Child of 1989 (art. 2).

[164] ICCPR, (n 112). [165] ICESCR, (n 111).

wish.[166] Article 1 of the Convention on the Elimination of Racial Discrimination (CERD) defines discrimination as any distinction, exclusion, restriction or preference based on race, colour, descent or national or ethnic origin which has the purpose or effect of nullifying or impairing the recognition, enjoyment or exercise, on an equal footing, of human rights and fundamental freedoms in the political, economic, social, cultural or any other field of public life.[167]

Article 7 of the Convention on the Elimination of All forms of Discrimination against Women (CEDAW) also provides for the elimination of discrimination against women in political and public life.[168] Article 5 encourages states to take measures to eliminate prejudices and stereotypes against women. Article 1 of the UDHR also provides that 'All human beings are born free and equal in dignity and rights.' Article 2 notes that 'everyone is entitled to all the rights and freedoms set forth in the Universal Declaration without distinction of any kind, such as race, colour, sex, language, religion, political or other opinion, national or social origin, property, birth or other status'. Article 4 provides that 'No one shall be held in slavery or servitude' while Article 7 declares unequivocally that 'All are equal before the law and are entitled without any discrimination to equal protection of the law.'[169]

Though not legally binding, the UN 'Protect, Respect and Remedy' Framework also provides detailed and persuasive explanation of how discriminatory trends can be avoided in project planning and execution.[170] It provides that business enterprises should ensure equality of opportunity and treatments, as recognized in international law, to ensure that all forms of discrimination are eliminated from business related activities.[171] The UN Framework, in paragraph 26, specifically notes

166 See United Nations document E/CN.4/52, Section V. See also A. Bayefsky, 'The Principle of Equality or Non-discrimination in International Law', *Human Rights Quarterly*, 11 (1990), 5.

167 See the International Convention on the Elimination of All Forms of Racial Discrimination, Mar. 7, 1966, 660 UNTS 195 (hereinafter CERD).

168 See Optional Protocol to the Convention on the Elimination of Discrimination Against Women (CEDAW) (Adopted 6 October 1999, entered into force 10 December 1999) 2131 UNTS 83, reprinted in 39 ILM 281 (2000).

169 See UDHR, (n 136).

170 United Nations, *Guiding Principles on Business and Human Rights: Implementing the United Nations 'Protect, Respect and Remedy'*, www.ohchr.org/Documents/Publications/GuidingPrinciplesBusinessHR_EN.pdf, accessed 12 December 2015.

171 'Guiding Principles on Business and Human Rights', *ibid.*

that 'States should take appropriate steps to ensure the effectiveness of domestic judicial mechanisms when addressing business-related human rights abuses, including considering ways to reduce legal, practical and other relevant barriers that could lead to a denial of access to remedy.' The Commentaries to this paragraph rightly identify that due to imbalance in power relations between the parties to business-related human rights claims; active discrimination; or unintended consequences of the way judicial mechanisms are designed and operate, individuals from groups or populations at heightened risk of vulnerability or marginalization often face additional cultural, social, physical and financial impediments to accessing, using and benefiting from justice mechanisms. The UN Framework notes that 'particular attention should be given to the rights and specific needs of such groups or populations at each stage of the remedial process: access, procedures and outcome.' The UN Framework provides normative guidance on the value of redesigning project approval frameworks to ensure that fundamental rights and freedoms of poor and vulnerable groups are protected, realized and fulfilled in the design, approval, finance and implementation of carbon projects.

The Conference of Parties to the UNFCCC (COP16) in the Cancun Decisions also recognised that gender equality and the effective participation of women and indigenous peoples are important for effective action on all aspects of climate change.[172] However, the COP did not provide concrete guidelines or frameworks on how issues of discrimination and inequity can be addressed in mitigation actions and projects.

To protect and fulfil non-discrimination obligations of international human rights law, host countries must be required to demonstrate site selection criteria, including detailed human rights impacts assessment reports on how the rights of vulnerable communities and groups have been taken into consideration at the design stages of the mitigation project. Host countries and parties to the international climate change regime in general must be encouraged to eliminate discriminatory practices that place the burden of climate change mitigation on a specific section of the society.

[172] International Shared Vision on Long-Term Cooperative Action, Cancun LCA Outcome, para. 7., UNFCCC, unfccc.int/files/meetings/cop_16/application/pdf/cop16_lca.pdf, accessed 15 April 2015.

2.3.7 *Loss of consultative/participatory rights*

One of the most reoccurring concerns with the implementation of above discussed CDM and REDD+ projects is that members of the public have not been provided adequate access to information (ATI) about project activities, most especially information on the short- and long-term impacts of projects on local communities. In addition, many projects have been designed, approved and implemented without broad consultation and participation of stakeholders and members of the public whose rights have been severely affected by such projects. There have also been concerns that project developers and national authorities fail to respect, and reflect indigenous people's rights and knowledge in the design of project activities and strategies. These concerns raise the need for greater recognition and protection of ATI and participatory rights in the international carbon finance architecture.

2.3.7.1 Right to information

As Aristotle once wrote, 'All human beings by nature desire to know'.[173] One of the key criticisms against the CDM is that many host communities are not provided adequate information on the likely impacts and benefits of projects. Most notable is the charge that states cloak project description documents in such secrecy that the public finds it hard to know how these projects might affect them in the short and long term.[174] As such, projects that should normally be lauded end up being opposed by the public who were excluded from the approval and implementation process. This raises the need for a public disclosure system that provides the public with up-to-date information on projects that could affect their lives.

Generally, under CDM Rules, project proponents are required to facilitate access to information by making information publicly available.[175] Project proponents are required to provide three categories of information: information that is related to a specific project; information contained in a database or registry; and information that is related to CDM governance, procedures or policy. While the rules stipulate that these categories of information must be made publicly available, the

[173] Aristotle, *Metaphysics*, Book VII, 4 (241B [1029b6–9]).

[174] See Onojiribholo, 'Kwale Chief Laments Plight of Communities', (n 36).

[175] Article 11 of the UNFCCC also requires a transparent system of governance over financial mechanisms through reporting and information disclosure requirement. See also Para. 27(h), CDM Modalities & Procedures, 3/CMP.1 (2005), cdm.unfccc.int/Reference/COPMOP/08a01.pdf, accessed 12 June 2015.

manner and time frame for making specified CDM documents 'publicly available' are not exhaustively elaborated.[176] This ambiguity with regard to the meaning and nature of 'publicly available' has been a major flaw in the CDM Rules.[177] The ambiguity is in the fact that information that appears after project registration perennially eludes the public as the CDM Rules do not establish mandatory requirements for project proponents to proactively disclosure project information post-registration and over the lifetime of a project. The rules do not also expatiate on the contents and elements of full and maximum disclosure.

Furthermore, while CDM Rules require states to provide information on an EIA, this requirement applies only when an EIA is necessary. According to the rules, an EIA would only be necessary 'if the impacts are considered significant by the project participants or the host party'.[178] What this implies is that it is the host country or the project developers that decide if an EIA is important. Current experience shows that most states have undermined the protection for the public by waiving aside the 'significant impact' requirement for a number of projects; thereby leading to a number of projects without an EIA.[179] This leaves the public at the mercy of huge projects that produce significant short- and long-term environmental effects.

There are some debates on the legal status of a right to information under international law. While some scholars identify the right to information as part of the right to expression and free speech, other scholars consider such views as attempts to expand the right to expression beyond its traditional limits.[180] This book aligns with the views that ATI, as enshrined in a number of core international human rights instruments, is part of the fundamental human right to expression, which includes the right to seek, receive and impart

[176] See Eddy and Wiser, 'Public Participation in the Clean Development Mechanism of the Kyoto Protocol', (n 159), 205.

[177] *Ibid.* [178] CDM Modalities & Procedures , (n 175).

[179] See D. Olawuyi, 'Fostering Accountability in Large Scale Environmental Projects: Lessons from CDM and REDD+ Projects', in J. Wouters, A. Ninio, T. Doherty and H. Cisse (eds.), *Improving Delivery in Development: The Role of Voice, Social Contract, and Accountability* (Washington, DC: World Bank Legal Review, The World Bank Group, 2015) 127–147.

[180] UNFPA, 'Core Concepts of a Human Rights Based Approach' (2010), www.unfpa.org/webdav/site/global/shared/documents/publications/2010/hrba/introduction.pdf, accessed 15 July 2014; see also F. Nguyen, 'Emerging Features of a Rights-Based Development Policy of UN, Development Cooperation & NGO Agencies' (2002) Discussion paper, OHCHR Asia-Pacific Regional Office, www.un.or.th/ohchr/issues/rba/Pdf/discussrba.pdf, accessed 12 October 2014.

information.[181] As Jagwanth rightly notes, 'given that rights are interdependent, in order to be able to exercise our rights more generally, people must be given access to information'.[182] Access to information (ATI) is defined in this book as the right to 'seek, receive and impart' information about carbon projects. Elements of ATI include maximum disclosure of project information, obligation to publish and disclose new information, establishing processes to facilitate timely access to information at a reasonable cost and increased public education and awareness. Information provides the necessary precondition that makes it possible for people to exercise and demand their rights.[183]

The right to seek, receive and impart information is recognized in Article 19(2) of the ICCPR which provides that everyone shall have the

[181] Several reports have been released by the Special Rapporteur on the Right to Freedom of Opinion and Expression that recognize ATI as part of the right to freedom of expression. Though not binding, these reports provide normative interpretations of the right to information as one of the most essential elements of freedom of speech and expression. According to the Special Rapporteur, 'The right to seek, receive and impart information imposes a positive obligation on states to ensure access to information, particularly with regard to information held by Government in all types of storage and retrieval systems. . .' See Report of the *Promotion and Protection of the Right to Freedom of Opinion and Expression*, UN Doc E/CN.4/1998/40, 28 January 1998, para. 1. See also Report of the Special Rapporteur, *Promotion and Protection of the Right to Freedom of Opinion and Expression*, UN Doc E/CN.4/1997/31, 4 February 1997; Report of the Special Rapporteur, *Promotion and Protection of the Right to Freedom of Opinion and Expression*, UN Doc E/CN.4/1995/31, 14 December 1995, para. 35. Joint Declaration of the UN Special Rapporteur on Freedom of Opinion and Expression, the OSCE Representative on Freedom of the Media and the OAS Special Rapporteur on Freedom of Expression of 26 November 1999, www.article19.org/pdfs/igo-documents/three-mandates-dec-1999.pdf, accessed 12 May 2015. See also Joint Declaration of the UN Special Rapporteur on Freedom of Opinion and Expression, the OSCE Representative on Freedom of the Media and the OAS Special Rapporteur on Freedom of Expression of 6 December 2004, www.unhchr.ch/huricane/huricane.nsf/0/9A56F80984C8BD5EC 1256F6B005C47F0?opendocument, accessed 12 May 2015; and the Joint Declaration of the UN Special Rapporteur on Freedom of Opinion and Expression, the OSCE Representative on Freedom of the Media and the OAS Special Rapporteur on Freedom of Expression of 19 December 2006, www.article19.org/pdfs/standards/four-mandates-dec-2006.pdf, accessed 12 May 2015.

[182] S. Jagwanth, 'The Right to Information as a Leverage Right', in R. Calland and A. Tilley (eds.), *The Right to Know, The Right to Live: Access to Information and Socio-Economic Justice* (Capetown: Open Democracy Advice Centre, 2002), 3–16; see also J. Klaaren, 'A Right to a Cell Phone? The Rightness of Access to Information' in Calland and Tilley (eds.), *The Right to Know, The Right to Live: Access to Information and Socio-Economic Justice*, *ibid.*, 17–26.

[183] P. Birkinshaw, 'Freedom of Information and Openness: Fundamental Human Rights', *Administrative Law Review*, 58 (2006), 177 at 210.

right to freedom of expression; this right shall include freedom *to seek, receive and impart information...*'[184] Article 19 of the UDHR also provides that 'everyone has the right to freedom of opinion and expression; this right includes the freedom to hold opinions without interference and *to seek, receive and impart information* and ideas through any media and regardless of frontiers'.[185] Similarly in 1998, the UN General Assembly adopted the Declaration on Human Rights Defenders. Though not legally binding, it provides in its Article 6 that the right to seek, obtain and receive information is fundamental to the effective promotion and protection of human rights.[186]

Regionally, Article 13 of the legally binding American Convention on Human Rights (ACHR) provides that 'everyone has the right to freedom of thought and expression. This right includes freedom to *seek, receive, and impart information* and ideas of all kinds, regardless of frontiers, either orally, in writing, in print, in the form of art, or through any other medium of one's choice'.[187] The Inter-American Court of Human Rights has interpreted Article 13 to include the freedom of information as 'a fundamental human right that is as important to a free society as freedom of expression'. This includes the right and freedom to seek, receive and impart information and ideas of all kinds.[188] The Special Rapporteur of the American Commission also noted that the right to access to official information is one of the cornerstones of representative democracy.[189] The Inter-American Court of Human Rights has upheld this right in *Marcel Claude Reyes et al v. Chile*, when it held that withholding information on a developmental project is a violation of the right to information of the petitioners.[190] The court noted that this right includes not only the right and freedom to express one's own thoughts, but also 'the right

[184] Emphasis added. See ICCPR, (n 112). [185] See UDHR, (n 136).

[186] *Declaration on the Right and Responsibility of Individuals, Groups and Organs of Society to Promote and Protect Universally Recognized Human Rights and Fundamental Freedoms* (Declaration on Human Rights Defenders), UN General Assembly Resolution 53/144 (8 March 1999), daccessdds.un.org/doc/UNDOC/GEN/N99/770/89/PDF/N9977089.pdf?OpenElement, accessed 12 May 2015.

[187] Emphasis added.

[188] *Compulsory Membership in an Association Prescribed by Law for the Practice of Journalism*, Advisory Opinion OC-5/85, 13 November 1985, para. 30.

[189] *Annual Report of the Inter-American Commission on Human Rights 1998, Volume III, Report of the Office of the Special Rapporteur for Freedom of Expression*, Chapter III.

[190] Case 12.108, Report No. 60/03, Inter-Am. C.H.R., OEA/Ser.L/V/II.118 Doc. 70 rev. 2 at 222 (2003). In that case non-governmental organizations (NGOs) and individuals, alleged that the Government of Chile violated their right to access to information under Article 13 of the American Convention on Human Rights when the Chilean

and freedom to seek, receive and impart information and ideas of all kind'.

Similarly, Article 10 of the European Convention on Human Rights (ECHR) provides that the right to freedom of expression includes freedom to hold opinions and to receive and impart information and ideas without interference by public authority. The Council of Europe also invited 'member states which have not yet done so to introduce a system of freedom of information, that is, access to government files. . .' In its Recommendation Council No.R (81) 19 on Access to Information Held by Public Authorities, the Council affirmed that everyone within the jurisdiction of a member state shall have the right to obtain, on request, information held by the public authorities other than legislative bodies and judicial authorities and that this shall apply subject only to such limitations and restrictions as are necessary in a democratic society for the protection of legitimate public interests (such as national security, public safety, public order, the economic wellbeing of the country, the prevention of crime, or for preventing the disclosure of information received in confidence), and for the protection of privacy and other legitimate private interests, having, however, due regard to the specific interest of an individual in information held by the public authorities which concerns him personally.[191]

The Council of Europe also provided perhaps what was the very first comprehensive attempt to link the right to seek, receive, and impart information to environmental matters – The Convention on Civil Liability for Damage Resulting from Activities Dangerous to the Environment, which provides for access to information and access to justice in environmental matters.[192] Under Articles 13 to 16, public

Committee on Foreign Investment failed to release information about a deforestation project known as 'Río Cóndor' which the petitioners tried to evaluate. The state argued that the requested information was confidential and that releasing it could discriminate against the investors. According to the state 'reservation of information in this type of company constitutes a cornerstone of constitutional economic guarantees and Chilean foreign investment policy' and that the release of said information would constitute arbitrary discrimination against the investors. The court upheld the right of the petitioners to seek, receive and impart information.

[191] Recommendation No. R (81) 19 of the Committee of Ministers of the Council of Europe on Access to Information Held by Public Authorities, adopted 25 November 1981, www.coe.int/t/e/legal_affairs/legal_co-operation/administrative_law_and_justice/texts_&_documents/Conv_Rec_Res/Recommendation(81)19.asp, accessed 12 May 2015.

[192] See Council of Europe, Convention on Civil Responsibility for Damage Resulting from Activities Dangerous to the Environment (Lugano, 21 June 1993). Speaking of environmental information, the Rio Declaration on Environment and Development, as far back

authorities are to provide access to information to any person upon request and without the person having to prove an interest. According to Article 14 any person shall, at their request and without having to prove an interest, have access to information relating to the environment held by public authorities. Article 15 provides that any person shall have access to information relating to the environment held by bodies with public responsibilities for the environment and under the control of a public authority. Access shall be given via the competent public administration or directly by the bodies themselves.[193]

In 1998, the United Nations Economic Commission for Europe (UNECE) adopted the Convention on Access to Information, Public Participation in Decision-making and Access to Justice in Environmental Matters (Aarhus Convention). Despite debates on the practical aspects of implementing some of its provisions, the Aarhus Convention, arguably, remains one of the most elaborate discussions of the right to information and participation in environmental matters.[194] Article 4 provides that each party shall ensure that public authorities, in response to a request for environmental information, make such information available to the public, within the framework of national legislation, including, where requested. Such information is to be made available as soon as possible and at the latest within one month after the request has been submitted, unless the volume and the complexity of the information justify an extension of this period up to two months after the request. The applicant shall be informed of any extension and of the reasons justifying it.[195]

as in 1992 recognized in its Principle 10 that environmental issues are best handled with the participation of all concerned citizens, at the relevant level. At the national level, each individual shall have appropriate access to information on hazardous materials and activities in their communities, and the opportunity to participate in decision-making processes. states shall facilitate and encourage public awareness and participation by making information widely available. Effective access to judicial and administrative proceedings, including redress and remedy, shall be provided. See also Agenda 21, the Blueprint for Sustainable Development, which was also adopted in Rio, it states that individuals, groups and organizations should have access to information relevant to environment and development held by national authorities, including information on products and activities that have or are likely to have a significant impact on the environment, and information protection measures.

193 *Ibid.*

194 See Convention on Access to Information, Public Participation in Decision–Making and Access to Justice in Environmental Matters, 2161 UNTS 447; 38 ILM 517 (1999) (Aarhus Convention). See also S. Wates, 'The Aarhus Convention: A Driving Force for Environmental Democracy', *Journal for European Environmental and Planning Law*, 2(1) (2005), 1–11.

195 Aarhus Convention, *ibid.*

Under this Convention, the grounds upon which a request for environmental information may be refused are: if the public authority to which the request is addressed does not hold the environmental information requested; if the request is manifestly unreasonable or formulated in too general a manner; or if the request concerns material in the course of completion or concerns internal communications of public authorities where such an exemption is provided for in national law or customary practice, taking into account the public interest served by disclosure.[196] The implication is that except when information is not available or when non-disclosure is in the overall interest of the public, all project information must be disclosed.

The right to seek, receive, and impart information provides a platform for individuals and communities to demand information on a project that could affect them. Governments therefore have responsibilities to provide or to compel project proponents to disclose to citizens pertinent information about development projects. As the African Commission on Human and Peoples' Rights held in *SERAC v. Federal Republic of Nigeria*, by colluding with Shell and by allowing the continuous oil spillage and gas flaring in the Niger Delta, the government of Nigeria had violated many rights of the Niger Delta people, including the right to access information on matters that concern their lives and well-being.[197]

The scope and contents of the right to information are elaborated in greater detail in other parts of this book. Two dimensions of the right to information are explored. First is the right of individuals to information to which access must be given upon request; this includes situations where citizens request additional or detailed information about a governmental policy or project. The right to information places a duty on

[196] See Art. 4 (4) of the Aarhus Convention, *ibid.*, on the grounds for withholding information in public interest.

[197] (2002) ACHPR/COMM/A044/1. This was a case instituted on behalf of the Ogoni people to challenge the Government of Nigeria on the impact of oil exploration in their community. They allege that the government has neither monitored operations of the oil companies nor required safety measures that are standard procedure within the industry; that the government has withheld from Ogoni Communities information on the dangers created by oil activities. Ogoni Communities have not been involved in the decisions affecting the development of Ogoniland and that the government has not required oil companies or its own agencies to produce basic health and environmental impact studies regarding hazardous operations and materials relating to oil production, despite the obvious health and environmental crisis in Ogoniland. They allege the violation of the rights to information under the African Charter. The commission held that the Nigerian government was in violation of several rights including the right to information by giving the green light to private actors, and the oil Companies in particular, to devastatingly affect the well-being of the Ogoni people.

governments to oblige such requested information. Secondly, the right to information includes the right to such information that must be published and disseminated *suo motu* (proactively) by public authorities. This places a duty on governments to run disclosure programmes that allow individuals to gain regular access to publicly held information. This could be by keeping a public database or by releasing periodic documents on governmental affairs. This second aspect is important especially in poor and less literate countries where some less privileged members of the society might not on their own volition request for public information. Putting the information out there through public disclosure programmes would ensure transparency and a level playing field.[198]

The right to seek, receive and impart information places a duty on states to lay down laws and policies that allow for maximum disclosure in the conduct of state affairs, except where public security and interests dictate otherwise.[199] A number of countries have enacted Freedom of Information (FOI) Acts, which enable this to happen.[200] However, the extent to which such laws are respected and information requests are complied with vary from country to country. In many developing countries where CDM activities take place, there have been tendencies not to comply with information requests on projects even when FOI Acts are

[198] See Article 19, 'A Model Freedom of Information Law', www.article19.org/pdfs/standards/modelfoilaw.pdf, accessed 12 May 2015; Article 19, 'The Johannesburg Principles: Principles on National Security, Freedom of Expression and Access to Information' (1996), www.article19.org/pdfs/standards/joburgprinciples. accessed 12 May 2012; Article 19, 'Who Wants to Forget? Truth and Access to Information about Past Human Rights Violations' (2000), www.article19.org/pdfs/publications/freedom-of-information-truth-commissions.pdf, accessed 12 May 2015; Centre for Policy Alternatives, Commonwealth Human Rights Initiative and Human Rights Commission of Pakistan, 'Global Trends on the Right to Information: A Survey of South Asia' (2001), www.article19.org/pdfs/publications/south-asia-foi-survey.pdf, accessed 12 May 2015.

[199] See *Gaskin v. The United Kingdom* (1989) 12 EHRR 36.

[200] As far back as 1766, the Government of Sweden enacted the *Freedom of the Press Act*, which allows Swedish citizens to access public data. This was followed by Finland many years later when it came up with the *Publicity of Documents Act* in 1951; Denmark's *Public Access Act* in 1970 and the Dutch *Administrative Transparency Act* of 1978. See M. Halstuk and B. Chamberlin, 'The Freedom of Information Act 1966–2006: A Retrospective on the Rise of Privacy Protection Over the Public Interest in Knowing What the Government's Up To', *Communication Law and Policy*, 11 (2006), 511. See also C.36, Laws of the United Kingdom 2000, www.legislation.gov.uk/ukpga/2000/36/contents, accessed 12 December 2014; Canada's Freedom of Information Act 1982; United States' Freedom of Information Act of 1966, New Zealand's Access to Information Act, Chapter A-1. See Birkinshaw, 'Freedom of Information and Openness', (n 183), 177, 179, 210.

present. For example in Nigeria, there were hopes that the passing of the Freedom of Information Act 2011 would end governmental secrecy and make public records and information more accessible. However, no FOI request has been granted till date.[201] This emphasizes the need for an international approach to ensuring that adequate information on mitigation projects are released. While we do not suggest here that international guidelines or threshold is a one time fix for all disclosure problems, it allows the CDM EB and other supervisory boards to block the approvals and registrations of projects that do not meet the necessary disclosure thresholds.

Inadequate access to project information creates tensions and resistance for carbon projects that with adequate information and awareness could have drawn support from environmental groups and individuals. Information disclosure increases the transparency of government decision-making, thereby reducing tensions and fostering public trust. Lack of transparency brings about and feeds the suspicion of corruption by the public even when it is absent. This places a perpetual burden on the government to defend its actions and policies thereby slowing down reforms and reducing the efficacy of mitigation programmes. Information disclosure leads to greater transparency as it allows citizens to understand governmental actions, programmes and projects; it also allows citizens to be carried along in the process of decision-making.[202]

Information disclosure fosters transparent public debates on governmental actions. It provides opportunities for citizens to oversee

[201] The Preamble to the Nigerian Act says it is enacted to make 'public records and information more freely available, provide for public access to public records and information, protect public records and information to the extent consistent with the public interest. . .' However, a number of journalists and human rights advocates have allegedly been arrested for requesting information under the FOI Act. Similarly, the Act is replete with more exception sections and clauses that tend to deny access to information than those that grant it, a situation which allows government agencies to deny information access. For example, while only two sections (sections 1 & 3) deal with access to information, ten sections (sections 7, 11,12,14,15,16,17,18, 19, and 28) contain exceptions that arguably deny the public access to information. Also, the time limit provided for in the law for granting or refusal of requests as provided for in section 4 of the Act is very short. The cost of requesting information is also very high. These problems were discussed in the case of *Nigerian Contract Monitoring Coalition v. The Power Holding Company of Nigeria (PHCN)* (2012) FHC/ABJ. See generally, 'Nigeria's Access to Information Law is Not Working' (4 December 2012), www.icirnigeria.org/index.php?id=28, accessed 12 January 2015; see also 'Nigeria: Reality of the Freedom of Information Act' (*Daily Trust*, 17 September 2012).

[202] L. Brandeis, 'What Publicity Can do' in *Other People's Money* chapter 5, 92 (1932), first published in *Harper's Weekly*, 20 December 1913.

governmental actions to ensure that human rights are respected.[203] It also keeps project proponents alert. Information disclosure supplies ready tools for environmental advocates to pin point and expose violations. It allows the public, and advocates, to serve as watchdogs for projects. Reducing secrecy allows members of the public to scrutinize the failures of the markets and to take them up accordingly. This prevents market forces from lowering sustainability standards and from trampling on the rights of individuals under the guise of secrecy.[204]

There is a need for the international climate change regime to establish a mandatory project disclosure threshold that stipulates project information that must be disclosed within the lifetime of a mitigation activity or project. This threshold is proposed and discussed in subsequent parts of this book. For example, people ought to be informed and updated about any project or action that might lead to human rights violations or any project that might lead to adverse environmental impacts.[205] Such information would be provided first by conducting an environmental impact assessment of the likely environmental effect of a proposed project and making public the findings of such an assessment. When the measure of the changes that a process of development might have on the environment are assessed, in order to establish how deleterious they might be, and these changes are communicated without secrecy, and people are given the chance to comment on these findings, then access to information will be guaranteed.

2.3.7.2 Participation

Apart from the lack of timely and adequate information on carbon projects, some of the discussed CDM projects were opposed for not providing adequate opportunities for stakeholders to participate in the planning and execution of projects.[206] This infringes the robust principles of participation and participatory rights that have been increasingly recognized in international law over the last decade. Participation is the right of the public to 'take part in' and to 'influence' decision-making processes on carbon projects. This goes beyond mere consultation. It involves the right

[203] C. Sunstein, 'Informational Regulation and Informational Standing: Akins and Beyond', *University of Pennsylvania Law Review*, 147 (1999), 613, 625.

[204] Jagwanth, 'The Right to Information as a Leverage Right', (n 182), 3–16.

[205] See J. Echeveria and J. Kaplan, *Poisonous Procedural Reforms: In Defense of Environmental Right to Know* (Georgetown University, 2002) 1–3.

[206] See International Rivers, 'Submission Regarding Human Rights to the CDM', (n 58).

of key stakeholders to influence and share control over projects, policies, decisions and resources that affect them.[207]

Scholars have questioned whether international law has fully embraced participatory rights; there are also questions about what participation truly entails; and whether the right to participate can be fully implemented.[208] Despite these debates, the doctrine of participation and inclusion has made its way into a number of core international human rights instruments. The legally binding ICCPR guarantees participatory rights in its Article 25(a), which stipulates that every citizen shall have the right and opportunity to participate, without distinction and without unreasonable restrictions, in the conduct of public affairs, directly or through freely chosen representatives.[209] Article 27 also provides that in states in which ethnic, religious or linguistic minorities exist, persons belonging to minorities shall not be denied the right, in community with the other members of their group, to enjoy their own culture, to profess and practise their own religion, or to use their own language.

Though not legally binding, the United Nations Declaration on the Right to Development prominently recognizes participation as a key requirement for development. This Declaration states that the right to development includes the right to participate in, contribute to, and enjoy economic, social, cultural and political development. It recognizes that there are three human rights standards that are particularly relevant to the full enjoyment of the right to development: the right to self-determination, sovereignty over natural resources and popular participation.[210]

The significance of the right to participation to the realization of development was later emphasized by the General Assembly in another resolution that stressed the importance of the adoption of measures to ensure the effective participation, as appropriate, of all the elements of society in the preparation and implementation of national economic and social development policies and of the mobilization of public opinion and

[207] The question of what constitutes 'stakeholders' and 'public' is addressed in greater detail in subsequent parts of this chapter. The World Bank, *The World Bank Participation Sourcebook* (Washington, DC: The World Bank, 1996), 3.

[208] See for example G. Fox, 'The Right to Political Participation in International Law', in G. Fox and B. Roth (eds.), *Democratic Governance and International Law* (Cambridge: Cambridge University Press, 2000) 48–55, Fox questioned whether the international community is prepared to accept a right to participation.

[209] ICCPR, (n 112).

[210] United Nations Declaration on the Rights to Development, GA Res. A/RES/41/128, 97th Plenary Meeting 41/128 (1986).

the dissemination of relevant information in the support of the principles and objectives of social progress and development.[211]

A more expansive provision on the right to take part and influence decision-making was included in the UNDRIP, which confers on indigenous peoples the right to choose the modalities of participation and the right to participate fully 'through representatives chosen by themselves in accordance with their own procedures'[212] and through procedures determined by them, in devising legislative and administrative measures that may affect them.[213] Article 18 provides that indigenous peoples have the right to participate in decision-making in matters, which would affect their rights, through representatives chosen by themselves in accordance with their own procedures, as well as to maintain and develop their own indigenous decision-making institutions. Article 32(2) further elaborates the duties of states to consult and cooperate in good faith with the indigenous peoples in order to obtain their free and informed consent prior to the approval of any project affecting their lands or territories and other resources, particularly in connection with the development, utilization or exploitation of mineral, water or other resources.[214] Though not legally binding, these provisions provide normative guidance on the importance of stakeholder participation, engagement and accommodation in the development and implementation of carbon projects. They also reinforce the core elements of participation to include freely taking part in decision-making processes and influencing final outcomes.

Similarly, Article 2 of the ILO Convention No. 169 concerning Indigenous and Tribal Peoples in Independent Countries provides that actions respecting indigenous peoples shall be developed with the participation of the peoples concerned. It provides that state parties must consult indigenous peoples and provide for their participation in formulating national and regional development plans that may affect them.[215] The rights include 'the right to participate in the use,

[211] UN Res. A/37/55 (1982).

[212] Art. 18, United Nations Declaration on the Rights of Indigenous Peoples, UN Doc/ A/ 61/L.67 (2007). UNDRIP was adopted by the General Assembly on 13 September 2007 by a majority of 144 states in favour, four votes against (Australia, Canada, New Zealand and the United States) and 11 abstentions. Since its adoption, Australia, New Zealand, Canada and the United States have all reversed their positions and now endorse the Declaration.

[213] *Ibid.*, Art. 19. [214] *Ibid.*

[215] International Labour Organization (ILO), Indigenous and Tribal Peoples Convention, C169, 27 June 1989, C169.

management and conservation of these resources'.[216] Article 30 requires governments to make known to the peoples concerned their rights and duties.

The United Nations Human Rights Council (UNHRC) has interpreted these provisions to include the rights of indigenous peoples to take part without any unreasonable restrictions and to influence decisions on projects that affect their linguistic, cultural and religious rights, land and resources, subsistence and participation rights, among others.[217] It recognizes the potential of indigenous peoples to enrich decision-making processes, to share perspectives and to participate as citizens and actors of change. Even though it specifically focuses on the rights of children, Article 12 of the Convention on the Rights of the Child also recognizes the rights of children to express their views freely in all matters affecting them, and that those views should be given due weight.[218]

In human right terms therefore, participation includes the right to *take part* in decision-making processes and *to influence* the decisions taken. It includes the right of individuals to take part in decision-making processes on matters affecting them and to give or withhold their consent to development activities that affect their human rights. According to the UNHRC, in *Ilmari Lansman et al v. Finland*, the essential thing is that the people have an adequate opportunity to ventilate their concerns fully

216 Art. 15, *ibid.*

217 See the case of *Mikmaq Tribal Society v. Canada* Communication No. 78/1980 (30 September 1980), UN Doc Supp. No. 40 a/39/40) at 200 (1984); where the Mikmaq Tribal Society alleged that they were excluded from Canada's political reform process in contravention of Article 25(a) of the ICCPR. The Court recognized that the right of citizens to *take part* in the conduct of public affairs without discrimination, but, however, held that even though inviting Representatives of the Mikmaq tribal society to meetings on Aboriginal matters was part of the right to take part in public affairs, in this case the right had not been violated as the group had not been subjected to any unreasonable restrictions. See Decisions, Para.6, Communication No. 205/1986. See also F. MacKay, 'A Briefing on Indigenous Peoples Rights and the United Nations Human Rights Committee' (2001) Forest Peoples Programme, online: Forest Peoples Programme, www.forestpeoples.org/documents/law_hr/unhrc_fpp_brief_dec01_eng.shtml#V_C, accessed 18 February 2015; D. Shelton, 'The UN Human Rights Committee's Decisions' (2005) Human Rights Dialogue: 'Cultural Rights', Carnegie Council for Ethics and International Affairs, www.cceia.org/resources/publications/dialogue/2_12/section_3/5151.html, accessed 18 February 2015.

218 It includes the right of a child to be heard in any judicial and administrative proceedings affecting the child, either directly, or through a representative or an appropriate body. See Convention on the Rights of the Child, UN GA Res. 44/25, Annex, 44 UN GAOR Supp. (No. 49) at 167, UN Doc A/44/49 (1989) (Adopted 20 November 1989, entered into force 2 September 1990).

about a proposed activity that may affect them, thus to determine their priorities and that those concerns are given due weight in the design and implementation of the project.[219]

2.3.7.3 Participation in environmental treaties

The need for active stakeholder participation in the design and implementation of activities that affect local environments and communities has also gained increased recognition in a number of international environmental instruments.[220] Though not legally binding, Agenda 21, adopted at the Rio Earth Summit in 1992 provides one of the early recognitions of participation as a prerequisite for sustainable development. It declares that

> one of the fundamental prerequisites for the achievement of sustainable development is broad public participation in decision-making. Furthermore, in the more specific context of environment and development, the need for new forms of participation has emerged. This includes the need of individuals, groups, and organizations to participate in environmental impact assessment procedures and to know about and participate in decisions, particularly those that potentially affect the communities in which they live and work.[221]

In 1992, the Convention on Transboundary Effects of Industrial Accidents became the first international environmental treaty to provide for the three procedural environmental rights: information, participation and access to remedies.[222] The Convention requires its parties to ensure that adequate information is given to the public in all areas capable of

219 Communication No. 511/1992, UN Doc CCPR/C/52/D/511/1992 (1994), UNHCHR, www.unhchr.ch/tbs/doc.nsf/(Symbol)/7e86ee6323192d2f802566e30034e775?Opendocument, accessed 17 February 2015. See also *Apirana Mahuika et al. v. New Zealand* Communication No. 547 /1993.

220 See the 1994 UN Convention to Combat Desertification in Countries Experiencing Serious Drought and/or Desertification, Particularly in Africa arts. 3(a), 10(e), and 19(1); 1971 Convention on Wetlands of International Importance Especially as Waterfowl Habitat 2 February 1971, 996 UNTS 245, 11 ILM 97 (1971) [Ramsar Convention].

221 See the preamble to chapter 23 of Agenda 21, approved by the UN Conference on Environment and Development on 13 June 1992: UN doc A/CONF.151/26 (vols. I–III) (1992). See also Principle 22 of the Rio Declaration, which states: Indigenous people and their communities and other local communities have a vital role in environmental management and development because of their knowledge and traditional practices. States should recognize and duly support their identity, culture and interests and enable their effective participation in the achievement of sustainable development.

222 United Nations Convention on Transboundary Effects of Industrial Accidents adopted 17 March 1992 in Helsinki 31 ILM 1330 (1992).

being affected by an industrial accident arising out of a hazardous activity. Article 9(3) provides that state parties shall provide natural and legal persons who are being or are capable of being adversely affected by the transboundary effects of an industrial accident with access to, and participation in relevant administrative and judicial proceedings, including the possibility of starting a legal action and appealing a decision affecting their rights, again equivalent to those available to persons within their own jurisdiction.[223]

The Convention on Biodiversity also contains detailed provisions on participation. Article 8(j) obliges each contracting party to, as far as possible, preserve and maintain indigenous knowledge, innovations and practices, and to promote their wider application with the approval and involvement of the holders of such knowledge, innovations and practices.[224] Article 1 of the 1993 North American Agreement on Environmental Cooperation provides that the objectives of the Agreement are to promote transparency and public participation in the development of environmental laws, regulations and policies.[225]

The Aarhus Convention also emphasizes the rights of the public to participate in environmental matters.[226] Article 3(9) provides that 'the public' shall have the opportunity to participate in decision-making 'without discrimination as to citizenship, nationality or domicile and, in the case of a legal person, without discrimination as to where it has its registered seat or an effective centre of its activities.' Article 6 adds that, public information and participation is required for all decisions on whether or not to permit industrial, agricultural and construction activities listed in Annex I to the convention. Article 6 (4) provides that each party shall provide for early public participation, when all options are open and effective public participation can take place. Article 7 also provides that each party shall make appropriate practical provisions for the public to participate during the preparation of plans and programmes relating to the environment, within a transparent and fair framework, having provided the necessary information to the public. According to the Convention, the public shall then be allowed to submit, in writing or, as appropriate, at a public hearing or inquiry with the applicant, any comments, information, analyses or opinions that it considers relevant to

223 *Ibid.*
224 Convention on Biodiversity (Adopted 5 June 1992, entered into force on 29 December 1993). See also Art. 14(1) (a).
225 North American Agreement on Environmental Cooperation (1993) 32 ILM 1480.
226 See Aarhus Convention, (n 194).

the proposed activity.[227] The opinion expressed by the public should as far as possible be taken into account in deciding on embarking on the proposed activity.[228] Aarhus therefore covers the rights of the public to take part in decision-making and to influence the final decision on whether or not an activity or project should move ahead.

Similarly, the Espoo Convention on Environmental Impact Assessment in a Transboundary Context guarantees non-discriminatory public participation in environmental impact procedures within the scope of the Convention.[229] The Convention requires parties to notify the public and to provide an opportunity for public participation in relevant environmental impact assessment procedures regarding proposed activities in any area likely to be affected by transboundary environmental harm. Article 2(6) provides that states shall 'provide an opportunity to the public in areas likely to be affected to participate in relevant impact assessment procedures regarding proposed activities and shall ensure that the opportunity provided to the public of the affected party is equivalent to that provided to the public of the Party of origin.' Article 3 states that the concerned states shall ensure that the public of the affected party in the areas likely to be affected be informed of, and be provided with, the opportunity for making comments or objections on the proposed activity, and for the transmittal of these comments or objections to the competent authority of the party of origin, either directly to this authority or, where appropriate, through the party of origin.[230]

Questions on what constitutes 'adequate participation'; specifically the normative elements of 'taking part' and 'influencing' decision-making processes, would be considered in greater detail in subsequent parts of this book. Arguably, the international climate change regime must move beyond establishing superficial and weak consultative procedures that fall short of robust human rights requirements on participation. For example, even though Article 6 of the UNFCCC provides that parties must promote and facilitate public access to information and public participation in addressing climate change and its effects, it fails to provide elaborate expansion on what constitutes good participation with respect to carbon projects.[231] It also fails to provide

[227] See Art. 6(7), and also Article 8 of Aarhus Convention, *ibid.* [228] Article 6(8), *ibid.*

[229] Convention on Environmental Impact Assessment in a Transboundary Context (Espoo Convention) (Adopted 25 February 1991, in force 27 June 1997), 30 ILM (1991) 802.

[230] Art. 3 para. 8, Espoo Convention, *ibid.*

[231] UN Framework Convention on Climate Change, 21 March 1994, 1771 UNTS 107, art. 6(a)(iii) (UNFCCC).

mandatory participatory procedures for project approvals. Similarly, Article 4 (1) (i) of the UNFCCC obliges parties to promote public awareness related to climate change and encourages the widest participation in this process, including that of non-governmental organizations. It fails to expatiate what 'widest participation' entails and how progress will be measured. Therefore, while the UNFCCC clearly recognizes the need for public participation in decision-making, many of those who stand to be directly affected by carbon projects, including indigenous peoples, perennially claim that their inputs have been ignored or have not been taken into account in project design and implementation.

Questions such as what constitutes wide participation, how and when should project proponents extend invitation to participate; what should the mode of participation be, what are the requirements for and frequency of consultative follow ups, and the requirements for the implementation of stakeholder views must be exhaustively answered. For example, CDM Rules have been criticized for failing to specify how or when stakeholder comments on projects should be solicited and the standards to be applied by the validating operational entities in deciding whether project participants satisfy the notice and comment provisions under the rules. Similarly, CDM Rules contain a narrow definition of stakeholders, which excludes important human rights gatekeepers from the participatory process created. Paragraph 1(e) of the CDM Rules define stakeholders to mean the public, including individuals, groups or communities affected, or likely to be affected, by the proposed clean development mechanism project activity.[232] The rules adopt a protective approach, which excludes non-governmental organizations and groups that even though may not be directly affected by a project; but could nonetheless 'have an interest' in a project.[233]

There is need for an expansive definition of stakeholders to include individuals, landowners, community leaders, corporations and other business organizations, and representatives from non-governmental organizations and interests groups affected or likely to be affected by, or having an interest in, the project. These are members of the public that should take part in and influence decision-making processes on

[232] Para. 1(e)., Modalities and Procedures for a Clean Development Mechanism as Defined in Article 12 of the Kyoto Protocol, FCCC/CP/2001/13/Add.2

[233] Eddy and Wiser, 'Public Participation in the Clean Development Mechanism of the Kyoto Protocol', (n 159), 210.

the design and approval of mitigation projects such as CDM projects.[234] A robust prior informed consent procedure must also be established to protect the rights of indigenous peoples and minority groups.[235]

Even though some studies point to the dangers of an expansive approach to participation,[236] broad-based participation by NGOs and human rights and environmental gatekeepers most times prove to be the only life line in less democratic countries where a large percentage of the population are either unsophisticated, unaware, disenchanted or simply uninterested due to many years of social exclusion, imbalance power relations and governmental oppression.[237] For example, the roles played by environmental groups such as ERA in raising awareness on environmental and human rights issues; and in monitoring the fulfilment of human rights obligations

[234] This definition follows the example of Art. 2(5) of the Aarhus Convention, which defines the 'public concerned' to mean the public affected or likely to be affected by, or having an interest in, the environmental decision-making; for the purposes of this definition, non-governmental organizations promoting environmental protection and meeting any requirements under national law shall be deemed to have an interest.

[235] World Commission on Dams (WCD), *Dams and Development: A New Framework for Decision-Making* (London and Sterling VA: Earthscan, 2000), 215, stating that 'Where projects affect indigenous and tribal peoples, such processes are guided by their free, prior and informed consent'. See also A. Johl and S. Duyck, 'Promoting Human Rights in the Future Climate Regime', *Ethics, Policy and Environment*, 15(3) (2012), 300.

[236] Some of the contentions include concerns as to whether pre-conceived standpoints and opinions will not stand in the way of real deliberation and real progress, and that some interests groups might delay or block the approvals of beneficial projects. For detailed discussions of the complexity of participation, and the limitations posed by government approaches to participation, see M. Lee, C Armeni, J. De Cendra, S. Chaytor, S. Lock, M. Maslin, C. Redgwell and Y. Rydin, 'Public Participation and Climate Change Infrastructure', *Journal of Environmental Law* 25(1) (2013), 33–62, where the authors discuss how untempered public participation might become a 'simple bureaucratic hurdle, frustrating for all concerned'. See also M. Lee, Y. Rydin, and S. Lock, 'Public Engagement in Decision-Making on Major Wind Energy Projects' *Journal of Environmental Law*, 27 (1) (2015), 139–150, discussing the dangers of government's excessive reliance on expert knowledge over lay constructions of knowledge by people who are directly affected by a project; M. Lee and C. Abbot, 'The Usual Suspects? Public Participation under the Aarhus Convention', *Modern Law Review*, 66 (2003), 80; C. Sunstein, 'Deliberative Trouble? Why Groups Go To Extremes', *Yale Law Journal*, 110 (2000), 71; J. Black, 'Proceduralizing Regulation: Part I', *Oxford Journal of Legal Studies*, 20 (2000), 597; and 'Proceduralizing Regulation: Part II', *Oxford Journal of Legal Studies*, 33 (2001), 21

[237] As Roth notes, the strength of NGOs such as Human Rights Watch in less democratic countries is not their rhetorical voice but their shaming methodology – their ability to investigate misconduct and expose it to public opprobrium. See Roth, 'Defending Economic, Social and Cultural Rights', (n 96), 63–73.

in Nigeria cannot be overemphasized.[238] Generally, absence of robust national planning laws, transparent environmental impact assessment processes, governmental accountability and an overall culture of human rights protection, as found in many established democracies, gives the right to participation a different impetus in developing countries- the epicenter of many of the concerns discussed in this book. Agreeably, there are legitimate concerns that untempered public participation process might act as a stumbling block to beneficial emission reduction projects, due to the tendencies of many interest groups to attend decision-making meetings with pre-conceived standpoints and opinions that could stagnate or frustrate decision-making processes. The importance of democratic checks and balances are, however, crucial, particularly in weaker democracies and countries with poor human rights records where there have been increased cases of governments colluding with project proponents to approve projects that affect the rights of host communities.

For example, one of the allegations against the Kwale project was that the Nigerian DNA only arranged a smokescreen EIA and consultation process after the project had already received host country approval.[239] The participatory process excluded many members of the Kwale community in the Niger Delta area of Nigeria who had no opportunity to take part in the process due to lack of information on the venue, date, time and details of the participatory meetings. Furthermore, the consultative

[238] For example, most of the communications to the Aarhus Compliance Committee regarding non-compliance with the provisions of the Convention by member states to the Convention were submitted by NGOs. See, e.g., Communication from Association Green Salvation (a Kazakh NGO) against Kazakhstan, ACCC/C/2004/02, 17 March 2004 (alleging that Kazakhstan constructed high-voltage power lines without notifying the residents along the proposed route regarding public hearings on environmental impact assessment of the project contrary to article 6 of the Convention); Communication from Ecopravo-Lviv (a Ukrainian NGO) against Ukraine, ACCC/C/2004/03 (alleging that Ukraine failed to provide for public participation in a decision-making process on state 'environmental expertiza' linked with the technical and economic evaluation of a proposed project to construct a navigation canal in the Danube Delta passing through an internationally recognized wetland contrary to article 6 of the Convention); and Communication by the Centre for Regional Development/Transparency International Armenia, the Sakharov Armenian Botanical Society (Armenian NGOs) against Armenia, ACCC/C/2004/08 (alleging that Armenia failed to provide access to information and public participation in decision-making on modification of land use designation and zoning as well as on the leasing of certain plots in an agricultural area of Dalma Orchads). The full texts of the Communications can be accessed at www.unece.org/env/pp/pubcom.htm accessed 12 January 2015.

[239] See Bassey, 'Foreword', (n 36); Carbon Trade Watch, 'Groups Slam Nigeria's Submission of Gas Flare Reductions for Carbon Credits', (n 36).

meetings were allegedly conducted in English language in a community where majority of the community speak and understand only the native Ukwuani-Aboh-Ndoni language.[240]

In the absence of robust protection at the national level to prevent exclusion and to allow for broad-based participation in the planning and design of mitigation projects, the responsibility to protect the rights of communities in host countries that are affected by mitigation projects must fall somewhere.[241] By building in project safeguards such as the requirements for host countries to demonstrate that a 'genuine participatory process' has been conducted, the international community will be able to monitor and ensure that projects designed to combat climate change are not creating long lasting human rights problems.[242] The way forward, arguably, lies in an international approach that builds in a necessary balance between the need for participation and the need to avoid extremes by NGOs. This balance can be achieved through a transparent process that allows NGOs with genuine complaints about a project the opportunities to access project information and take part in the decision-making processes on such projects.

2.3.7.4 Project review/complaints procedure and access to justice

The importance of providing stakeholders opportunities to seek redress or review projects that violate human rights is very crucial. The idea of access to justice has gained increased recognition in a number of core international human rights instruments. Access to justice is the ability of the public to seek and obtain quick, effective and fair remedy through formal and informal institutions.[243]

240 See Onojiribholo, 'Kwale Chief Laments Plight of Communities', *ibid.*, also Allen et al., 'Niger Delta Oil Flares, Broken Laws, Pollution and Oppression', (n 36), chp. 4.

241 Bassey argues that with the spate at which CDM projects affect the rights of communities in Africa, Africa may be 'cooked' before the international community wakes up to the reality of investigating the dangers of carbon markets. N. Bassey, *To Cook a Continent* (Oxford: Pambazuka Books, 2011). See also T. McVeigh, 'Biofuels Land Grab in Kenya's Tana Delta Fuels Talk of War' (*Guardian*, 2 July 2011), www.guardian.co.uk/world/2011/jul/02/biofuelsFlandFgrabFkenyaFdelta, accessed 12 December 2014.

242 The elements of a 'genuine participatory process' are considered in Chapter 5.

243 See Report of the United Nations High Commissioner for Human Rights (A/HRC/27/25) para. 9, which discusses access to justice as the right of individuals to obtain a quick, effective and fair response, to prevent or solve disputes, to control the abuse of power through transparent processes and to have affordable and accountable mechanisms available.

One of the strongest weaknesses observable in the international climate change regime and rules applicable to the CDM is the failure to establish complaint mechanisms and procedures for stakeholders or private individuals whose consultative rights or human rights in general have been infringed by projects to seek redress, to challenge the approval of such projects or to seek the review of already approved projects. For example, the Kyoto Compliance Committee only examines questions of implementation with obligations under the Kyoto Protocol, which are mainly issues relating to estimation, reporting, and verification of emissions and credits of Annex I parties and overall compliance by parties with their emission reduction obligations.[244] Since the Kyoto Protocol itself does not expressly include human rights obligations or respect for human rights in the design and execution of climate change responses, the Kyoto Compliance Committee does not have the mandate to consider compliance with human rights. As such, issues relating to how policy measures aimed at achieving emission reductions affect human rights fall outside the mandate and expertise of the enforcement or facilitative branches of the compliance committee.[245] This gap must be addressed

[244] Some of the overall weaknesses of the Kyoto Compliance Mechanism and Procedures with respect to monitoring and verifying emission reductions have been exhaustively identified elsewhere. See M. Doelle, 'Early Experience with the Kyoto Compliance System: Possible Lessons for MEA Compliance System Design', law.dal.ca/Files/MEL_Institute/Doelle_Kyoto_Compliance_Final_Draft.pdf, accessed 12 July 2015; S. Oberthür and R. Lefeber, 'Holding Countries to Account: The Kyoto Protocol's Compliance System Revisited after Four Years of Experience', *Climate Law*, 1 (2010), 133–158; J. Brunnée, 'A Fine Balance: Facilitation and Enforcement in the Design of a Compliance for the Kyoto Protocol', *Tulane Environmental Law Journal*, 13 (2000), 223. See also C. Breidenich and D. Bodansky, *Measurement, Reporting and Verification in a Post-2012 Climate Agreement* (Pew Centeron Global Climate Change, 2009), www.pewclimate.org/docUploads/mrv-report.pdf, accessed 12 January 2015; P.R. Kalas and A. Herwig, 'Dispute Resolution under the Kyoto Protocol', *Ecology Law Quarterly*, 27 (2000), 53; D. Victor, 'Enforcing International Law: Implications for an Effective Global Warming Regime', *Duke Environmental Law and Policy Forum*, 10 (1999), 147.

[245] Procedurally, a question of implementation can be brought before the Compliance Committee either by a party to the Kyoto Protocol or by an Expert Review Team (ERT). The Bureau of the Committee then determines whether it comes within the jurisdiction of the Enforcement Branch (EB) or the Facilitative Branch (FB), or both. Once the EB receives a question of implementation from the Bureau it conducts a preliminary review of the issue raised and makes a determination whether to proceed. A human rights question will most likely stop at this stage as it falls outside the jurisdiction of the committee. For issues coming within its jurisdiction, the party under investigation is informed of the decision of the EB to proceed. Such party has the right to request a hearing and make written submissions. The EB will usually hear from the party, the ERT, any other party, and any independent experts it feels are needed to resolve the issue raised. The EB can also request specific information from the party

if the COP is truly serious about its Cancun Decision that 'Parties should in all climate change actions respect human rights.' It is not enough to simply include human rights paragraphs or safeguards in the international climate change regime. It must be accompanied by a reform in governance structures to establish robust systems to monitor and enforce compliance with human rights.

This aside, the Compliance Committee does not have powers to receive compliance complaints from the public. Compliance issues can only be referred either by a party to the Kyoto Protocol or by an Expert Review Team (ERT).[246] Private individuals or members of the public do not have access to request a review or a question of implementation under the current Compliance Rules. This further narrows the scope of the committee and reduces the impacts it could make with respect to investigating questions of implementation or complaints by members of the public.

Apart from the Compliance Committee, the CDM EB and DOEs do not have the mandate to consider projects from a human rights standpoint.[247] Rather, both bodies examine the technical requirements of projects and whether projects have met national requirements of the host country. For example, the CDM EB in response to allegations of fraud and procedural impunities brought against DOEs established a complaint procedure in 2010. This procedure would allow the public to bring forward complaints about fraud or unethical behaviour by

under investigation, and can consider submissions from non-parties. Timelines are set for the major steps in the process. After the hearing, the EB makes a preliminary finding as to whether the party is in compliance. The party has an opportunity to comment on the preliminary finding. If it does not, the preliminary finding stands as the final decision of the EB. If the party submits comments on the preliminary finding, the EB issues a final decision in light of the comments filed. The EB has to give reasons for its decisions. A finding of non-compliance will result in a range of consequences depending on the nature of the violation. A key part of the process is the preparation of a compliance plan within three months of the determination of non-compliance, with regular updates thereafter on the implementation of the compliance plan. Key substantive requirements for the compliance plan are set out in Section XV (6) of the Compliance Procedures. For a comprehensive overview of the process, see Oberthür and Lefeber, 'Holding Countries to Account', (n 244); see also M. Doelle, 'Early Experience with the Kyoto Compliance System: Possible Lessons for MEA Compliance System Design', law.dal.ca/Files/MEL_Institute/Doelle_Kyoto_Compliance_Final_Draft.pdf, accessed 23 June 2015.

246 See Annexes to Decisions 4/CMP.2 and 4/CMP.4 on the Compliance Committee, FCCC/KP/CMP/2006/Add.1 (2 March 2007), 17, and FCCC/KP/CMP/2008/11/Add.1 (19 March 2009), 14 (hereinafter 'Rules of Procedure').

247 See The CDM Executive Board, cdm.unfccc.int/EB/index.html, accessed 12 February 2015.

a DOE or to challenge a DOE that has not complied with its own accredited system and/or the CDM requirements.[248] However, despite the potentials and promises of this new complaint procedure, its efficacy has been hindered by the fact that respect for human rights is not one of the CDM requirements, as such, DOEs just like the CDM EB do not have powers to consider human rights impacts of a project. This gap makes it impossible for individuals to approach the executive board to review projects that affect their human rights, leaving states as the only option for redress.

Another loophole is the fact that individuals are not recognized under the CDM Rules and cannot request a project review. Currently, only governments or three CDM EB members can request a review of projects under the CDM Rules.[249] As expected, states that have approved projects would most unlikely instigate such review processes. Stakeholders whose rights are violated do not currently have any way to request a review of a CDM project prior to registration. The Kyoto Protocol also does not confer the CDM EB with the authority to refuse a project based on human rights complaints or with the discretion to hear appeals from members of the public whose rights might be affected by a project even in cases such as the Aguan gas project where several petitions were received that indicated significant infringements of human rights.

These gaps create a one-way mechanism in which the decision of the State is final and most often rubber stamped by the international supervisory body. This has arguably entrenched a culture of 'approvals', which freezes out the common citizen whose fundamental human rights may be repressed, by the home state and the repression endorsed by the UNFCCC. Grievance mechanisms support the identification of any adverse human rights impact as part of the human rights due diligence on a project; they also make it possible for grievances, once identified, to be addressed and for any adverse impact to be remediated early and directly by project proponents, thereby preventing harm from being compounded and grievances from escalating.[250] As new project information emerges, new human rights issues could also emerge. As such, it is not enough to only provide updated information on projects, there must also be a project

[248] CDM Executive Board, 'Appendix 3: Handling of Complaints against an AE/DOE', in Procedure for Accrediting Operational Entities by the Executive Board of the Clean Development Mechanism (CDM) cdm.unfccc.int/Reference/Procedures/accr_proc01.pdf, accessed 1 April 2015.

[249] CDM Modalities & Procedures, 3/CMP.1, (n 232).

[250] *Ibid.*

review dispute resolution platform for stakeholders to seek a review of projects and to address any human rights concerns that might arise. A review mechanism complements wider stakeholder engagement as it provides opportunities for stakeholders to raise emerging issues that were not discussed during the pre-approval consultations.

A number of international human rights instruments guarantee access to justice as a fundamental human right. Though not legally binding, Article 8 of the UDHR provides that 'everyone has the right to an effective remedy by the competent national tribunals for acts violating the fundamental rights granted him by the constitution or by law'.[251] The right to remedy includes the right not to be denied fair representation. Article 10 also provides that everyone is entitled in full equality to a fair and public hearing by an independent and impartial tribunal, in the determination of his rights and obligations. The ICCPR also discusses access to justice as a human right, specifically with respect to criminal trials.[252]

Similarly in December 2012, the International Commission of Jurists (ICJ) released a Declaration on Access to Justice and Right to a Remedy in International Human Rights Systems. Though not legally binding, this declaration identifies that the universal realization of human rights requires meaningful access to justice for all persons and that it is the responsibility of states and other human rights duty bearers to act to remove barriers to the full access to justice. The ICJ also noted that without the availability to rights holders of avenues to seek and obtain effective remedies and reparations for violations, human rights guarantees may be illusory and go unrealized.[253] Regionally, Article 7 of the AfCHPR specifically provides that 'every individual shall have the right to have his cause heard'.[254] Article 6 of the European Convention on Human Rights (ECHR) goes further to provide that 'in the determination of his civil rights and obligations. . .everyone is entitled to a fair and

[251] Universal Declaration on Human and Peoples Rights, Adopted on 10 December 1948) GA Res. 217A (III), UN Doc A/810 at 71 (1948).

[252] ICCPR, (n 112).

[253] International Commission of Jurists (ICJ) is an international non-governmental organization consisting of a group of 60 eminent jurists (judges and lawyers) from across the world that seeks to promote human rights and the rule of law. It is not part of the UN systems, neither is it related to the International Court of Justice (ICJ). See www.icj.org/about/, accessed 12 January 2015.

[254] African Charter Human and Peoples Rights (Adopted 27 June 1981, entered into force 21 October 1981) OAU Doc. CAB/LEG/67/3 Rev. 5, reprinted in 21 ILM 58 (1982).

public hearing within a reasonable time by an independent and impartial tribunal established by law'.[255]

Virtually all international human rights treaties establish individual complaint procedures that allow private citizens to bring complaints before an international quasi-judicial committee for human rights abuses.[256] The complaint procedures under these treaties are designed to be uncomplicated and accessible to the layperson, such that any member of the public whose rights are threatened and is unfamiliar with legal and technical terms could bring a complaint before the bodies concerned.

Such is the reform that is needed to revitalize the international climate governance architecture. Considering the fact that climate change concerns have peculiarities that might not be easily addressed before a human rights body, especially due to the fact that many of these human rights instruments do not specifically address climate change issues, there is a need for the UNFCCC to build in a project review mechanism, and a project appeal panel, that would allow members of the public whose rights are directly or indirectly threatened by climate change to seek and obtain redress. A number of international environmental law treaties already have such robust grievance redress mechanisms that could

[255] European Convention for the Protection of Human Rights and Fundamental Freedoms (Signed 4 November 1950, entered in to force 3 September 1953, 213 UNTS 221, ETS 5).

[256] At the international level, individual complaint mechanisms are found under eight human rights conventions: in the First Optional Protocol to the International Covenant on Civil and Political Rights (ICCPR); the new Optional Protocol to the ICESCR (which opened for signature and ratification in March 2009), Article 22 of the Convention against Torture (CAT) talks about the competence of the Committee to receive and consider communications from or on behalf of individuals subject to state jurisdiction who claim to be victims of a violation by a state party of the provisions of the Convention; same as the Optional Protocol to the CEDAW; Article 14 CERD, Article 77 CMW, the Optional Protocol to CRPD and in Article 31 of the International Convention for the Protection of All Persons from Enforced Disappearance (As of February 2010, 81 states have signed, and eighteen have ratified. It will come into force when ratified by 20 states). The African Charter on Human and Peoples Rights also allows individuals and NGOs to submit complaints. Article 56 states that communications can be submitted by private individuals, non-governmental organizations and various other entities and the petitioner does not have to be the victim. Under the European system, an individual complaint mechanism is found under Article 34 of the European Convention for the Protection of Human Rights and Fundamental Freedoms. At the Inter-American level, Article 44 of the American Convention on Human Rights allows petitions to be brought unconditionally before its supervisory body, the Inter-American Commission on Human Rights; unconditionally meaning that no separate acceptance by the state of the individual complaint procedure is required. The petitioner under this system does not have to be the victim but the petition must be submitted to the Commission within six months after the local remedies have been exhausted.

provide lessons to the UNFCCC on how best to establish an international complaint mechanism.[257] For example, the Council of Europe's Convention on Civil Liability for Damage Resulting from Activities Dangerous to the Environment provides access to justice and adequate compensation for damage resulting from activities dangerous to the environment and also provides for means of prevention and reinstatement for victims of environmental damage.[258] It provides broad standing for environmental organizations to seek the prohibition of an unlawful activity that poses a grave threat of damage to the environment and to seek orders against operators in order to prevent or mitigate damage. Article 14 provides remedies for any person who considers that his request for information has been unreasonably refused or ignored or has been inadequately answered by a public authority. Such persons are entitled to seek a judicial or administrative review of the decision, in accordance with the relevant legal system. Article 16 provides that a person suffering damage may at any time request the court to order an operator to provide specific information in so far as this is necessary to establish the existence of a claim for compensation under the Convention.

The Aarhus Convention also establishes a robust framework that allows individuals and members of the public to demand the enforcement and review of an infringement of any of the procedural rights guaranteed under the Convention.[259] Article 9 (1) of the Aarhus Convention provides that 'any person' who has requested information under Article 4 can challenge decisions. Article 4 allows 'any member of the public' to request information, while Article 4(2) defines the 'public' as natural or legal persons, and their associations, organizations or groups. This grants standing to a wide variety of the society falling within the definitions of the public. This provision opens the door to NGOs and members of the public to challenge any action that violates the established procedural rights in the Convention. Despite the implementation challenges of this expansive approach, it allows members of the public to

[257] Council of Europe, Convention on Civil Responsibility for Damage Resulting from Activities Dangerous to the Environment (Lugano, 21 June 1993) 21.VI.1993. 18 ETS 150.

[258] See Art 1, *ibid.*

[259] See Aarhus Convention; See C. Redgwell, 'Access to Environmental Justice', in F. Francioni (ed.), *Access to Justice as a Human Right* (Oxford: Oxford University Press, 2007), 153–175. Redgwell identifies the landmark contributions of the Aarhus Convention as the only environmental convention that significantly addresses the obstacles to environmental justice.

challenge actions of public bodies that infringe procedural rights to information, participation and access to justice.[260] The Aarhus Convention also establishes one of the more forward-looking compliance committees. The Aarhus Compliance Committee allows member of the public to send communications to it concerning a party's compliance with the convention. Article 15 of the Aarhus Convention establishes the compliance committee to be a forum for reviewing compliance with the provisions of the Convention. This provides a vanguard framework through which the public can take governments to task for violating procedural human rights and environmental laws. Based on the fact that the Aarhus Convention is opened to accession by non-ECE countries, it possesses stronger potentials to serve as a global framework for strengthening access to justice, participation and access to information in environmental matters globally. Despite some concerns on its practical implementations, the Aarhus framework can serve as model for the improvement of the international climate regime to also allow the public to enforce human rights obligations.[261]

Elements of access to justice are explored in greater detail in Chapter 5 of this book. Generally, there is a need to establish formal grievance mechanisms through which members of the public could seek redress for human rights violations that could result from a planned project. There is also a need for a project review board or panel that could examine new evidence of human rights violations by a registered project. These bodies would allow the public to block the approval and registration of a project that could affect their human rights and provide the opportunity to seek a post-registration review of an approved project in the light of new facts or evidence. We also discuss the need for compliance mechanisms under the international climate change regime to measure and monitor human rights compliance in the execution of mitigation projects. This might be by expanding the current mandates of the international climate change

[260] While Aarhus is well regarded for providing broad legal standing and access to justice provisions, its implementation has been problematic. A recent study shows that almost all EU countries have problems in at least one of the essential elements for the implementation of Article 9 provisions mostly in the areas of transparency, length of procedures and cost of legal action. See J. Darpo, 'Effective Justice? Synthesis Report of the Study on the Implementation of Articles 9.3 and 9.4 of the Aarhus Convention in Seventeen of the Member States of the European Union', (2012), ec.europa.eu/environment/aarhus/pdf/2012_access_justice_report.pdf, accessed 23 January 2015. See also 'The Aarhus Convention: How Are Its Access to Justice Provisions being implemented?' A Report on a Conference organized by the European Commission in Brussels, 2 June 2008.

[261] See Aarhus Convention.

institutions and supervisory bodies to empower them to monitor human rights compliance or by establishing a separate human rights compliance monitoring branch as a third branch of the Compliance Committee.

However, such efforts must begin with identifying human rights as part of the international climate change regime. Before we speak of compliance, the international climate change regime must first mainstream the obligation to protect and respect human rights as part of efforts to combat climate change. This would provide legal basis to expand and reform climate change governance structures. When such structures are established, they can then monitor how countries are respecting human rights in climate change actions. Such governance structures would also create synergy with national efforts by encouraging member states to monitor approved projects and report human rights compliance as part of the periodic reporting requirements which parties currently have under the Kyoto Protocol.[262]

2.4 Conclusion: bringing human rights to the table

To eliminate human rights risks in the design, approval, finance and implementation of carbon projects, human rights obligations of both governments and private actors, specifically those obligations relating to rights to information, equity and non-discrimination, public participation in decision-making and access to justice, must be centrally recognized in the climate governance architecture. Generally, the international climate change regime must expand the mandates and authorities of supervisory bodies for carbon projects to enable them to address and respond to human rights petitions and concerns resulting from carbon projects. Placing the decision to execute a project in the hands of national authorities alone, without strategic oversight from international supervisory bodies under the international climate change regime, has arguably resulted in a culture of 'approvals' under which questionable projects, even those with documented human rights impacts, are approved by host governments and registered for carbon credits.[263]

[262] The UNFCCC Reporting Guidelines on Annual Inventories require Annex I parties, to provide by 15 April each year, annual national GHG inventories, including a National Inventory Report (NIR) on the institutional structures and quality assurance and control procedures adopted by parties in meeting emission reduction commitments. See Articles 4 and 12 of the Convention; see also The UNFCCC Reporting Guidelines on Annual Inventories FCCC/SBSTA/2006/9 (18 August 2006).

[263] See Eddy and Wiser, 'Public Participation in the Clean Development Mechanism of the Kyoto Protocol', (159), 203.

The international climate governance architecture must be reformed and enhanced to better deal with identified gaps and loopholes that exacerbate human rights risks in carbon projects and climate-related activities. The way forward lies in an approach that mainstreams human rights safeguards into the processes and procedures for the financing, approval and registration of carbon projects.[264] The human rights mainstream paradigm has emerged within the last decade as an integrated approach for promoting human rights ideas at the decision-making level. By mainstreaming human rights standards into project finance, approval and registration requirements, project proponents, host countries and stakeholders in project lifecycle would have greater responsibility to anticipate and address human rights risks in carbon projects, at planning stages, in order to reduce the perennial conflicts between human rights and carbon projects.[265]

However, the human rights mainstream paradigm is not without conceptual and theoretical debates. Questions continue to emerge regarding the legal and theoretical basis for human rights mainstreaming; whether mainstreaming human rights would not further exacerbate the concerns on human rights proliferation and the devaluation of the human rights currency; whether human rights mainstreaming could effectively tackle market-driven inequalities; and whether human rights mainstreaming would add anything new to the extant international climate change regime.

The next chapter evaluates these and other debates on the human rights mainstream paradigm.

[264] See E. Meijer and J. Werksman, 'Keeping It Clean: Safeguarding the Environmental Integrity of the Clean Development Mechanism' in D. Freestone and C. Streck, (eds.), *Legal Aspects of Implementing the Kyoto Protocol* (Oxford University Press, 2005), 191, 201; The World Bank, *Development and Climate Change: World Development Report 2010* (Washington, DC: The World Bank, 2010); Oxfam International, *Climate Wrongs and Human Rights: Putting People at the Heart of Climate-Change Policy*, Oxfam Briefing Paper 117 (Oxfam International, 2008); See P. Baer, T. Athanasiou, and S. Kartha, *The Right to Development in a Climate Constrained World: The Greenhouse Development Rights Framework* (Heinrich Böll Foundation, 2007); E. Page, *Climate Change, Justice and Future Generations* (Cheltenham: Edward Elgar, 2006), 132–160.

[265] See S. Jodoin, 'Human Rights and Climate Change Governance' (Center for International Sustainable Development Law, Legal Working Paper, December 2010); see also D. Olawuyi, 'Towards a Transparent and Accountable Clean Development Mechanism: Legal and Institutional Imperatives', *Nordic Environmental Law Journal*, 2 (2) (2012), 33–52.

PART II

Mainstreaming human rights in the international climate change regime

3

The concept of mainstreaming in international law

3.1 Introduction

As Sartre rightly observed, human beings are by nature 'stalkers of meaning', the entire task of law centres on meanings and reasons.[1] An enquiry into the practical implications of adopting a human rights-based approach to carbon finance must begin with a detailed understanding of its legal, moral, philosophical, ethical and economic basis and value. This is the focus of this chapter. This chapter analyses and discusses the theoretical underpinnings, legal basis and practical value of the human rights-based approach to carbon finance.

The human rights-based approach to carbon finance has a strong legal and theoretical basis. It builds on human rights obligations already owed by states to their citizens and to the international community, and provides legal frameworks through which policymakers can mainstream and integrate these obligations into decision-making processes. This idea of mainstreaming human rights safeguards in the design, approval, finance and implementation of carbon projects is the crux of the human rights-based approach to carbon finance. This chapter evaluates the concept of human rights mainstreaming as theoretical anchor for the human rights-based approach to carbon finance. It discusses how the principle of systemic integration in international law; the inter-doctrinal legal transplant/cross-fertilization theory; participatory development theory; and the full belly thesis provide normative foundations and philosophical supports for the idea of integrating human rights safeguards and norms into carbon finance.

This chapter will also evaluate the practical and analytical value of mainstreaming human rights safeguards into the international carbon finance architecture. Theoretical questions such as the definition question, novelty or duplication question, neo-liberalism critique, determinacy and the conflict of rights debates will be examined in order to understand if and how a rights-based approach to carbon finance may

[1] J.P. Sartre, 'Merleau-Ponty Vivant', *Les Temps Modernes* (184–54), 316.

aid efforts to harmonize human rights and climate change obligations and institutions in international law.

3.2 Human rights mainstreaming: a background

Over the last decade, the idea of mainstreaming has emerged under different genres of public international law as a legal strategy for achieving a coherent and expansive understanding and application of the law. It is a paradigm that seeks to understand the interrelationships between diverse societal strata, such as the arts, business, education, environment, faith, finance, government, labour, gender, media, public safety, public health and science.[2] Even though the concept of mainstreaming is not new to international law, it has recently gained increased recognition as a multifaceted and action-oriented approach to responding to social justice concerns.[3] A plenitude of literature has been generated on gender mainstreaming, anti-corruption mainstreaming, health mainstreaming, environmental mainstreaming and human rights mainstreaming.[4] Even

[2] See J. Soussan, *Making Mainstreaming Work* (Stockholm: Stockholm Environment Institute, 2007), 2, www.unpei.org/sites/default/files/PDF/influencingpolicyprocesses/Making-mainstreaming-work-marine-coastal.pdf, accessed 12 December 2015; see also World Justice Project, *Mainstreaming the Rule of Law*, (2011), worldjusticeproject.org/mainstreaming/, accessed 12 December 2014.

[3] For recent developments on legal mainstreaming, see World Justice Project, *Mainstreaming the Rule of Law, ibid.*

[4] See N. Kabeer, *Mainstreaming Gender in Social Protection for the Informal Economy* (Commonwealth, 2008); A. Brooke, 'Gender Mainstreaming and Women's Human Rights', Prepared for the International Studies Association Annual Convention, Honolulu, Hawaii, 1–5 March 2005; H. Charlesworth, 'Not Waving but Drowning: Gender Mainstreaming and Human Rights in the United Nations', *Harvard Human Rights Journal*, 18 (2005), 1–18; C. Bacchi, 'Gender Mainstreaming: A New Vision, More of the Same, or Backlash?', *Dialogue: Academy of the Social Sciences*, 20(2) (2001), 16–20. 'Environmental Mainstreaming' is the informed inclusion of relevant environmental concerns into the decisions of institutions that drive national and sectoral development policy, rules, plans, investment and action. See B. Clayton and S. Bass, *The Challenges of Environmental Mainstreaming* (London: International Institute for Environment and Development, 2009); World Bank, *Mainstreaming Climate Change Adaptation in Irrigated Agriculture Project* (Washington, DC: World Bank, 2008); M. Kok and H. de Connick, 'Widening the Scope of Policies to Address Climate Change: Directions for Mainstreaming', *Environmental Science & Policy*, 10(7–8) (2007), 587–599; Meta Management and Overseas Development Institute, *Ideas and Experiences in Mainstreaming Environment and Water* (Netherlands Ministry of Foreign Affairs / Development Cooperation, 2006); F. Seymour, C. Maurer, and R. Quiroga, *Environmental Mainstreaming: Applications in the Context of Modernization of the State, Social Development, Competitiveness, and Regional Integration* (Inter-American Development Bank, 2005); See also T. Clark and F. Crepeau, 'Mainstreaming Refugee

though these different models of mainstreaming have distinctive applications and conceptual connotations that are unique to them, the critical concepts and theoretical underpinnings are arguably the same. Mainstreaming advocates the adoption of harmonized and multi-pronged strategies to engage the expertise of policymakers in designing laws, policies and regulations that reflect pressing socio-justice issues.[5] This book focuses on human rights mainstreaming. It discusses the potentials and pitfalls of mainstreaming human rights principles into the international climate change regime, particularly the approval and execution of carbon projects.

Human rights mainstreaming advocates the harmonization or integration of human rights norms and practices with social justice issues such as development, environmental protection, health, education, business, safety, government and labour.[6] It is a process-based strategy aimed at integrating human rights norms at all levels of political, legislative and regulatory decision-making. Mainstreaming human rights means making sure that human rights is integrated and considered at all stages of development and delivery of actions, programmes and policies, and by everyone involved in such programme delivery.[7]

The origin of the human rights mainstreaming movement can be traced to the 1997 United Nations Programme for Reform, launched by the United Nations Secretary General.[8] The report designates human rights as a crosscutting issue in each of the four substantive fields covered

Rights: The 1951 Refugee Convention and International Human Rights Law', *Netherlands Quarterly on Human Rights*, 17 (1999), 389–390.

5 See World Justice Project, *Mainstreaming the Rule of Law*, (n 2); see also Y. Ghai and J. Cotterel, 'The Rule of Law and Access to Justice', in Y. Ghai and J. Cotterel (eds.), *Marginalized Communities and Access to Justice* (Abingdon: Routledge-Cavendish, 2009).

6 New Zealand Agency for International Development (NZAID), *Human Rights Mainstreaming* (Wellington: NZAID, 2009) 4, www.aid.govt.nz/sites/default/files/Reference%20Guide%20on%20Mainstreaming%20Human%20Rights.pdf, accessed 12 December 2015; I. Bynoe and S. Spencer, *Mainstreaming Human Rights in Whitehall and Westminster* (London: Institute of Public Policy Research, 1999); E. Petersmann, 'From "Negative" to "Positive" Integration in the WTO: Time for "Mainstreaming Human Rights" into WTO Law', *Common Market Law Review*, 37 (2000), 1363–1382; T. Pajuste, *Mainstreaming Human Rights in the Context of the European Security and Defence Policy* (Erik Castrén Institute, 2008), 10.

7 NZAID, *Human Rights Mainstreaming, ibid.*

8 United Nations, *Renewing the United Nations: A Programme for Reform*, Report of the Secretary General, (A/51/1950), 14 July 1997, 87.

by the United Nations.[9] The reform programme called on all entities of the United Nations to 'mainstream human rights norms' into their respective mandates and activities.[10] The report recommends the extension of human rights activities by the integration of human rights into all principal UN activities and programmes. It gives a mandate to UN agencies to fully integrate human rights into their areas of activities.[11]

This report was followed by the 1998 Report of the Secretary General to the UN Economic and Social Council which also recommended *inter alia* the: (i) adoption of a 'human rights-based approach' to activities carried out within the respective mandates of components of the United Nations system; (ii) development of programmes/projects addressing specific human rights issues; and (iii) reorientation of existing programmes as a means of focusing adequate attention on human rights concerns.[12]

Ever since this reform programme was launched, the idea of mainstreaming human rights has found significant support in the activities of most UN agencies, the World Bank, regional bodies such as the European Union, African Development Bank, Asian Development bank and even national governments and agencies at the domestic level.[13] For example,

[9] These include peace and security; development cooperation; economic and social affairs; and humanitarian affairs. *Ibid.*

[10] For an excellent and detailed account of the history of human rights mainstreaming, see A. Frankovits, *Mainstreaming Human Rights, The Human Rights-Based Approach and the United Nations System* (Paris: UNESCO, 2005) 15–30, unesdoc.unesco.org/images/0014/001469/146999e.pdf, accessed 10 December 2015.

[11] United Nations, *Renewing the United Nations*, (n 8).

[12] See UN/UNDOC/GEN/N97/A/51/950/1997. See UNESCO, 'Guidelines for Human Rights-Based Programming Approach' Executive Directive 1998–004; see also J. Theis, *Evolution and Future of Rights-based Approaches* (Sweden: Save the Children, 2003), 10.

[13] See T. Greiber, M. Janki, M. Orellana, A. Savaresi-Hartmann and D. Shelton, *Conservation with Justice: A Rights-Based Approach* (Gland: International Union for Conservation of Nature, 2009); Frankovits, *Mainstreaming Human Rights*, (n 10); C. Nyamu-Musembi, and A. Cornwall, 'What is the "Rights-Based Approach" all about? Perspectives from International Development Agencies' (Brighton: Institute of Development Studies Working Paper, 2004), 234, www.ids.ac.uk/files/dmfile/Wp234.pdf, accessed 10 December 2015; K. Svadlenak-Gomez, *Human Rights and Conservation: Integrating Human Rights in Conservation Programming* (Bronx, NY: World Conservation Society and Washington, DC: Translinks, USAID, 2007), 48; I. VeneKlasen, *Rights-Based Approaches and Beyond: Challenges of Linking Rights and Participation* (UK: Institute of Development Studies Working Paper, 2004), 235; J. Theis, 'Promoting Rights-based Approaches: Experiences and Ideas from Asia and the Pacific'; B.I. Hamm, 'A Human Rights Approach to Development', *Human Rights Quarterly*, 23 (2001), 1005–1031; G. Castillo and M. Brouwer, 'Reflections on Integrating a Rights Based Approach in Environment and Development', in *Conservation and Human Rights* (Gland: IUCN Commission on Environment, Economic and Social Policy (CEESP), 2007), cmsdata.iucn.org/downloads/pm15.pdf, accessed 15 May 2015.

in response to the UN Secretary's Reform Programme, the United Nations Educational, Scientific and Cultural Organization (UNESCO) in 2003 adopted its Strategy on Human Rights to 'strengthen. . .the promotion and protection of human rights through the application of a coherent and organization-wide vision of human rights'.[14] The UNESCO strategy proposed the development of a 'phased plan to mainstream human rights into all of UNESCO's programmes and activities on the basis of human rights instruments and the conclusions of treaty monitoring bodies'.[15] The need to mainstream human rights in decision-making at all levels of governance has also been identified as top priority by the United Nations High Commissioner for Human Rights.[16]

In 2005, world leaders at the United Nations World Summit resolved to integrate the promotion and protection of human rights into national policies and to develop and support human rights mainstreaming strategies.[17] According to the summit outcome:

> We resolve to *integrate* the promotion and protection of human rights into national policies and to support the *further mainstreaming of human rights* throughout the United Nations system, as well as closer cooperation between the Office of the United Nations High Commissioner for Human Rights and all relevant United Nations bodies.[18]

In 2009, the United Nations Development Group (UNDG) launched what is arguably the most comprehensive human rights mainstreaming plan within the international development circle – the UNDG Human Rights Mainstreaming Programme (UNDG-HRM).[19] The UNDG-HRM

[14] UNESCO, *Strategy on Human Rights*, Adopted by the 32nd UNESCO General Conference 2003 (Document 32 C/57), unesdoc.unesco.org/images/0013/001316/131627e.pdf, accessed 23 February 2015.

[15] *Ibid.*

[16] The Commissioner called for the reform of the Human Rights secretariat so as to be the centre of mainstreaming human rights within the UN systems. See UN Office of the High Commissioner for 'Human Rights, Mainstreaming Human Rights', www.unhchr.ch/development/mainstreaming-01.html, accessed 23 January 2015.

[17] See UN World Summit, *Outcome Document* (2005), daccess-dds-ny.un.org/doc/UNDOC/GEN/N05/487/60/PDF/N0548760.pdf?OpenElement, accessed 12 January 2015.

[18] Emphasis added. See Para. 126, UN World Summit, *ibid.*

[19] The UNDG was established in 1997 as a coalition of over 32 UN agencies, funds, programmes, and offices, plus five observers that play a role in development. The group meets at least three times yearly to decide on issues related to country level coordination to achieve the internationally agreed development goals, including the Millennium Development Goals (MDGs). The UNDG seeks to coordinate, harmonize and align the 32 UN agencies, departments, and programmes that are involved in international

aims to mainstream human rights norms and standards into the work and activities of all UN agencies and programmes. It provides platform for a coordinated and coherent UN system-wide approach in the integration of human rights principles and international standards into operational activities and development programmes. In 2011, the UNDG-HRM released a comprehensive action plan on how human rights mainstreaming could be achieved within UN systems.[20] The plan seeks to strengthen coherence in human rights mainstreaming policies and practices across UN systems; enhance the adoption of a human rights-based approach (HRBA) within UN systems and teams; enhance UN system-wide knowledge codification and sharing on HRM; and strengthen capacity, collaboration and policy dialogue on human rights mainstreaming across the United Nations.[21]

Human rights mainstreaming has also received enormous attention from legal scholars and commentators.[22] McCrudden, a leading

development. The group's common objective is to deliver more coherent, effective and efficient programmes within the UN system. See UNDG, www.undg.org/index.cfm?P=2, accessed 20 January 2015. The United Nations Environment Programme (UNEP) is a part of the UNDG. For a comprehensive list of members see UNDG Members, www.undg.org/index.cfm?P=1503, accessed 20 January 2015.

[20] The UNDG-HRM identifies four broad priorities:

1. Promoting a coordinated and coherent UN system-wide approach towards the integration of human rights principles and international standards into UN operational activities for development;
2. Providing coherent and coordinated support to Resident Coordinators and UN country teams in mainstreaming human rights;
3. Developing a coherent UN-system wide approach, through cooperation and collaboration among UN agencies, to providing support towards strengthening national human rights protection systems at the request of governments; and
4. Contributing to the integration of human rights issues in the overall UNDG advocacy on development agenda and global issues.

See UNDG, *UNDG Human Rights Mainstreaming Mechanism: Operational Plan 2011–2013* (November 2011), undg.org/docs/12173/UNDG-HRM%20OperationalPlan%20Nov%202011.pdf, accessed 23 January 2015.

[21] See 'Priority 1: Promoting a coordinated & coherent UN system wide approach to human rights integration' in UNDG, *UNDG Human Rights Mainstreaming Mechanism: Operational Plan 2011–2013, ibid.*

[22] See Petersmann, 'From "Negative" to "Positive" Integration in the WTO', (n 6), 1363–1382; Pajuste, *Mainstreaming Human Rights in the Context of the European Security and Defence Policy*, (n 6), 10; F. Beveridge and S. Nott, Mainstreaming: A Case for Optimism and Cynicism, *Feminist Legal Studies*, 10(3), (2002), 299–311; J. Squires, 'Is Mainstreaming Transformative? Theorizing Mainstreaming in the Context of Diversity and Deliberation', *Social Politics: International Studies in Gender, State and Society*, 12(3) 2005), 366–388; Theis, *Evolution and Future of Rights*, (n 12), 10; P. Alston,

commentator on human rights mainstreaming, defines human rights mainstreaming as the reorganization, improvement, development and evaluation of policy processes, so that a human rights perspective is incorporated in all policies at all levels and at all stages, by the actors normally involved in policymaking.[23] Clark defines mainstreaming as a social justice-led approach to policymaking in which human rights principles, strategies and practices are integrated into the work of government and other public bodies.[24] It has been rightly framed as a new, better revolutionary strategy and a dominant policy framework aimed at reducing the gap between policy intentions and actual results, and as 'a wonder drug' for equal opportunities in international law.[25] Harvey describes mainstreaming as the task for the century and as a revolutionary strategy for an effective enforcement and implementation of international human rights standards. An underlying theme in these definitions is that human rights mainstreaming provides a practical framework through which human rights principles and standards can be integrated and reflected in decision-making processes to ensure coherence, harmonization and effectiveness.

By mainstreaming human rights norms in regulatory and legislative decision-making processes, countries can better protect, fulfil and promote the realization of the current robust principles of international human rights law that came about after several years of campaigns against social exclusions and injustices.[26] According to the

'Resisting the Merger and Acquisition of Human Rights By Trade Law: A Reply To Petersmann', *European Journal of International Law*, 13(8) (2002), 815; M. Koskenniemi, 'Human Rights Mainstreaming as a Strategy for Institutional Power', *Humanity*, 1(1) (2010), 47.

23 J.C. McCrudden, 'Mainstreaming Human Rights', in Colin Harvey (ed.), *Human Rights in the Community: Rights as Agents for Change* (Oxford: Hart Publishing, 2005); J.C. McCrudden, 'Mainstreaming Equality in Northern Ireland 1998–2004: A Review of Issues Concerning the Operation of the Equality Duty in Section 75 of the Northern Ireland Act 1998', in E. McLaughlin and N. Faris (eds.), *Section 75 Equality Duty: An Operational Review* (North Ireland Office, 2004).

24 See Clark and Crepeau, 'Mainstreaming Refugee Rights', (n 4), 389–390.

25 Beveridge and Nott, 'Mainstreaming: A Case for Optimism and Cynicism', (n 22), 299–311.

26 Human rights are a product of philosophical debates that have raged for over two thousand years. The debates have focused on a search for moral standards of political organization and behaviour that are independent of the contemporary society. For an excellent overview of the history of human rights, see J. Shestack, 'The Philosophical Foundations of Human Rights', in J. Symonides (ed.), *Human Rights Concepts and Standards* (Burlington, VT: Ashgate, 2000), 31–56; A. Heard, 'Human Rights: Chimeras in Sheep's Clothing?' (Simon Fraser University, 1997), www.sfu.ca/~aheard/intro.html,

International Covenant on Civil and Political Rights (ICCPR), 'the ideal of free human beings enjoying civil and political freedom and freedom from fear and want can only be achieved if conditions are created whereby everyone may enjoy his civil and political rights, as well as his social, economic and cultural rights.'[27] Human rights mainstreaming guides policymakers to ensure that conditions that favour the realization of human rights are put in place in the design and implementation of development activities. As the United Nations High Commissioner for Human Rights rightly noted:

> Mainstreaming human rights in the UN system is not just a mandate for me as the High Commissioner, but a common endeavour for all UN agencies and staff... Over the last decade, there has been significant progress in mainstreaming human rights in the work of the UN system. An increasing number of UN agencies are not only integrating human rights into their internal policies, but also actively advocating for human rights through their mandated work – on the rights of people affected by HIV, on the rights to food, education, health and shelter, and on the fundamental rights of workers. . .*human rights are no longer add-ons to a long list of other priorities*, but a foundation on which they can build the UN's coherence and comparative advantages.[28]

Over the last decade, the international community has increasingly recognized the need to ensure that respect for human rights is incorporated into efforts aimed at addressing climate change.[29] The aim and importance of this new drive is to ensure that human rights are not only recognized when there is a problem; instead, respect for human rights must underpin and form part of the rules of the game in the design of the international legal regime on climate change.

accessed 12 December 2014. See also J. Manique, 'Universal and Inalienable Human Rights: A Search for Foundations', *Human Rights Quarterly*, 12 (1990), 465–485.

[27] See Preamble to the International Covenant on Civil and Political Rights, Dec. 19, 1966, 999 UNTS 171 (hereinafter ICCPR).

[28] Statement by N. Pillay, United Nations High Commissioner for Human Rights at the Launch of the United Nations Development Group Human Rights Mainstreaming Mechanism (UNDG-HRM) Multi-Donor Trust Fund (27 November 2011), www.undg.org/docs/12174/Statement%20by%20High%20Commissioner%20for%20Human%20Rights%20Ms.%20Navi%20Pillay.pdf, accessed 23 January 2015.

[29] Emphasis added. See also a recent report of the International Bar Association, which called for the integration of human rights into climate change mitigation and adaptation efforts to ensure climate change justice. International Bar Association, *Achieving Justice and Human Rights in an Era of Climate Disruption* (Climate Change Justice and Human Rights Task Force Report July 2014).

3.2.1 Tenets of human rights mainstreaming

A fundamental purpose of human rights mainstreaming is to develop 'a human rights culture' in international, regional and national governance.[30] Under this paradigm, human rights issues would not only come to the table when there is a violation, infringement or problem. Rather, human rights would always be placed at the table when governmental decisions are made. Under this paradigm, human rights will not only be invoked when there is a violation or concern, rather respect for human rights will run throughout the implementation and delivery of climate actions and projects, and it will be understood that projects and actions have failed if they do not guarantee and protect the human rights of host communities.

The importance of fostering a 'human rights culture' in governance has been extensively canvassed. As Blackburn rightly argued, 'human rights should form part of the "rules of the game" under which the system of politics and government is conducted'.[31] Decisions such as building a new airport, constructing a new road or sports complex, exploiting natural resources building a pipeline and constructing a hydroelectric power plant, ought to be made with full appraisal of their likely human rights consequences, and with the full participation and inclusion of citizens whose rights would be affected in the decision-making process. Creating a human rights culture would represent a radical shift from a needs-based approach, under which human rights are treated as 'add ons', to an approach that enshrines a human rights culture at the heart of policymaking. Under this model, human rights are not relegated to the background; instead, human rights issues are placed at the fore of policy-making, actions and resource allocation.[32]

[30] According to the United Nations, the crux of human rights mainstreaming is to integrate human rights in all development efforts undertaken by UN Agencies, organizations, funds and programmes, in order to ensure that development is participatory, equitable, sustainable and accountable. See Human Rights Council, 'Draft Concept Note: High Level Panel Discussion on Human Rights Mainstreaming' (1 March 2013), Palais des Nations, Geneva, Switzerland); M. Wewerinke and C. Doebbler, 'Exploring the Legal Basis of a Human Rights Approach to Climate Change', *Chinese Journal of International Law*, 10 (2011), 141–160; Bynoe and Spencer, *Mainstreaming Human Rights in Whitehall and Westminster*, (n 6); see also UNDG, *UNDG Human Rights Mainstreaming Mechanism: Operational Plan 2011–2013*, (n 20).

[31] R. Blackburn, *Towards a Constitutional Bill of Rights for the United Kingdom* (London: Frances Pinter Publishers Ltd., 1999), xxxii.

[32] *Ibid.*

The human rights mainstream paradigm presents an opportunity for the public to demand from governments and private actors the respect for and the enforcement of human rights in emission reduction schemes. This approach would enlighten the public about their rights and provide them with practical opportunities to enforce them. It will also enable the public to demand that governments, public authorities and private actors, ensure that they comply with human rights standards when designing projects.[33] Members of the public will be provided with the information, education and legal assistance required to demand the enforcement of their human rights. Generally, under this paradigm, citizens are not mere observers or objects of decision-making; rather, they are active participants, subjects and in fact the most important stakeholders in decision-making processes. The public are empowered to understand their rights to act as active gatekeepers in ensuring that decisions, policies or governmental projects that can affect their human rights do not see the light of the day. As Baros rightly canvassed:

> a new type of relationship between the individual and state is needed. A full and participatory citizenship would have to be an ideological framework within which the proposed improvements should take place. Greater awareness of legal and political entitlements by citizens and meaningful participation in political life is a precondition for developing a partnership with the government and therefore contributing to a more positive human rights culture.[34]

Human rights mainstreaming is not all about giving individuals a stick with which to beat the state, or a voice to compete with state apparatus; rather, it seeks to create a society in which rights and responsibilities are properly balanced.[35] It promotes the establishment of institutions that promote effective relationships and correlations between rights and responsibilities. The whole idea is not only about empowering the individual with rights against governmental excesses. Rather, mainstreaming presents what Hunt describes as the 'communitarian conception of a human rights culture' whereby rights and responsibilities are properly

[33] *Ibid.*; see also C. Harvey, 'Talking about Human Rights', *European Human Rights Law Review*, 5 (2004), 500–516.

[34] M. Baros, 'A New Human Rights Culture and Deliberate Constitutionalisation', *Hertfordshire Law Journal*, 5(1) (2007), 115.

[35] J. Straw, Home Secretary, 'Building a Human Rights Culture', an address delivered at a Civil Service College Seminar, 9 December 1999.

balanced.[36] Citizens understand their rights and freedoms and at the same time understand their obligations and duties.

The human rights mainstream paradigm is a move from individualistic libertarian conceptions of rights that fail to emphasize the linkages between human rights and duties that are enshrined in international human rights instruments. This paradigm seeks to establish a culture of mutual respect for human rights between citizens and the state, and between citizens themselves. Such community-based method of promoting human rights finds expression in international human rights instruments. For example, the preamble to the ICCPR notes that the 'individual, having duties to other individuals and to the community in which he belongs, is under a responsibility to strive for the promotion and observance of the rights recognized in the present Covenant'.[37] Article 29 of the Universal Declaration of Human Rights (UDHR) states that 'everyone has duties to the Community in which alone the free and full development of his personality is possible'.[38] The African Charter on Human and Peoples Rights (AfCHPR) specifically notes that an individual has duties and obligations both to other individuals and to the state. Human rights mainstreaming allows for a balanced mix of rights and duties. Through balanced policy designs, mainstreaming manifests itself in equal measures: the human rights of citizens and their responsibilities to other individuals, the community and to the state. It promotes a human rights culture that is not all about a collection of individuals and interest groups who see human rights as a free good and primarily as a means of enforcing their rights against national authorities without regard to the interests of other individuals and/or the wider community.[39]

Human rights mainstreaming also advocates protection of human rights through multifaceted, multi-dimensional and cross-disciplinary

[36] M. Hunt, 'The Human Rights Act and Legal Culture: The Judiciary and the Legal Profession', *Journal of Law and Society*, 26(1) (1999), 90; see also R. Bellamy, 'Constitutive Citizenship versus Constitutional Rights: Republican Reflections of the EU Charter and the Human Rights Act', in T. Campbell, K. Ewing and A. Tomkins, (eds.), *Sceptical Essays on Human Rights* (Oxford: Oxford University Press, 2001), 15–39.

[37] ICCPR, (n 27).

[38] See Universal Declaration on Human and Peoples Rights, (Adopted on 10 December, 1948) GA Res. 217A (III), UN Doc A/810 at 71 (1948) (UDHR).

[39] See United Nations, 'Concept Note: High Level Panel on Human Rights Mainstreaming', (1 March 2013), www.ohchr.org/.../panel-on-mainstreaming2013_concept_note.docx, accessed 12 April 2015; see also Home Office notes on 'The Human Rights Culture', www.nationalarchives.gov.uk/ERORecords/HO/421/2/hract/hrcult.htm, accessed 12 May 2015.

efforts.[40] The fundamental postulation of human rights mainstreaming is that human rights are relevant in all aspects of human endeavour, including environment, arts, business, education, engineering, faith, labour, media, military, public health, public safety and law enforcement. The mainstream paradigm advocates holistic and harmonized responses to emerging societal issues, rather than a fragmented, sectoral or demand-based approach. The problem of fragmentation of international law has been well theorized.[41] Human rights obligations are scattered in a plenitude of treaties and conventions, while there are also countless other international treaties that speak to a broad range of issues ranging from the environment, terrorism, trade, agriculture, tourism, medicine, aviation, war, humanitarian intervention, health, food, water, gender, housing, banking, religion, family and culture to mention but a few. Some of these treaties are, for the most part, overlapping, counterproductive or duplicative, raising the need for a coordinated and harmonized approach to international law.[42] Human rights mainstreaming provides a normative framework that allows policymakers to integrate and implement their obligations under diverse instruments in a coherent and reinforcing manner.

An integrated approach, as typified by the human rights mainstream paradigm, represents a shift from the 'one pipe-at-a-time' approach to

[40] See M. Freeman, *Human Rights: An Interdisciplinary Approach*, 2nd edn. (Cambridge, UK and Malden, MA: Polity Press, 2011), 8–10, see also 89.

[41] See M. Young, 'Climate Change Law and Regime Interaction', *Carbon and Climate Law Review*, 5 (2011), 147, 147–151; M. Young (ed.), *Regime Interaction in International Law: Facing Fragmentation* (Cambridge: Cambridge University Press, 2012); S. Singh, 'The Potential of International Law: Fragmentation and Ethics', *Leiden Journal of International Law*, 24(1) (2011), 23–43; M. Koskenniemi and P. Leino, 'Fragmentation of International Law? Postmodern Anxieties', *Leiden Journal of International Law*, 15(3) (2002), 553–579; J. Pauwelyn, 'Bridging Fragmentation and Unity: International Law as a Universe of Inter-Connected Islands', *Michigan Journal of International Law*, 5 (2004), 903; M. Koskenniemi, 'Fragmentation of International Law: Difficulties Arising from the Diversification and Expansion of International Law', Report of the Study Group of the International Law Commission, Finalized by M. Koskenniemi, A/CN.4/L.682.

[42] See H. Van Asselt, F. Sindico, and M. Mehling, 'Global Climate Change and the Fragmentation of International Law', *Law and Policy*, 30 (2008), 423, where the authors describe how the climate change regime conflicts with pre-existing international law regimes particularly the biodiversity regime. They argue that an understanding of the institutional aspects of these relationships may allow enhanced political cooperation and coordination. See also M. Doelle, 'Linking the Kyoto Protocol and Other Multilateral Environmental Agreements: From Fragmentation to Integration?' *Journal of Environmental Law and Practice*, 14 (2004), 75–104.

a forward-looking, harmonized and holistic approach. Such an approach is important because socio-justice issues are crosscutting and often react with one another; the root causes of societal injustices are also similar. For example, the causes of poverty, environmental pollution, health issues, food scarcity and lack of housing can all be traced to the inability of governments to recognize, protect and fulfil basic rights of the public, such as right to life and right to education. As such, a human rights mainstream approach recognizes that international law obligations interact with one another, and that if they must all be respected and fulfilled, there is a need for an 'integrated approach' which takes into account the cross-cutting issues in different treaty regimes. It promotes a coherence of international obligations with respect to human rights and other issues through a harmonized approach.[43] Unlike traditional piecemeal approaches, the mainstream approach makes it possible for the underlying problems to be identified and causes tackled, rather than merely treating the symptom on a 'as it happens' basis. The fundamental idea is to foster a human rights culture and to spread the human rights message by ensuring that all governmental actions integrate human rights principles.

Despite the fact that human rights mainstreaming has been espoused and developed by a plethora of UN agencies, regional bodies and state governments, the theoretical basis for the human rights mainstreaming concept itself has not been significantly probed in existing literatures. Much scholarly attention has not been devoted to asking the question: what is the legal and theoretical basis for human rights mainstreaming? This has led to the charge that human rights mainstreaming has a fuzzy core and weak theoretical underpinnings.[44] As some commentators have argued, 'the somewhat vague and non-specific character of the concept of mainstreaming has probably aided this rapid ascendancy; everyone understands the general idea but no one is sure what it requires in practice'.[45]

[43] According to the United Nations, human rights mainstreaming aims to promote a coordinated and coherent UN system-wide approach towards the integration of human rights principles and international standards; and to develop a coherent UN-system wide approach, through cooperation and collaboration among UN agencies. See 'Priorities of the UNDG-HRM', www.undg.org/index.cfm?P=1452, accessed 12 March 2015.

[44] See Koskenniemi, 'Human Rights Mainstreaming as a Strategy', (n 22), 47.

[45] *Ibid.*; see also G. Oberleitner, *Global Human Rights Institutions* (Cambridge, UK; Malden, MA: Polity Press, 2007), 45; Beveridge and Nott, 'Mainstreaming: A Case for Optimism and Cynicism', (n 22), 299–311.

The value of the mainstreaming approach, would, arguably, be manifest if its scope and content are clarified. There is a need to further the conceptual understanding of the human mainstreaming paradigm by providing strong legal and theoretical analyses of the potential benefits and problems associated with its application within the carbon finance context.[46]

3.3 Legal and theoretical basis for human rights mainstreaming

This section analyses the legal and theoretical underpinnings for the human rights mainstreaming paradigm. This section discusses how the principle of systemic integration in international law, the interdoctrinal legal transplant/cross-fertilization theory, the participatory development theory and the full belly thesis, all provide strong theoretical justifications and basis for mainstreaming human rights norms in climate governance.

3.3.1 Principle of systemic integration in international law

One of the central tenets of human rights mainstreaming is the idea of integration, that is, integrating or harmonizing human rights obligations and norms into policymaking through a coordinated or holistic approach.[47] Arguably, the importance of the principles of systemic integration and coherent interpretation of international law obligations laid down in Article 31(3)(c) of the Vienna Convention on the Law of Treaties can be invoked to emphasize that integrating obligations established in climate change treaties with obligations in human rights instruments is a mandatory requirement of treaty interpretation and implementation under international law.[48]

[46] See Oberleitner, *Global Human Rights Institutions*, *ibid.*, see also McCrudden, 'Mainstreaming Human Rights'.

[47] See United Nations, 'Concept Note (n 39).

[48] See C. Mclachlan, 'The Principle of Systemic Integration and Article 31(3)(C) Of The Vienna Convention', *International and Comparative Law Quarterly*, 54 (2005), 279; see also Combacau and Sur, who refer to it as the 'Principe d' Integration', in J. Combacau and S. Sur, *Droit International Public* (Paris: Montchrestien, 2004), 175. See the Report of the International Law Commission which describes Article 31(3)(c) of the Vienna Convention on the Law of Treaties as a ' a mandatory part of the interpretation' that covers all the sources of international law, including custom, general principles, and, where applicable, other treaties. International Law Commission (ILC), *Fragmentation of International Law: Difficulties arising from the Diversification and Expansion of International Law*, Report of the Study Group of the International Law Commission, 13 April 2006, A/CN.4/L.682, at para. 426. See also WTO Panel in *EC – Measures Affecting*

Article 31 of the Vienna Convention establishes general principles and rules for interpreting international treaties. It provides that one of the factors to be taken into account while interpreting a treaty includes 'any relevant rules of international law' applicable in the relations between the parties.[49] It is the idea that states must take into account relevant rules or obligations under international law while interpreting or implementing specific treaty obligations. The principle underpins the significance of reading international law instruments in mutually supportive ways.[50]

The main crux of the principle of systemic integration is the holistic interpretation and harmonious reading of rules, from external sources that are relevant to the norm under interpretation, so as to take into account the broader normative environment or legal matrix.[51] It also speaks of a presumption that states, when creating new rules of international law, do not intend or aim to violate their obligations under other, pre-existing rules but rather intend to operate within a coherent international law framework.[52] Generally, international obligations must be interpreted with reference to their normative environment or existing international law obligations.[53]

What this principle advocates therefore is that a treaty must be applied and interpreted coherently with existing general obligations, rules or principles of international law.[54] As Hanqin rightly noted, Article 31 (3) (c) operates like a 'master key' to the house of international law. In case there is

the Approval and Marketing of BiotechProducts (7 February 2006), WT/DS291–293/INTERIM, 300, para. 7.69, which confirmed the application of Art. 31 (3) to treaty interpretation. See also Wewerinke and Doebbler, 'Exploring the Legal Basis of a Human Rights Approach', (n 30), 141–160.

49 Vienna Convention on the Law of Treaties (VCLT), 22 May 1969, 1155 UNTS 331.

50 International Law Commission (ILC), *Fragmentation of International Law.*

51 See V. Tzevelekos 'The Use of Article 31(3)(c) of the VCLT in the Case Law of the ECtHR: An Effective Anti-Fragmentation Tool or a Selective Loophole for the Reinforcement of Human Rights Teleology? Between Evolution and Systemic Integration', *Michigan Journal of International Law*, 31 (2010), 621, at 631.

52 See B. Simma, 'Universality of International Law from the Perspective of a Practitioner', *European Journal of International Law*, 20(2) (2009), 265–297; see also P.-M. Dupuy, 'Unification Rather than Fragmentation of International Law? The Case of International Investment Law and Human Rights Law', in P.-M. Dupuy, F. Francioni and E.U. Petersmann (eds.), *Human Rights in International Investment Arbitration* (Oxford: Oxford University Press, 2009), 45–62; R. Jennings and A. Watts, *Oppenheim's International Law*, 9th edn. (Oxford: Oxford University Press, 1992), 1275.

53 International Law Commission (ILC), *Fragmentation of International Law.*

54 R. Gardiner, *Treaty Interpretation* (Oxford: Oxford University Press, 2008), 11; O. Elias and P. Merkouris (eds.), *Treaty Interpretation and the Vienna Convention on the Law of Treaties: 30 years On* (Leiden: Martinus Nijhoff, 2010), 4.

a systemic problem – a conflict, an inconsistency, an intersection or an overlap between two or more international law obligations or norms – and no other interpretative means provides a resolution, then recourse may always be had to Article 31 (3) (c) in order to proceed in a reasoned way such that they produce effects which accord with existing rules of international law.[55] Even though it is a principle of treaty interpretation, it provides a basis for harmonizing and coordinating international law obligations such that an attempt to perform one obligation does not lead to conflict with or the violation of another.[56]

The principle of systemic integration is not new to international law. Perennially, there have been questions on how different treaty provisions that receive their force and validity from different sources, and set up rights and obligations that exist alongside rights and obligations established by other treaty provisions and rules of customary international law, can be harmonized to make them operate as parts of a coherent and meaningful whole. This gave rise to the reasoning that treaty provisions must be 'applied and interpreted against the background of the general principles of international law', as pointed out by Lord McNair as far back as 1961.[57] Despite the fact that the idea was codified in the 1969 Vienna Convention, it remained in obscurity for an extended period of time.[58] However, this principle has gained increased prominence within the last years following the decision of the International Court of Justice (ICJ) in the *Oil Platform case.*[59]

[55] Xue Hanqin coined this term during the ILC debate on the significance of art. 31(3)(c). ILC, see Final Report of the Study Group on Fragmentation of International Law: Difficulties Arising from the Diversification and Expansion of International Law, para. 420, UN Doc A/CN.4/L.682 (13 April 2006) (prepared by M. Koskenniemi), ILC Para. 420.

[56] See 2009 Arbitral Award in *Phoenix Action Ltd v. Czech Republic*, where it was held that investment protection 'should not be granted to investments made in violation of the most fundamental rules of protection of human rights. . .'. *Phoenix Action Ltd v. Czech Republic*, ICSID Case No ARB/06/5, 15 April 2009, para. 78. See also Dupuy, 'Unification Rather than Fragmentation of International Law?', (n 52).

[57] Lord A.D. McNair, *The Law of Treaties*, 2nd edn. (Oxford: Clarendon, Press 1961), 466.

[58] For example, Thirlway mentioned in 1991 that he doubted '. . .whether this subparagraph will be of any assistance in the task of treaty interpretation'. See H. Thirlway, 'The Law and Procedure of the International Court of Justice 1960 – 1989: Part Three', *British Yearbook of International Law*, 62 (1991), 3 at 58. See also Mclachlan, 'The Principle of Systemic Integration', (n 48), 279 where the author noted that this principle has for so long remained one of the most neglected principles of international law.

[59] Case Concerning Oil Platforms (Islamic Republic of Iran *v.* United States of America), Judgment on Preliminary Objections of 12 December 1996, ICJ Rep. 1996, 803, Separate Opinion of Judge Buergenthal, para. 22.

In the *Oil Platform* case, the ICJ was called upon to interpret two provisions of the 1955 Treaty of Amity, Economic Relations and Consular Rights between Iran and the United States. In this case, Iran instituted proceedings against the United States arguing that attack on, and destruction of, three Iranian offshore oil production complexes by the US Navy constituted a fundamental breach of various provisions of the treaty as well as of international law. The United States argued that it was not in breach of its obligations under the treaty. In interpreting obligations of both countries under the treaty, Judge Simma identified the need for a holistic interpretation of the treaty, with reference to other relevant rules of international law as stipulated by Article 31(3)(c), which according to the court, is a 'sound and undisputed principle as far as treaty interpretation is concerned'.[60] The court held *inter alia*:

> Moreover, under the general rules of treaty interpretation, as reflected in the 1969 Vienna Convention on the Law of Treaties, interpretation must take into account 'any relevant rules of international law applicable in the relations between the parties' (Article 31, paragraph 3(c)). The Court cannot accept that Article XX, paragraph 1(d), of the 1955 Treaty was intended to operate wholly independently of the relevant rules of international law on the use of force, so as to be capable of being successfully invoked, even in the limited context of a claim for breach of the Treaty, in relation to an unlawful use of force. The application of the relevant rules of international law relating to this question thus forms an integral part of the task of interpretation entrusted to the Court by. . .the 1955 Treaty.[61]

The decision of the Court acknowledged the importance of a coherent interpretation and application of international law instruments, and the idea that treaty obligations are not intended to operate wholly independently of other relevant rules of international law. However, the Court did not provide further guidance as to when and how it should be applied. This was one of the issues considered by the International Law Commission Study Group that was established in 2002 to deal with the issue of fragmentation in international law, specifically the question on how best to integrate different regimes of international law to avoid fragmentation and multiplicity.[62]

[60] Case Concerning Oil Platforms (Islamic Republic of Iran *v.* United States of America), Judgment on Preliminary Objections of 12 December 1996, ICJ Rep. 1996, 803.

[61] *Ibid.*, para. 41.

[62] The debate on fragmentation of international law continues. See J. Trachtman, 'Fragmentation and Coherence in International Law', *Tufts University Legal Research Paper Series* (2011), 2; as Outi Korhonen puts it, it is a question on 'how a synthetic order, which is both common enough to produce cohesion and pluralistic enough not to reduce the various cultural differences, can be achieved without succumbing to either hegemony

The final report of the Study Group discussed Article 31(3)(c) as a possible response to the issue of fragmentation in international law. The report suggested that all relevant rules of international law must be considered when interpreting treaties. The report also seems to have laid to rest perennial arguments on the meaning of 'relevant rules', when it stated that relevant rules within the meaning of this provision cover all sources of international law.[63] The ILC Study Group noted the applicability of all general principles of international law in the interpretation of treaties and points to a need to take into account the normative environment more widely while interpreting treaties.[64]

Despite debates on the legal status and practical application of Article 31(3) (c) as a solution to fragmentation, this book aligns with the views that the principle of systemic integration, as enshrined in the 1969 Vienna Convention, is part of customary international law that could bring about the harmonization and integration of international law obligations that are scattered in different treaties to make them appear as part of a coherent and meaningful whole.[65]

or unmanageable fragmentation'. See O. Korhonen, *International Law Situated: An Analysis of the Lawyer's Stance towards Culture, History and Community* (The Hague: Kluwer, 2000) at 42; D. Pulkowski, 'Narratives of Fragmentation International Law between Unity and Multiplicity' (2005), www.esil-sedi.eu/sites/default/files/Pulkowski_0.PDF, accessed 21 March 2015.

63 However, it is still disputed whether parties relate to all parties to the treaties or specifically to parties in dispute. The WTO panel in the *EC – Approval and Marketing of Biotech Products* case stated that it applied to all the parties to the treaties. The main question relates to the lack of congruence in the membership of important international conventions, for example when a country is part of the ICCPR but not party to the UNFCCC. One can argue that the interpretation of this would vary from one context to another. One cannot generalize; hence there is a need to base conclusions on the text of the specific issue at stake. For example, with respect to climate change and human rights, a review of membership shows that over 95 per cent of the countries that are parties to the Kyoto Protocol are also parties to the core international human rights treaties. This congruence reinforces the view that all parties to the Kyoto Protocol have obligations to respect existing rules of international human rights law while designing emission reduction obligations under the Kyoto Protocol. See ILC, *Fragmentation of International Law*, at para. 426. See B. Simma and T. Kill, 'Harmonizing Investment Protection and International Human Rights: First Steps Towards a Methodology', in C. Binder, U. Kriebaum, A. Reinisch, and S. Wittich (eds.), *International Investment Law for the 21st Century: Essays in Honour of Christoph Schreuer* (Oxford: Oxford University Press, 2009), 678–707.

64 See International Law Commission (ILC), *Fragmentation of International Law*, at para. 415.

65 International courts and tribunals have consistently found that Article 31 of the VCLT is part of customary international law. See Case Concerning Oil Platforms (Islamic Republic of Iran *v.* United States of America), Judgment on Preliminary Objections of 12 December 1996, ICJ Rep. 1996, 803, para. 23; *Georges Pinson* case (*France/United*

The next question is whether international human rights falls within the broad meaning of 'relevant rules of international law' that must be considered when interpreting treaties. This question has been exhaustively answered by Simma and Kill. They developed a tripod test on elements or requirements for a rule or principle to qualify for consideration under Article 31(3)(c): there must be a rule; the rule must be 'relevant': and the rule must be 'applicable in the relations between the parties'. They subjected international human rights law to these tests and concluded that human rights are part of the relevant rules of international law.[66] I align with the extensive methodology adopted by Simma and Kill in arguing that the 'relevant rules' within the meaning of this provision are norms that have been recognized in core sources of international law, which include international human rights. This would require that human rights obligations of a country must be harmonized or integrated with its other obligations under any other international law instrument, for example emission reduction obligations under the international climate regime.

Even though the systemic integration principle is specifically a treaty interpretation principle, it is a principle that could bring about coherence and harmonization of crosscutting international law

Mexican States) Award of 13 April 1928, UNRIAA, vol. V, p. 422. (holding that a treaty must be deemed 'to refer to such principles of international law, for all questions which it does not itself resolve expressly and in a different way'). See also *Southern Bluefin Tuna* case (2000) (holding that it is a commonplace of international law and State practice for more than one treaty to bear upon a particular dispute. . . There is frequently a parallelism of treaties. . . The universal range of international legal obligations benefits from a process of accretion and accumulation); see also Japan – Taxes on Alcoholic Beverages, Appellate Body Report adopted on 1 November 1996, WTO, WT/DS8, 10,11/AB/R (Japan – Alcohol); Arbitral Award of 31 July 1989 (Guinea-Bissau *v.* Senegal), Judgment of 12 November 1991, ICJ Rep. 1991, 53; Maritime Delimitation and Territorial Questions between Qatar and Bahrain (Qatar *v.* Bahrain), Judgment of 15 February 1995, ICJ Rep. 1995, 18, para. 33; Territorial Dispute (Libyan Arab Jamahiriya *v.* Chad), Judgment of 3 February 1994, ICJ Rep. 1994, 6, para. 41; *Territorial Dispute case (Libyan Arab Jamahiriya/Chad) I.C.J. Reports 1994* p. 6; *Kasikili/Sedudu Island case (Botswana/Namibia) I.C.J. Reports 1999* p. 1059; *LaGrand case (Germany v. United States of America) I.C.J. Reports 2001* p. 501, para. 99. See also *Golder v. the United Kingdom*, Judgment of 21 February 1975, ECHR Series A (1975) No. 18, p. 14, para. 2+ 9; *Restrictions to the Death Penalty Cases*, Judgment of 8 September 1983, Advisory opinion, Int-Am CHR, OC-3/83, ILR, vol. 70 p. 449 and, for example, *United States – Standards for Reformulated and Conventional Gasoline* (29 April 1996) WT/DS2/AB/R, DSR 1996:I, p. 16.

66 See Simma and Kill, 'Harmonizing Investment Protection and International Human Rights', (n 63), 678–707.

obligations.[67] Applying this principle to the international climate change regime reinforces the notion that emission reduction obligations under the international climate change regime are not intended to operate wholly independently of other relevant rules of international law, including human rights. Treaty obligations under the international climate change regime must be applied and interpreted against the background of the general principles of international law, which include international human rights obligations. As such, projects, actions and measures designed to meet obligations under climate change agreements must not violate obligations under other international law regimes but rather must operate coherently as part of the international legal order.[68]

Considering the fact that many of the parties to global climate change treaties are also parties to the core international human rights treaties that establish human rights that are infringed by carbon projects, the coherent interpretative approach provides an avenue to understand why countries must endeavour to respect human rights while combating climate change. The systemic interpretation and international coherence approach in the Vienna Law of Treaties could provide a basis for countries to understand that human rights obligations are part of the 'relevant rules of international law' that must be taken into consideration or mainstreamed into the design of carbon projects and emission reduction schemes. Coherent interpretation and application of treaty obligations requires a mutually reinforcing approach that enables a country to meet its climate change obligations without violating its human rights obligations, and other obligations it may have, under international law. Human rights mainstreaming, specifically the human rights-based approach, provides a methodological approach through which human right obligations can be incorporated into climate change actions, projects and governance structures.[69]

[67] See Van Asselt, et al., 'Global Climate Change and the Fragmentation of International Law', (n 42), 423; S. McInerney-Lankford, 'Human Rights and Climate Change: Perspectives on the Value Added of Human Rights Law', (2012); Mclachlan, 'The Principle of Systemic Integration', (n 48), 279; see also Simma and Kill, 'Harmonizing Investment Protection and International Human Rights', (n 63); see also Dupuy, 'Unification Rather than Fragmentation of International Law?', (n 52), 45–62.

[68] See S. Stephenson, 'Jobs, Justice, Climate: Conflicting State Obligations in the International Human Rights and Climate Change Regimes', *Ottawa Law Review*, 42 (2010), 155–180.

[69] See Wewerinke and Doebbler, 'Exploring the Legal Basis of a Human Rights Approach', (n 30), 141–160.

Human rights mainstreaming provides an avenue to package a collection of international obligations under both human rights treaties and the international climate change regime into one box to enhance implementation.[70] Through a rights-based approach to carbon finance, carbon projects that violate human rights would not be funded, approved or registered, thereby ensuring that countries meet their obligations to respect, protect and fulfil human rights in their efforts to mitigate climate change. Similarly, it would provide an avenue to draw upon human rights obligations to underpin and inform the design and execution of emission reduction goals provided in climate change agreements, for example, the aspirations of the COP 16 in Cancun that '[P]arties should in all climate change actions respect human rights'. Human rights mainstreaming allows countries to integrate their obligations to respect human rights and to take action with regard to climate change through a holistic framework. This would also ensure that communities, private actors and governments will mitigate sources of climate change, and provide for adaptation to its effects, in a manner that respects human rights.

3.3.2 Legal transplant theory/inter-doctrinal legal transplants

The legal transplant theory, arguably, provides another theoretical justification on the value and importance of mainstreaming human rights norms in the design, approval, finance and implementation of carbon projects. While legal transplants in the traditional sense refer to the transfer of established laws and the borrowing of legal materials from one legal system or jurisdiction to another,[71] contemporary legal transplant has evolved as a basis for moving best practices, norms and legal rules from one area of law to another.[72]

Modern ideas of cross-fertilization challenge traditional comparative law, which, according to Ewald, have failed to recognize new possibilities

[70] See S. McInerney-Lankford, 'Human Rights and Climate Change: Perspectives on the Value Added', *Yale Journal of International Law*, 37: 2, (2012).

[71] For an excellent and detailed description of the idea of legal transplants, see D. Nelken and J. Feest (eds.), *Adapting Legal Cultures* (Oxford: Hart Publishing, 2001).

[72] Modern theorists have developed colourful vocabularies to describe the movement of legal models between different genres of law. Concepts such as mobility of law, grafting, implantation, cross-fertilization, cross-pollination, engulfment, emulation, infiltration, infusion, digestion, salad bowl mixture, melting pot and transposition have been used to describe the inter-doctrinal transplant idea. See E. Orucu, 'Family Trees for Legal Systems: Towards a Contemporary Approach', in M. van Hoecke (ed.), *Epistemology and Methodology of Comparative Law* (Oxford: Hart 2004) see chapter 18.

in a globalized world where unification and internationalization of laws are paramount.[73] Contemporary legal transplant theories provide theoretical justification for the transfer of best practices and the evolution of mutually reinforcing partnerships between two areas of law, for example human rights law and climate change law.[74] Through contemporary legal transplant and cross-fertilization, best practices, standards and norms could be adopted or moved from human rights law to strengthen and underpin the design of climate change actions, responses and projects, and vice versa. Human rights mainstreaming involves forging legal partnerships between two inherently distinct legal communities and treaty bodies: human rights and climate change communities.

The idea of moving legal norms and practices between two areas of law is, however, not without debate.[75] While some scholars accentuate the importance of such legal diffusions, others question whether the diffusion of laws and legal systems between legal families are possible or desirable. Framed simply, it is the question whether it is logical and desirable to contemplate the movement or borrowing of legal materials, practices or norms from human rights law to climate change law. Different schools of thought have emerged on the idea of

[73] See H. Spamann, 'Contemporary Legal Transplants: Legal Families and the Diffusion of Law', *Brigham Young University Law Review*, 18, (2009); E. Rock and M. Wachter, 'Dangerous Liaisons: Corporate Law, Trust Law and Inter-doctrinal Legal Transplants', *Northwestern University Law Review*, 96(2), (2002), 652; W. Ewald, 'Comparative Jurisprudence (I): What Was It Like to Try a Rat', *Pennsylvania Law Review*, 143 (1995), 2111–2112; see also P. Monateri, 'The "Weak" Law: Contaminations and Legal Cultures', in P. Monateri (ed.), *The Weak Law: Italian National Reports to the XVth International Congress of Comparative Law, Bristol 1998* (Milano: Giuffrè editore, 1998), 84; R. Sacco, 'Legal Formants: A Dynamic Approach to Comparative Law (Instalment I of II)', *American Journal of Comparative Law*, 1, (1991), 39.

[74] See Rock and Wachter, 'Dangerous Liaisons', *ibid.*; N. Burrows, 'European Community: The Mega Mix', in E. Örücü, E. Attwooll, and S. Coyle (eds.), *Studies in Legal Systems: Mixed and Mixing* (The Hague: Kluwer Law International, 1996), 297–312.

[75] See M. Chen-Wishart, 'Legal Transplant and Undue Influence: Lost in Translation or a Working Misunderstanding', *International and Comparative Law Quarterly*, 62(1) (2013), where the author argued that there is generally no yes or no answer or grand theories to explain the desirability of legal transplants, instead the answer owes to a careful examination of the nature of the transplanted law and the relationship between the formal and informal legal orders of the originating and the recipient systems raised by a particular transplant; see also Spamann, 'Contemporary Legal Transplants', (n 73), 1813–1877; Rock and Wachter, 'Dangerous Liaisons', (n 73), 652; N. Zhu, 'A Case Study of Legal Transplant: The Possibility of Efficient Breach in China', *Georgetown Journal of International Law*, 36 (2005), 1145.

inter-doctrinal legal transplants.[76] As such, there is no straightforward answer in the literature, on the question whether diffusion of laws between bodies of law are possible or desirable. This book does not intend to delve extensively into these debates. Instead, it aligns with the views that borrowing human rights principles, best practices and governance structures to govern climate change responses and projects are desirable and possible. This must, however, be done with careful attention to context.[77]

A good example is the excellent analogy drawn by Rock and Watcher on the dangers of inter-doctrinal legal transplants. While demonstrating the importance of context, Rock and Watcher gave the example of transplanting the idea of 'duty of care' from trusts and agency laws to become the corporate law on 'duty of loyalty'. They note that the toxicity of this transplant lies in the dangers of adopting a market concept in a firm context. It fails to recognize that the market/firm dichotomy is real and critical. The boundaries include the governance structures, the difference between third-party enforcements of obligations in markets and the non-legal self-governance structures within firms. According to them, in moving unselfconsciously across these boundaries, the seeds of mischief and confusions were sown. Even though this transplant by the Delaware parliament has a happy ending, it is a cautionary tale to the fact that transplanting legal concepts without paying attention to the underlying context can cause great mischief.[78]

Rock and Watcher's analogy rightly elucidates the need to examine issues of compatibility, conflict and the spirit of the law when engaging in inter-doctrinal legal transplants between human rights and climate change. The spirit of the law refers to the circumstances, reasons and

[76] These schools of thought can be traced to the initial standpoints on traditional legal transplants. See P. Legrand, 'Impossibility of Legal Transplants', *Maastricht Journal of European and Comparative Law*, 4 (1997), 111; P. Legrand, 'European Legal Systems are not Converging', *International and Comparative Law Quarterly*, 79 (1996).

[77] See Rock and Wachter, 'Dangerous Liaisons', (n 73), 651, where the authors argued that while some transplants are possible, others could be toxic.

> . . .When a legal concept is taken from one context and incorporated into a fundamentally different setting, it may carry with it, implications that do not serve anyone's interests. . . Borrowing concepts and transplanting legal rules is critical to the legal enterprise, but in doing so, one must be extremely sensitive to the differences in the context lest the borrowing becomes toxic.

See also C. Harlow, 'Book Review of J. Beatson and T. Tridimas (eds.), *New Directions in European Public Law* (Hart Publishing, 1998)', *Legal Studies*, 18 (1998), 558.

[78] See Rock and Wachter, 'Dangerous Liaisons', *ibid.*, 652.

context within which a law is made.[79] It is important to pause and determine, what the intentions are of negotiators or lawmakers in making the law or rule that we seek to borrow. It would be counterproductive to borrow a law or rule that would not fit into its new abode. For example, could human rights be effectively applied to market-based instruments, are there compartible institutions for implementing borrowed rules or principles, and are there normative or epistemic barriers to a successful transplant of the rule or structure. These questions would influence the success or otherwise of the transplant. A law reformer must thus ask whether the contexts within which the transplant is negotiated are similar, mergeable and realistic.[80] As Obiora rightly noted, the need to reflect and respect the varying context and diversity while maintaining a universalist claim cannot be overemphasized.[81]

It is also necessary to put in place governance structures or mechanisms for the follow-up and review of laws transplanted between two areas of law. Here the basic question is how well are the laws and institutions mixing? Are there institutions in place in the new destination for the practical implementation of these laws and how would the issue of a conflict of laws and institutions be dealt with? If the laws and institutions are not mixing or are not bringing about the desired results, then the transplant has failed and must be aborted.

An effective mainstream would, arguably, be dependent on three factors: organizational ideology and procedures; appropriate legal frameworks; and the existence and capacity of institutions and structures. The success or failure of any inter-doctrinal legal transfer between human rights and environmental law would depend in most part on the legal and institutional structures put in place to make the transplant successful. There is a need for a radical reform of laws and institutions in preparation for the reception of laws. It is not enough to theorize on transplants without following up with institutions. If human rights mainstreaming is to be effective, it must spearhead a transformative reform of policymaking procedures and a radical integration and reconfiguration

[79] See B. Montesquieu, *The Spirit of the Laws* (University of California Press, 1977). See also E Orucu, 'Mixed and Mixing Systems, A Conceptual Search' in E. Örücü, E. Attwooll, and S. Coyle (eds.), *Studies in Legal Systems: Mixed and Mixing*, 1st edn. (The Hague: Kluwer Law International, 1996), 344.

[80] See M. Chen-Wishart, 'Legal Transplant and Undue Influence', (n 75).

[81] L. Obiora, 'Toward an Auspicious Reconciliation of International and Comparative Analyses', *American Journal of Comparative Law*, 46 (1998), 669.

of power relationships. This would also include capacity development and training, enhancing structure and institutional capacity to reflect the changes and equipping personnel with practical strategies for implementing the reforms.

While the inter-doctrinal transplant theory provides a theoretical basis for transferring human rights principles and institutions into the international legal regime on climate change, such transfers must be considered side by side with the question of appropriate structures to make the integration successful. The question of providing the appropriate institutions that would make the transfers smooth should not be neglected, as this would go a long way to determine the success or failure of integrating or mainstreaming human rights principles into the international climate change regime.

A greater part of this book will examine and address these questions. A special attention will be paid to evaluating strategies and governance structures required to overcome the problem of ineffective transplant of human rights norms in the carbon finance context.

3.3.3 Participatory Development Theory

The Participatory Development Theory (PDT) is another pertinent theoretical foundation for the human rights mainstream paradigm.[82] Human rights mainstreaming and the human rights-based approach advocate the integration of human rights principles and safeguards in the design of carbon projects and climate change responses. One of the most important international human rights safeguard is the right to participate in decision-making processes. Participation is the right of the public to *take part* in decision-making processes and to *influence* the decisions taken. Built on a belief that the public can be trusted to shape their own future, participatory development uses local decision-making and capacities to steer and define the nature of an intervention or policy responses. This theory provides a basis for mainstreaming human rights rules and structures that allow the public to influence

[82] See R. Jennings, 'Participatory Development as New Paradigm: The Transition of Development Professionalism', (October 2000), www.usaid.gov/our_work/crosscutting_programmes/transition_initiatives/pubs/ptdv1000.pdf, accessed 12 May 2015. See also G. Pring and S. Noe, 'The Emerging International Law of Public Participation Affecting Global Mining, Energy, and Resources Development', in D. Zillman, N. Donald, A. Lucas, and G. Pring (eds.), *Human Rights in Natural Resource Development* (Oxford: Oxford University Press, 2002), 11.

and take part in decision-making processes with respect to carbon projects.

The idea of participation has its theoretical root in the PDT.[83] PDT is based on the tenets of equity and self-determination. Its fundamental assumption is that local people have a right to determine their own future and to take part in decisions that affect their lives.[84] The right to self-determination is not just about meeting people's needs; it is about creating an atmosphere whereby people can effectively identify and address their own needs. As Barton rightly notes, 'if you are ever going to get to a system of the people, by the people, and for the people then you'd better engage the people as early as you can'.[85]

Even though public participation began to gain attention as early as the 1970s, it gained increased recognition as a potent mechanism for good governance by the early 1990s.[86] PDT emerged as a response to traditional approaches to development that have been blamed for escalating the development problems in the developing and least developed countries.[87] The top-down development approach of the early 1960s

[83] Many other theories can be brought to bear on participation; these include neo-liberalism, multi-stakeholder theory, liberal democracy, rational elitism, pluralism, civic republicanism, and deliberative democracy, among others. Participatory development theory, however, is arguably the most important to the environmental and natural resource management discourse. See Pring and Noé, 'The Emerging International Law of Public Participation', 13; R. Chambers, *Rural Appraisal: Rapid, Relaxed and Participatory* (Brighton: Institute of Development Studies, 1992), 28–29; See also G. Mohan and K. Stokke, 'Participatory Development and Empowerment: The Dangers of Localism', *Third World Quarterly*, 21(2) (2000), 247–268; A. Agrawal and C. Gibson, 'Enchantment and Disenchantment: The Role of Community in Natural Resource Conservation', *World Development*, 27(4) (1999), 629–639; C. Kumar, 'Revisiting "Community" in Community-Based Natural Resource Management', *Community Development Journal*, 40(3) (2005), 275–285; and W. Meynen and M. Doornbos, 'Decentralizing Natural Resource Management: A Recipe for Sustainability and Equity?', *European Journal of Development Research*, 16(1) (2004), 235.

[84] B. Barton, 'Underlying Concepts and Theoretical Issues in Public Participation in Resources Development', in Zillman, Donald, Lucas, and Pring (eds.), *Human Rights in Natural Resource Development*, 81.

[85] *Ibid.*

[86] B. Richardson and J. Razzaque, 'Public Participation in Environmental Decision-making', in B. Richardson and S. Woods (eds.), *Environmental Law for Sustainability* (Portland, OR: Hart Publishing, 2006), 168. They describe how citizens' protests and activism by NGOs in the black communities in the United States and in indigenous communities in Canada, Australia and New Zealand brought about legal recognition for participatory development.

[87] Many dimensions of participation have been identified in literatures. According to Cohen and Uphoff, when speaking of participation the following questions must be asked: i) what kind of participation? Is it participation in decision-making, participation in implementation, participation in benefits (or harmful consequences), or participation

has been criticized for stifling the growth of human rights by neglecting the rights of citizens to be involved in the conduct of public affairs and the rights of individuals to be heard.[88] This approach failed to meet the needs of the poor mainly because 'the decisions were made by experts far removed from the people and their needs, and implemented through structures intended to be more responsive to central direction than local reality'.[89] As a result, outsiders who knew little about and paid little attention to the indigenes formulated projects designed to empower the indigenes. As Mtandawire rightly notes, this was a period when development failed to adequately address the processes of social exclusion as large sections of the populations were marginalized.[90]

These criticisms led to calls for an approach that respects, recognizes and mainstreams human rights of people to participate in public affairs and in decision-making.[91] What followed was a flurry of views and activities on participatory development. PDT gained exponential recognition as an approach that addresses the flawed assumptions of traditional approaches. As Barton observes, 'if there is a political evolution going on throughout the world, it is what might be called the

in evaluation? ii) who is participating in it (local residents, local leaders, government personnel, and foreign personnel)? and iii) how is participation occurring (basis of participation, form of participation, extent of participation, and effect of participation)? The HRBA is mainly concerned with participation in the decision-making process. J. Cohen and N. Uphoff, 'Participation's Place in Rural Development: Seeking Clarity through Specificity', *World Development*, 8 (1980), 214.

[88] See J. Lemmer, 'Cleaning up Development: EIA in Two of the World's Largest and Most Rapidly Developing Countries', *Georgetown International Environmental Law Review*, 19 (2007), 275; Y. Atal, 'Involving the Poor: The Many Question Marks', in Y. Atal and D. Ogen, (eds.), *Poverty and Participation in Civil Society* (Copenhagen: UNESCO, 1997); D. Korten, 'Rural Development Programming: The Learning Process Approach' in D.C. Korten and R. Klauss (eds.), *People-Centered Development: Contributions toward Theory and Planning Framework* (West Hartford: Kumarian Press, 1984), 183–184.

[89] See J. Hartz-Karp and M. Briand, 'Institutionalizing Deliberative Democracy', *Journal of Public Affairs*, 9 (2009), 125; D. Korten and F. Alfonso (eds.), *Bureaucracy and the Poor: Closing the Gap* (West Hartford: Kumarian Press, 1983), 2; Atal, 'Involving the Poor', *ibid.*; Korten, 'Rural Development Programming', *ibid.*

[90] T. Mtandawire, 'Bringing Development Back In', Keynote lecture at Dublin City University, 24 March 2006, Dublin City University, www.dcu.ie/~cis/Mkandawire.html, accessed 12 March 2014.

[91] See T. Webler, "Right" Discourse in Citizen Participation: An Evaluative Yardstick' in R. Ortwin, T. Webler and P. Wiedemann (eds.), *Fairness and Competence in Citizen Participation: Evaluating Models for Environmental Discourse* (Boston, MA: Kluwer Academic Publishers, 1995), 35; J. Ebbesson, 'Participatory and Procedural Rights in Environmental Matters: State of Play', (2009) High Level Expert Meeting on the New Future of Human Rights and Environment: Moving the Global Agenda Forward Co-organized by UNEP and OHCHR Nairobi, 30 November–1 December 2009.

participation explosion'.[92] PDT condemns top-down approaches to development that treat people as abstract concepts or as 'statistical figures to be moved up and down in the charts of progress'.[93] It recommends the adoption of 'people first' and 'people centred' principles to development and planning.[94] PDT seeks to assign to an individual the role not of subject but of an actor who participates in defining the goals and directing the processes of achieving them.[95] Its main thrust is the rejection of 'pre-packaged solutions' that fail to take cognizance of the socio-cultural reality and opinion of members of the society whose interests would be affected by a governmental decision.[96] It can be broadly conceived to embrace the idea that all stakeholders should take part in and influence regulatory and legislative decision-making processes related to carbon projects.

Participation could play an important role in improving the effectiveness and sustainability of carbon projects.[97] When host communities for carbon projects play active roles in project design and in the decision-making processes, the potentials or possibilities for oppositions, protests and conflicts are largely reduced. Participation provides policymakers with more complete information about details that might not be known by policymakers; for example, the complexity of the problems, how local circumstances, demographics and culture can have an impact on projects. When this information is gathered from

[92] See Barton, 'Underlying Concepts and Theoretical Issues', (n 84), 11 (notes that 'to keep calling it an explosion is hardly accurate now').

[93] See Atal and Ogen (eds.), *Poverty and Participation in Civil Society*, (n 88).

[94] K. Bottriell and M. Cordonier Segger, *The Principle of Public Participation and Access to Information and Justice* (Montreal: CISDL, 2005) 1.

[95] Barton, 'Underlying Concepts and Theoretical Issues', (n 84), 81.

[96] The concept of participation was first used in the late 1950s, and in the context of the Third World. At that time, social activists began to criticize the prevailing technocratic and top-down development approaches and to call for the participation of intended beneficiaries in development projects. By the 1970s, international development agencies had recognized the failure of pre-packaged solutions to deliver solutions to the problems of poverty. J Richards, *Paradigm for the development of South African prisoners in the post-apartheid dispensation* (thesis submitted to the Department of Development Studies in partial fulfilment of the requirements for the award of D.Litt Et Phil of the Rands Afrikaans University (RAU), South Africa 2003), 112, RAU, etd.rau.ac.za/theses/available/etd-10062004-104219/, assessed 12 May 2015; Atal, 'Involving the Poor', (n 88), 16.

[97] See generally R. Chambers, *Whose Reality Counts? Putting the First Last* (London: Intermediate Technology Publications, 1997); G. Gran, *Development by People: Citizen Construction of a Just World* (New York: Praeger 1983), 154; Korten, 'Rural Development Programming', (n 88), 183–184.

individuals who know it best, it allows for high-quality decisions to be taken.[98]

Participation also engenders public trust towards the government. It can increase the involvement of people in project design and implementation, thereby bringing them closer to the government. When people feel that the process was fair and that their input was taken into consideration, they will be motivated to take part in further decision-making processes.[99] They therefore view governments as partners in progress rather than as an abstract entity or policy overlords.

As Chambers, one of the main proponents of the participatory development theory, notes, there is a need for reversal of values and methods to one that puts the last (i.e. local people) first.[100] He calls for a paradigm shift in development theory, a reversal of values and methods so that the rural poor can be put at the centre of development.[101] To do better requires a reversal of this 'normal' trend - to an approach that puts people first. Even though Chambers' theory on participation has been extensively criticized, it remains a holistic summary of the idea of public participation, that is, a people-centred approach to development[102] - one that includes 'a sense of involvement' within a participatory and deliberative democracy where people become the subjects of their own world rather than objects in other people's world.[103] PDT provides a theoretical

98 As Macias rightly noted: 'The main substantive rationale for public participation is that it promotes better outcomes. Environmental decision-making is improved by a range of opinions, concerns, information and types of knowledge. When deliberation takes place, information and technologies will presumably be better applied.' R. Macias, 'Public Participation in Energy and Natural Resources Development: A Theory and Criteria for Evaluation', *Canadian Institute for Resources Law* (2010), 10.

99 *Ibid.* See also M. Reed, 'Stakeholder Participation for Environmental Management: A Literature Review', *Biological Conservation*, 141 (2008), 2417.

100 Chambers, *Whose Reality Counts?*, (n 97); R Chambers, *Challenging the Professions: Frontiers for Rural Development* (London: Intermediate Technology Publications, 1993); Chambers, *Rural Appraisal*, (n 83), 28–29.

101 Chambers, *Whose Reality Counts?* (n 97).

102 For detailed and extensive critique of Chambers, see I. Kapoor, 'The Devil's in the Theory: A Critical Appraisal of Robert Chambers' work on Participatory Development', *Third World Quarterly*, 23(1) (2002), 107–117; Agrawal and Gibson, 'Enchantment and Disenchantment', (n 83), 629–639; Kumar, 'Revisiting "Community" in Community-Based Natural Resource Management, (n 83), 275–285; Meynen and Doornbos, 'Decentralizing Natural Resource Management', (n 83), 235; Mohan and Stokke, 'Participatory Development and Empowerment', (n 83), 247–268.

103 For an excellent overview of the concept of participation in resource management, see C. Nwapi, 'A Legislative Proposal for Public Participation in Oil and Gas Decision-Making in Nigeria', *Journal of African Law*, 54 (2010), 184–211.

anchor on the need to develop appropriate legal and governance structures that promote the full participation of stakeholders in decision-making. Elements of a good participatory process are discussed in greater detail in Chapter 5 of this book.

3.3.4 *Full belly thesis*

One of the most important aspects of human rights mainstreaming is the need to respect, protect and fulfil *all* human rights in project planning and design. Rights infringed by carbon projects are commonly social, economic and cultural rights which some scholars still consider as secondary and second generation rights. Rather than bifurcating rights, human rights mainstreaming considers all human rights as equally important and necessary, whether they are civil and political or whether they are social, economic and cultural. The full belly thesis is a theoretical anchor for the position that social, economic and cultural rights are important and vital and must be respected and protected with equal vigour.

The full belly thesis was first propounded by Rhonda Howard in her watershed publication in 1983.[104] In it, Howard argued that economic rights to basic needs are more important than civil and political human rights and that 'civil and political rights can wait until basic economic needs are secured'.[105] She argues for the need to think of civil and political rights as a means to achieve a stable societal order, which is necessary for the society itself to exist in the first place. She noted that all human beings need a certain sense of dignity and autonomy. Achieving this requires the provision and guarantee of basic human needs such as the need for food, freedom from poverty, environmental protection and provision of adequate health resources. A full belly requires the active participation of people in the development and fulfilment of basic social and economic needs.[106] As Acheampong, a former president of Ghana, once noted, 'one man one vote is meaningless unless accompanied by the principle of one man, one bread'.[107] Summarily, the full belly thesis is that 'a man's belly must be full before he can indulge in the luxury of worrying about his political freedom'.[108] Without a belly full of food

[104] R. Howard, 'The Full Belly Thesis: Should Economic Rights Take Priority over Civil and Political Rights? Evidence from Sub-Saharan Africa', *Human Rights Quarterly*, 5 (1983), 469.

[105] *Ibid.* [106] *Ibid.*, 470.

[107] Amnesty International, *Background Paper on Ghana* (Mimeo, 1994).

[108] Howard, 'The Full belly Thesis', (n 104), 469.

and other vital needs, human rights, dignity, self-respect, or personal freedoms are meaningless.[109]

Other scholars have applied and built on the full belly thesis. For example, Gwam has classified environmental protection as one of the prerequisites of a full belly. He argues that civil and political rights such as the right to life are meaningless in the presence of environmental pollution such as the toxic waste dumping in Koko, Delta State, Nigeria and in Abidjan, Cote d'Ivoire. Gwam reckons that a clean environment is a sine qua non for the enjoyment of civil and political rights and that without a clean environment man may not live to enjoy the other recognized forms of rights, such as the right to life and the right to education.[110] Gwam calls for a participatory approach to development, which recognizes environmental protection as a prerequisite for the protection and fulfilment of fundamental human rights.[111]

Howard's thesis is particularly important as it made an early contribution to debates on the hierarchy of human rights and the place of economic and social rights in international law. These debates have been extensively dealt with elsewhere and need not be rehashed.[112] The full belly thesis takes a functionalist approach to debates on how attempts to respect, promote and fulfil human rights could fail if they do not reflect justice perspectives or take into consideration basic social, economic and environmental needs. Howard illustrates this with developments in Nigeria, Kenya and other sub-Saharan African countries where human rights mean nothing to so many people who cannot afford the basic necessities of life. She calls for an approach to development that recognizes that subsistence and security for people is a basic human right. According to Howard, 'in economic terms it means the right to adequate nutrition and a minimum standard of health care'.[113] She calls for an integrated approach to understanding human rights. Howard's approach rightly emphasizes food security, good health,

[109] *Ibid.* [110] C. Gwam, *Toxic Wastes and Human Rights* (UK: AuthorHouse, 2010), 105.
[111] *Ibid.*, 105.
[112] See O. De Schutter, *International Human Rights Law: Cases, Materials, Commentary* (Cambridge: Cambridge University Press, 2010), 553; See also H. Steiner, P. Alston and R. Goodman, *International Human Rights in Context: Law, Politics and Morals*, 3rd edn (Oxford: Oxford University Press, 2008), 280; N. Aryeh, 'Social and Economic Rights: A Critique', *Human Rights Brief*, 13(2) (2006), 1; R. Plant, 'Social and Economic Rights Revisited', *Kings College Law Journal*, 14 (2003), 13.
[113] Howard, 'The Full Belly Thesis', (n 104), 469.

empowerment, participatory development and environmental protection as prerequisites for the protection of human rights.[114]

Howard's thesis that human rights are meaningless if they do not act as catalysts for providing the basic human needs of individuals to social security and protection is apt. Environmental problems carry serious and far-reaching effects which could lead to death, displacement, change in life pattern and loss of ability to meet basic needs, and when these happen, they deprive people of the ability to enjoy other recognized human rights. There is thus a need for an approach that recognizes the linkages between the basic need for environmental protection and the protection and fulfilment of human rights. To achieve a full belly, it is pertinent to develop an understanding of how existing human rights instruments can be employed to achieve environmental protection goals. There is a need to focus on a holistic approach that mainstreams human rights norms into environmental planning to avoid situations whereby governmental policies or projects, for example in the areas of climate change mitigation, impact or threaten subsistence, culture and traditional land use rights.

The importance of harmonizing environmental policies and actions with fundamental human rights has gained increased attention over the last decades. As Shelton notes, it is necessary to examine how existing human rights are capable of responding to environmental degradation.[115] The United Nations Environment Programme has also emphasized the need for countries to 'transform current legal systems to incorporate issues of environmental rights within national and international agreements and to establish a legal framework for partnerships between environment and human rights'.[116] The full belly thesis underscores the need to adopt an integrated approach that recognizes cross-cutting issues in human rights and climate change.[117] A holistic approach

[114] *Ibid.* See also Gwam, *Toxic Wastes and Human Rights*, (n 110), 106.

[115] See D. Shelton, 'A Rights Based Approach to Climate Change Mitigation', in Greiber, Janki, Orellana, Savaresi-Hartmann and Shelton, *Conservation with Justice*, (n 13), 5.

[116] UNEP, Declaration of the Fourth Global Civil Society Forum held from 1 to 2 February 2003 at UNEP Headquarters, Nairobi, Kenya, www.unep.org/civil_society/PDF_docs/4thGCSF_Recommendations_GC22_Agenda.pdf, accessed 15 October 2014.

[117] See C. Sunderland, T. Greiber and G. Oviedo (eds.) *Rights-Based Approaches: Exploring Issues and Opportunities for Conservation* (Center for International Forestry Research (CIFOR) and IUCN, 2009); IUCN, *Conservation with Justice: A Rights-Based Approach*, 3; S. Vanderheiden, *Atmospheric Justice Political Theory of Climate Change* (Oxford University Press, 2008), 192–202.

to carbon finance would strike a partnership between climate change governance and human rights.[118] This approach will emphasize mainstreaming human rights norms, standards and principles into the various stages of mitigation responses, including policymaking, planning, funding, implementation, monitoring and evaluation of carbon projects, so as to ensure that the interests of the public are protected at all times.

The full belly thesis advocates a shift from a need-based approach to an approach that requires governments and planning agencies to consider the impact of a particular project on the enjoyment of human rights. To achieve a full belly, human rights must be harmonized and integrated into legislation, policies and project activities, thereby giving concerned communities and individuals a platform to demand enforcement. By mainstreaming human rights principles such as participation, access to justice and access to information into the international climate change regime we can best tackle sources of societal exclusions and prevent them.

Furthermore, by adopting a holistic approach that integrates all human rights into climate actions, policies and projects, social, economic and cultural rights become more visible and meaningful. Without incorporating human rights issues into mitigation strategies and projects, human rights might remain delusional or a merely utopian concept. It is only when human rights are brought to the table in decision and policymaking, that genuine opportunities would be provided to everyone, including the marginalized and vulnerable communities. This would enable such groups to have a say on the desirability or otherwise of projects that affect their environment or human rights – thereby making it possible to achieve Howard's perceptions of a full belly.

The above theoretical principles and norms reinforce the importance and value of integrating human rights requirements to the design, approval, finance and implementation of carbon projects. They are, however, far from exhaustive. Despite increased calls on the need for mainstreaming human rights norms into the climate change regime, some scholars question the practicability of this approach. This is mainly

[118] See S. Kas, 'Integrated Justice: Human Rights, Climate Change, and Poverty', *Transnational Law & Contemporary Problems*, 18(1) (2009), 115–138; J. Carlson, 'Reflections on a Problem of Climate Justice: Climate Change and the Rights of States in a Minimalist International Legal Order', *Transnational Law & Contemporary Problems*, 18(1) (2009), 45–68.

because many of the practical implications, content and requirements of implementing this approach in practice have not been exhaustively communicated. Some of the key outstanding theoretical questions on the paradoxes and potentials of human rights mainstreaming are examined in the next section.

3.4 Theoretical critiques of the human rights mainstream paradigm

Despite optimism in the literature on the value and potentials of the human rights mainstream paradigm, its scope and content have been criticized by a number of scholars. Most importantly, human rights mainstreaming has been criticized for its 'fuzzy' core.[119] Ambiguities surrounding human rights mainstreaming are arguably tied to the fact that key concepts central to the human rights mainstreaming movement have not been effectively communicated.[120] For example, it becomes hard to speak convincingly about mainstreaming human rights norms into the international climate change regime without a conceptual clarification of what human rights mainstreaming itself does and does not entail. Consequently, mainstreaming most times emerges as a monologue, rather than a dialogue with policymakers.[121] There is therefore a need to define and clarify key concepts associated with mainstreaming human rights in a specific context.

There is also a need to analyse both the theoretical and practical implications of a commitment to mainstream. There is need for a greater understanding of the potential benefits and problems associated with its application within a specific context, for example in the context

[119] See Oberleitner, *Global Human Rights Institutions*, (n 45), 45; Alston, 'Resisting the Merger and Acquisition of Human Rights', (n 22), 815. For example, Koskenniemi argues that:

> Mainstreaming should be understood as a project for seizing institutional power that is profoundly ambiguous in its effects. There is no reason to be either 'for' or 'against' mainstreaming without a clear sense of what priorities it is intended or likely to support and what foreseeable effect it might have on the allocation of resources among human groups.

See Koskenniemi, 'Human Rights Mainstreaming as a Strategy for Institutional Power', (n 22), 47.

[120] See Oberleitner, *Global Human Rights Institutions ibid.*, 2–3; Beveridge and Nott, 'Mainstreaming: A Case for Optimism and Cynicism', (n 22), 299–311.

[121] See C. Bacchi and J. Eveline, 'Mainstreaming and Neoliberalism: A Contested Relationship', *Journal of Public, Foreign and Global Policy*, 22 : 2, (2004), 118.

of climate change governance.[122] This is because the implications, challenges and strategies for human mainstreaming would vary from one context to another. This section sheds more light on key contested issues in the human rights mainstreaming literature. Four key debates are examined and discussed: proliferation debates, the novelty or duplication question, neo-liberalism critique and conflict of rights.

3.4.1 *Proliferation debates*

In the literature on human rights mainstreaming, one key debate is that applying human rights to broader areas of moral or political discourse could rob the human rights language of its primordial importance. Attempts to apply human rights language to policymaking have often been contested by traditional human rights theorists as attempts to weaken the concept of human rights. As Tasioulas notes, 'discourse of human rights has acquired in recent times. . .the status of an ethical *lingua franca*'.[123] Galtung and Wirag described this as the failure of mankind to distinguish between human rights and human needs.[124] They point to the geometric rise in appeals for human rights recognition under international law, even at times for trivial claims. Examples of proposals waiting for international recognition as human rights include the right to sexual pleasure;[125] the right to be free from poverty;[126] a right to commit suicide;[127] a right to

[122] See Wewerinke and Doebbler, 'Exploring the Legal Basis of a Human Rights Approach', (n 30), 141–160; McInerney-Lankford, 'Human Rights and Climate Change: Perspectives on the Value Added', (n 70).

[123] J. Tasioulas, 'The Moral Reality of Human Rights', in T. Pogge (ed)., *Freedom from Poverty as a Human Right: Who Owes What to the Very Poor?*, (Oxford: Oxford University Press, 2007), 75.

[124] J. Galtung and A. Wirak, 'On the Relationship between Human Rights and Human Needs', (1978) UNESCO Doc. SS-78/CONF.630/4, 48; See also J. Galtung and A. Wirak, 'Human Needs and Human Rights: A Theoretical Approach', *Bulletin of Peace Proposals*, 8 (1977), 251.

[125] *Sexual Rights are Fundamental and Universal Rights* (A Declaration Adopted by the World Association of Sexual Health in Hong Kong at the 14th World Congress on Sexology, 26 August 1999), www.assert.org.au/downloads/DecSexualRights.pdf, accessed 23 March 2015. According to the declaration 'Sexual rights are universal human rights based on the inherent freedom, dignity, and equality of all human beings. Since health is a fundamental human right, so must sexual health be a basic human right'. Para. 5 reads that 'Sexual pleasure, including autoeroticism, is a source of physical, psychological, intellectual and spiritual well-being' hence a fundamental right.

[126] See Pogge (ed.), *Freedom from Poverty as a Human Right*, (n 123).

[127] This includes a right to think about suicide and a right to commit suicide. See L. Stevens, 'Suicide: A Civil Right' (2006), www.antipsychiatry.org/suicide.htm, accessed

sleep;[128] the right not to be exposed to an excessively and unnecessarily heavy, degrading, dirty or boring job;[129] the right of individuals to permanent peace;[130] and a right to tourism.[131] Scholars have criticized these claims as attempts to further devalue the human rights language, impugn the integrity and credibility of the human rights tradition and erode the legitimacy of existing human rights.[132] The crux of the arguments is that linking every moral claim to a human right can lead to a human rights proliferation, thereby weakening the meaning of human rights and the resolve of enforcers to respect the entire human rights system.[133] As Wellman rightly argues:

> Just as inflation gradually reduces the real value of one's savings because one can now purchase fewer goods and services with the same amount of money, so also the rights inflation in political discourse has devalued any and every public appeal to rights.[134]

The recognition of every moral claim in the human rights lexicon will agreeably render the rights language useless and ordinary. However, not all human rights claims are trivial and frivolous. Asserting this will amount to over generalization. There are some genuine aspirations that deserve closer attention, particularly because of their relevance and

12 March 2014. This debate was intensified by the US case of *Cruzan v. Missouri* 497 US 261 (1990) where the court declared that an individual has a right to die. According to the court, people have a right of self-determination, which includes the right to refuse medical treatment, and a right to choose to die. See L. Perl, *Cruzan v. Missouri: A Right to Die* (New York: Marshall Cavendish, 2007), 11.

128 See 'Essays on the Right to Sleep', sleepinghumanrights.blogspot.com/, accessed 10 January 2015.

129 See Galtung and Wirak, 'On the Relationship between Human Rights and Human Needs', (n 124), 48.

130 *Ibid.*

131 See B. George and V. Varghese, 'Human Rights in Tourism: Conceptualization and Stakeholder's Perspectives', (2009), ejbo.jyu.fi/pdf/ejbo_vol12_no2_pages_40–48.pdf, accessed 24 March 2014.

132 Alston notes that the lists of rights are growing as if they are brought up by magic. He argues for an *appellations controlees* system or a quality control system through which the long list of 'would be' rights can be examined and pruned. See P. Alston, 'Conjuring Up Human Rights: A Proposal for Quality Control', *American Journal of International Law*, 78 (1984), 615.

133 See N. Ferreira, 'The Expanding Realm of Human Rights', *Res Publica*, 14 (2008), 57. Ferreira notes that the danger in the sudden proliferation of human rights is that 'when a concept is used in so many different circumstances; its meaning might become confused'.

134 See Wellman, *The Proliferation of Rights*, (Westview Press, 1999), 177–178.

importance to human liberty and freedom.[135] The challenge then is how to balance between protecting the human rights tradition and giving recognition to true and emerging threats to human existence. This book aligns with Wellman's views that 'the appeal to rights ought to be an invitation to reason with one another. . .the solution is not to abandon the language of rights in political discourse but to use it with more restraint.'[136]

Climate change represents one such fundamental concern that could benefit from the human rights language. Overwhelming scientific studies have shown that if climate change is not slowed down or mitigated through drastic cuts in emissions, it will threaten human existence and make living on planet earth impossible.[137] Similarly, recent studies have shown that if efforts and actions aimed at mitigating climate change are not harmonized with human rights, such measures might render the protection and fulfilment of human rights delusional.[138] For example, of what use is the right to life to citizens whose health and lives are threatened by extreme weather events caused by climate change or the right to property to citizens who have been forcefully displaced from their homes due to carbon projects? These show that climate change poses both direct and indirect threats to human rights. As such, there is a need to harmonize emission reduction schemes and measures with human rights in order to achieve robust and coherent international legal protection.

Similarly, recognizing these linkages has the potential of giving climate change mitigation the moral urgency it deserves; it could also go a long way in emphasizing the seriousness of its direct and indirect threats to human existence and the need to act fast. For example, recognizing how many lives are lost due to carbon projects, how many individuals are displaced from their homes, how many jobs are lost and how

135 This argument was extensively explored in D. Olawuyi, 'Adopting the Language of Rights for Environmental Protection: Legal and Theoretical Justifications', *Canadian Journal of Globalization*, 3(1) (2012), 35.

136 See Wellman, *The Proliferation of Rights*, (n 134), 180.

137 Intergovernmental Panel on Climate Change (IPCC) (2007) *Climate Change 2007: Synthesis Report. A Contribution of Working Groups I, II, and III to the Fourth Assessment Report of the Intergovernmental Panel on Climate Change* (Cambridge University Press, 2007).

138 See E. Cameron, 'Human Rights and Climate Change: Moving from an Intrinsic to an Instrumental Approach', *Georgia Journal of International and Comparative Law*, 38 (2010), 699, 673; N. Roht-Arriaza, 'Human Rights in the Climate Change Regime', *Journal of Human Rights and the Environment*, 1(2) (2010), 211 at 215–219; Olawuyi, 'Adopting the Language of Rights for Environmental Protection', (n 135), 35.

international human rights are threatened could strike the right chords with human rights gatekeepers and bring these concerns to the attention of the global community. Due to perceptions, human rights violations tend to attract more outrage and condemnation in many developing countries compared to environmental pollution.

Mainstreaming human rights norms into decision-making would place additional weight and responsibility on decision-makers to examine not only the economic aspects of a project but also its impact on the people.[139] This paradigm places the people as priorities in decision-making processes, thereby empowering them to demand enforcement and to play active roles in decision-making processes.

Adopting human rights norms in the international climate change regime would place humans at the centre of climate change mitigation. It would also give human rights protection precedence over emission reduction goals. The current approaches to mitigation that tend to place emission reduction, economic benefits and carbon credits ahead of negative impacts of large emission reduction schemes on host communities need to be replaced with a more transparent, accountable and rights-based framework anchored on the respect for human rights.

Similarly, linking climate change mitigation to human rights would provide a substantive basis for non-governmental organizations (NGOs), media, individuals and other constituencies beyond environmentalists to 'name and shame' projects, policies and governmental actions that do not reflect human rights considerations.[140] Since human rights are considered sacrosanct in many countries, exposing how carbon projects threaten human rights could be an effective way to draw more attention to the linkages between climate change mitigation and human rights. This could encourage governments to reflect on human rights while designing climate change responses and vice versa.[141]

Thirdly, incorporating human rights norms into the international climate change regime has the potential of providing individuals with a legal framework through which they could directly hold national authorities accountable before international complaints bodies for direct and indirect impacts of carbon projects. While a number of core

[139] See Cameron, 'Human Rights and Climate Change', *ibid.*

[140] Johl and Duyck, 'Promoting Human Rights in the Future Climate Regime', *Ethics, Policy and Environment*, 15(3) (2012), 300; McInerney-Lankford, 'Human Rights and Climate Change: Perspectives on the Value Added', 10.

[141] E. Posner, 'Climate Change and International Human Rights Litigation: A Critical Appraisal', *University of Pennsylvania Law Review*, 155 (2007), 1925–1945.

international human rights treaties establish individual complaints mechanisms that allow citizens to challenge governmental decisions, the current international climate change regime fails to do so. Both the Kyoto Protocol and the UNFCCC do not provide human rights recognition or protection for individuals. Mainstreaming human rights standards in the international climate change regime could provide opportunities to address this gap. For example, the Aarhus Convention is a major environmental treaty that has leveraged on the experiences of human rights treaty bodies by establishing a complaint mechanism that is accessible and open to individuals.[142] It establishes procedural human rights such as access to information, participation in the decision-making process and access to justice and also provides tools that could empower individuals to demand protection against policies that may affect them. Mainstreaming these procedural human rights in the international climate change regime could provide robust platforms for individuals and communities to block or appeal for a review of projects that infringe on the enjoyment of existing human rights. Mainstreaming human rights norms in climate change mitigation efforts may provide an opportunity to restructure climate change governance institutions and structures such that similar grievance mechanisms that allow individuals to claim their human rights can be established within the international climate change regime.[143]

Fourthly, mainstreaming human rights norms in the international climate change regime could provide a holistic and coherent approach for implementing carbon projects without creating human rights problems. Current market-based approaches to mitigation arguably tend to resolve the problem of climate change by creating other concerns, making the cure to climate change to appear worse than the problem itself. For example, many industrialized countries have focused on business-as-usual projects and 'emission credits at all cost' policies rather than projects that are truly efficient and sustainable in the long term. Such trade-offs have been the main reason for the ineffectiveness of the Kyoto

[142] UNECE, Aarhus Convention on Access to Information, Public Participation in Decision-Making and Access to Justice in Environmental Matters (Adopted 25 June 1998, entered into force 30 October, 2001) No. 37770, treaties.un.org/doc/Publication/MTDSG/Volume%20II/Chapter%20XXVII/XXVII-13.en.pdf, accessed 19 March 2015 (Aarhus Convention).

[143] See N. Roht-Arriaza, 'First, Do No Harm': Human Rights and Efforts to Combat Climate Change', *Georgia Journal of International and Comparative Law*, 38 (2009–2010), 593, 595; O. Padersen, 'The Janus Head of Human Rights and Climate Change: Adaptation and Mitigation', *Nordic Journal of International Law*, 80 (2011), 403.

Protocol. Balancing mitigation objectives with human rights obligations would place citizens at the centre of mitigation measures and projects. It would place an obligation on governments not only to ask how many CERs a project could generate but to also ascertain how the project could affect host communities in the long run; how many individuals could be displaced as a result of this project; and what measures could have been put in place to address such concerns. This balanced approach would not only ensure that responses to climate change will conform to existing human rights obligations and principles but that they also have the potential to strengthen the effectiveness and long-term success of mitigation measures.

Lastly, mainstreaming human rights norms in the international climate change regime would offer a more coordinated, detailed and well-documented approach for dealing with the linkages between climate change and human rights. Currently, these linkages have developed in a fragmented manner across national and regional courts and in a plenitude of declarations and resolutions. This has not fostered a broad understanding of the linkages; it has also resulted in some courts failing to recognize these linkages. A rights-based reform of the international legal regime on climate change would codify the linkages; foster a common understanding on the direct and indirect effects of climate change on human rights; clarify how the linkages could be better reflected in mitigation plans; and provide clear guidelines on thresholds that carbon projects must meet before they can be registered or approved as sustainable.

3.4.2 Novelty or duplication – redundancy debates

There have been arguments that human rights mainstreaming would bring nothing new to the processes and institutions of climate change.[144] Some theorists have also noted that mainstreaming human rights might be unnecessary, redundant and duplicative as existing human rights treaties and multilateral environmental agreements are already sufficient for shaping climate change responses and for dealing with any area of

[144] See E. Morgera, 'No Need to Reinvent the Wheel for a Human Rights-Based Approach to Tackling Climate Change: The Contribution of International Biodiversity Law', in E. Hollo, K. Kulovesi and M. Mehling (eds.), *Climate Change and the Law: A Global Perspective* (Netherlands: Springer-Verlag London Ltd., 2012). See also C. Gearty, 'Do Human Rights Help or Hinder Environmental Protection?', *Journal of Human Rights and the Environment*, 1 (2010), 7–22.

intersection between climate change and human rights. The question therefore is 'what is new about human rights mainstreaming?'

Human rights mainstreaming is based on the need to fully integrate and harmonize separate, and sometimes scattered, developments in human rights with developments in international climate change in order to develop more coherent, robust and holistic responses.[145] There are currently over nine hundred international legal instruments that are either primarily directed to international human rights issues or contain important provisions on them. There are also several international agreements, conventions, protocols, decisions and modalities that already speak to the issue of climate change under international law.[146] However, even though many areas of international law are relevant to the problems raised by carbon projects, they were not created with the challenge of climate change in view and are, arguably, not well suited to address human rights risks in carbon projects.[147]

Due to the complex nature of issues involved in carbon projects, it is important to design customized and holistic responses to address the problem of exclusions and lack of coherent legal protection, in a manner not currently envisaged by extant legal regimes. Human rights mainstreaming takes an integrative look at establishing a legal partnership between human rights and climate change such that obligations under human rights instruments and climate change agreements are harmonized. It is an approach that seeks to recognize the interactions between human rights and climate change and to promote coherence in the design of actions and responses such that an attempt to combat climate change would not lead to human rights violations and vice versa. The novelty of this approach is that instead of establishing new or conflicting regimes, it seeks to harmonize the existing obligations of project proponents and project host countries under human rights and the international climate change regime to ensure compliance with both sets of obligations.

Establishing legal partnerships between human rights and climate change would enhance the works of the respective human rights, climate

[145] See Van Asselt, Sindico and Mehling, 'Global Climate Change and the Fragmentation of International Law', (n 42), 423; see also E. Weiss, 'International Environmental Law: Contemporary Issues and the Emergence of a New World Order', *Georgetown Law Journal*, 81 (1993), 675–684, 702–710.

[146] See for example the Conference of the Parties to the Framework Convention on Climate Change, *Kyoto Protocol*, 10 December 1997, UN Doc FCCC/CP/1997/L.7/add.1, 37 ILM 22 (1998) [*Kyoto Protocol*].

[147] See Young, 'Climate Change Law and Regime Interaction', (n 41), 147–151.

change institutions and legal regimes and provide a chance to maximize resource allocation, avoid duplication of roles and also eliminate counterproductive measures. Also where enforcement options are unavailable or missing in one area, the other area might strengthen implementation. By establishing a complaint mechanism under the international climate change regime through which members of the public can ask for a review of any project that affects human rights, the international climate change regime would serve as an alternative window of redress for human rights violations with respect to carbon projects.

3.4.3 Neo-liberalism and human rights mainstreaming

The idea of introducing human rights norms into market or economic instruments has been hugely contested.[148] This debate is centred on the question whether command and control-type instruments, such as human rights, should be applied to market systems.[149] As Beveridge *et al.* rightly noted, 'one of the greatest concerns regarding mainstreaming is the degree to which mainstreaming can tackle the tendency of

148 M. Childs, 'Privatising the Atmosphere: A Solution or Dangerous Con?' *Ephemera Journal*, 12(1/2) (2012), 12–18; Center of Concern, 'Why is a Human Rights Approach Needed in Financial Regulation?' (Center of Concern, 2011); See also the Alston/Petersmann exchange on this subject, Petersmann, 'From "Negative" to "Positive" Integration in the WTO', (n 6), 1363–1382; Alston, 'Resisting the Merger and Acquisition of Human Rights', (n 22), 815; E. Petersmann, 'Taking Human Dignity, Poverty and Empowerment of Individuals More Seriously: Rejoinder to Alston', *European Journal of International Law*, 13 (2002), 845–851. See also R. Howse, 'Human Rights in the WTO: Whose Rights, What Humanity? Comment on Petersmann', *European Journal of International Law*, 13 (2002), 652.

149 Fletcher has argued that the growing effort to address anthropogenic climate change through carbon markets and other financial mechanisms constitutes a form of 'disaster capitalism'. See R. Fletcher, 'Capitalizing on Chaos: Climate Change and Disaster Capitalism', *Ephemera Journal*, 12 (1/2) (2012), 97; see also M. Dowell-Jones, and D. Kinley, 'Minding the Gap: Global Finance and Human Rights', *Ethics and International Affairs*, 25(2) (2011), 183–210; L. Lohmann and S. Böhm, 'Critiquing Carbon Markets: A Conversation', *Ephemera Journal*, 12(1/2) (2012), 81–96; Bacchi and Eveline 'Mainstreaming and Neoliberalism', (n 121), 118; S. Scott, 'Is the Crisis of Climate Change a Crisis for International Law: Is International Law too Democratic, too Capitalist and too Fearful to Cope with the Crisis of Climate Change?', *Australian International Law Journal*, 14 (2007), 31–44; T. Hayward, 'Human Rights Versus Emissions Rights: Climate Justice and the Equitable Distribution of Ecological Space', *Ethics and International Affairs*, 21(4) (2007), 431–450. For an older but still pertinent perspective, see J. Coleman, *Markets, Morals and the Law* (Cambridge: Cambridge University Press, 1988).

markets to replicate and aggravate inequalities in society'.[150] Also known as the free market ideology, neo-liberalism advocates free trade, open markets and the transfer of market control and economic decision-making from public institutions to the private sector.[151] It proposes that regulations, rules and governmental interventions in markets must be kept to a bare minimum to enhance the functionality and self-sustaining nature of markets.[152] Under this model, everyone cannot win; there would inevitably be winners and losers.[153]

Neo-liberalists are therefore quick to reject the idea of mainstreaming human rights norms in market-based instruments such as the CDM or REDD+.[154] The crux of the argument is that market-based mechanisms are incentive-based mechanisms that should not be diluted by conduct-based instruments such as human rights. The neo-liberalistic standpoint therefore is that mainstreaming human rights norms in incentive-based regimes, such as the climate regime, would reduce their effectiveness and might defeat the entire flexibility purpose that markets are designed to serve. Furthermore, it has been argued that applying the language of human rights to cooperative mechanisms such as emissions trading and the CDM would distort the competitive and incentive-based nature of market systems.[155]

Another neo-liberalist argument is that flexibility of costs cannot be side lined in an effective regulatory system. As Weiner notes, incentive instruments would be more cost effective than conduct instruments and fixed performance standards such as human rights instruments. This is because market-based instruments serve as incentives for developed

150 Beveridge and Nott, 'Mainstreaming: A Case for Optimism and Cynicism', 308; see also Dowell-Jones and Kinley, 'Minding the Gap', *ibid.*

151 Neoliberalism is said to have developed from the *laissez faire* ideas of Adam Smith that the government should not intervene in economic matters. See J. Campbell and O. Pedersen (eds.), *The Rise of Neoliberalism and Institutional Analysis* (Princeton, NJ: Princeton University Press, 2001). See D. Thorson, 'What is Neoliberalism', (2009) University of Oslo Working Paper Series, folk.uio.no/daget/neoliberalism2.pdf, accessed 12 January 2014; See also R. Plant, *The Neoliberal State* (Oxford: Oxford University Press, 2009); Campbell and Pedersen (eds.), *The Rise of Neoliberalism and Institutional Analysis*, *ibid.*, 288.

152 D. Harvey, *A Brief History of Neoliberalism* (Oxford: Oxford University Press, 2007), 2.

153 S. George, 'A Short History of Neoliberalism', (1999), a paper presented at a Conference on Economic Sovereignty in a Globalizing World (Bangkok, 24–26 March 1999).

154 See Bacchi and Eveline 'Mainstreaming and Neoliberalism', (n 121), 118; Lohmann and Böhm, 'Critiquing Carbon Markets', (n 149); See also D. Levy, 'Can Capitalism Survive Climate Change', *Ephemera Journal*, 12(1/2) (2012), 253.

155 See Fletcher, 'Capitalizing on Chaos', (n 149), 97; see also Childs, 'Privatising the Atmosphere', (n 148), 12–18.

countries to achieve steady progress in environmental protection at the least cost and with much flexibility. He maintained that such an incentive approach is cost effective, fair and has the essential attributes of dynamic innovation, low administrative cost and can be morally justified.[156] Portney also argues that the issue of costs cannot be left aside in policymaking. For Portney, the question should be efficiency and not fairness.[157]

These arguments are in contrast with views of human rights mainstreaming theorists that social justice issues such as the environment are far too important to be left to market distortions.[158] Petersmann has been one of the most consistent proponents of an approach that mainstreams human rights in world trade and global markets.[159] He argues that human rights could enhance the effectiveness of market economies; empower citizens to demand a correction of market injustices; and also limit the strong governmental influences and control on markets. For example, the right to information can act as a basis for citizens to demand market information such as market prices from governments, which would enable them to make more informed decisions. He calls for the transformation of market freedoms into fundamental human rights and the integration of human rights norms into the laws and treaties of worldwide organizations.[160] Furthermore, Petersmann calls for a decentralized 'market governance mechanism' which treats citizens as legal subjects rather than mere objects.[161]

[156] See J. Weiner, 'Global Environmental Regulation: Instrument Choice in Legal Context', *Yale Law Journal*, 108 (1999), 691–692.

[157] P. Portney, 'Counting the Cost: The Growing Role of Economics in Environmental Decision-Making', *Environment*, 2 (1998), 14.

[158] See Roht-Arriaza, 'First, Do No Harm', (n 143), 593, 595; S. Kravchenko, 'Right to Carbon or Right to Life: Human Rights Approaches to Climate Change', *Vermont Journal of Environmental Law*, 9 (2008), 522.

[159] See Petersmann, 'From "Negative" to "Positive" Integration in the WTO', (n 6), 1363–1382; E. Petersmann, 'Time for United Nations "Global Compact" for Integrating Human Rights into the Law of Worldwide Organizations: Lessons from European Integration", *European Journal of International Law*, 13 (2002), 626; Petersmann, 'Taking Human Dignity, Poverty and Empowerment of Individuals More Seriously', (n 148), 845–851; E. Petersmann, 'Constitutionalism and WTO Law: From a State-Centered Approach towards a Human Rights Approach in International Economic Law', in D.L.M. Kennedy and J.D. Southwick (eds.), *The Political Economy of International Trade Law* (Cambridge: Cambridge University Press, 2002), 32–67; E. Petersmann and Ernst-Ulrich, 'Human Rights and the Law of the World Trade Organization', *Journal of World Trade* (2003), 241–281.

[160] Petersmann, 'Time for a United Nations "Global Compact"', 641.

[161] Petersmann, 'Taking Human Dignity, Poverty and Empowerment of Individuals More Seriously', (n 148), 845–851.

Similarly, there have been calls for a new form of leadership that addresses manipulations and injustices of market systems, specifically tendencies of market participants to achieve emission reductions at all cost.[162] This is the concern that markets fail to condemn actors who play outside well-accepted rules and norms; so far, they meet up with market rules. For example, condoning the purchase and sale of the 'right to pollute' fails to condemn the actor who causes increased pollution and thereby undermines ethical norms. As Rees argued, prevailing policy and economic approaches that focus almost exclusively on continuous material growth and profitability are not working.[163] Klein calls this 'the shock doctrine'. She argued that free market policies have played a big role in perpetuating massive human rights violations since the mid-1970s. She noted that most of these policies are often unwillingly imposed upon people immediately after a major catastrophe since the national psyche is still in a state of shock and hence unable to mount effective resistance. Klein refers to the idea of market-based instruments as 'disaster capitalism'.[164]

Furthermore, markets have been criticized for fostering discriminations and inequities. For example, Silayan has analysed the inequalities in project distribution under the CDM.[165] Since investors in the markets seek to pursue an optimal investment portfolio and to maximize profits and minimize cost, they tend to achieve results at all costs, including by fostering inequities in host countries.[166] Apart from the issue of concentrating projects in a few selected countries, host countries, in a desperate bid to attract carbon investments, are also forced to dance to the tune of the investors or lose out. This has led to issues such as governments repressing and oppressing its own citizens just to retain projects and satisfy investors.[167] In this, the only winners would be those countries that lower sustainability standards, increase investment incentives and

[162] I. Khan, 'Foreword: It's not just the economy, it's a human rights crisis' in *Amnesty International Human Rights Report 2009*, report2009.amnesty.org/en/introduction, accessed 23 April 2015.

[163] W. Rees, 'Foreword: Global Change and the Social Construction of Human Rights', in L. Westra (ed.), *Human Rights: The Common and the Collective* (Vancouver, BC: University of British Columbia Press, 2011), xiv.

[164] See N. Klein, *The Shock Doctrine: The Rise of Disaster Capitalism* (New York: Metropolitan Books 2007), 9–10; see also N. Klein, 'Disaster Capitalism: The New Economy of Catastrophe' (2007), www.democruptcy.com/disaster-capitalism/, accessed 12 May 2015.

[165] See A. Silayan, 'Equitable Distribution of CDM Projects Among Developing Countries', *HWWA-Institut für Wirtschaftsforschung-Hamburg (HWWA) Report*, 255 (2006), 56–57.

[166] *Ibid.* [167] *Ibid.*, 56.

turn a blind eye to human rights standards when they stand in the way of economic gain.[168] Countries that are perceived as not providing the right atmosphere for carbon investments to proceed become losers in the market, as they would be left behind. As Silayan notes, the end result may be an overall negative effect for developing nations as a whole.[169]

Sokona has also been critical of neo-liberalism, as emphasized by markets, as an unfair approach to environmental protection as it imposes strenuous conditions on developing countries, particularly African countries. It also fosters human rights abuses and social repressions.[170] Haya was also of the view that market-based mechanisms will lead to further societal injustices in developing countries with already abysmal human rights records.[171] She maintained that most industrialized countries are likely to capitalize on the nature of the markets to propose 'free-rider' and 'business-as-usual projects' that only worsen environmental and human rights situations of developing countries. Haya argues that leaving solutions to the problems of climate change in the hands of markets would be counter-productive in the long run.[172] Shestack argues that markets have profit minimization motivations under which the well-being of the individual may be sacrificed for aggregate market interests; this would have the effect of leaving liberty and rights vulnerable to contingencies and risks.[173]

Despite the critiques of carbon markets, Weiner insists that market mechanisms do not in any way represent immoral means of achieving environmental protection. He opined that while critics of incentives may feel the environment is too important to leave to markets, the better view is that environmental protection is too important to leave out of market. According to Weiner, if the moralist's claim is that the person who causes more pollution is blameworthy and should be condemned, then ironically the person who causes more pollution is the moralist who opposes the more effective instruments.[174]

168 *Ibid.* See also R. Saner 'Hype or Reality: Can the CDM trigger FDI', *European Climate Platform*, 2 (2005).

169 See Silayan, 'Equitable Distribution of CDM Projects Among Developing Countries', (n 165), 58; see also R. Timmons and B. Parks, *A Climate of Injustice: Global Inequality, North-South Politics, and Climate Policy* (Cambridge, MA: Massachusetts Institute of Technology Press, 2007), 11.

170 Y. Sokona (ed.), *CDM- What Prospects For Africa* (Dakar, Senegal: ENDA 2005), 1.

171 B. Haya, *Damming the CDM: Why Big Hydro is Ruining the Clean Development Mechanism* (Berkeley, CA: International Rivers Network, 2002), 1.

172 *Ibid.*, 2–3.

173 See Shestack, 'The Philosophical Foundations of Human Rights', (n 26), 44–45.

174 See Weiner, 'Global Environmental Regulation, (n 156), 677 at 691–692.

The overall arguments put forward by market enthusiasts are not strengthened by human rights petitions and protests that have trailed current market systems. Issues such as the transfer of outdated and inefficient technologies in exchange for emission credits, governmental repressions, killing and maiming of citizens to force through carbon projects; lack of fair and transparent opportunities for participation by citizens in project planning and implementation; siting of projects in poor and vulnerable communities; lack of governmental accountability on projects, to the absence of judicial and quasi-judicial remedies for victims of the above-mentioned problems have plagued the current system. The international community would have failed in its role as gatekeeper and protector of human rights and dignity if nothing is done to curtail the excesses of the current carbon markets. The loss of human lives cannot be quantified in terms of costs, be they monetary or in the form of emission credits. It would be difficult to appraise the cost effectiveness of markets if individuals continue to pay the higher prices of displacements from homes and at times death.

Generally, markets are not the biggest problems. Arguably, the biggest problem is an unregulated or ineffectively governed market, that is, a market without adequate rules that prevent oppression and injustice. For example, in terms of climate change, the adoption of market mechanisms is not by any stretch the main problem. In fact, the history of the process that led to the Kyoto Protocol shows that but for the market mechanisms a global agreement on climate change could not have been reached.[175] This is because flexibility and cost abatement offered by markets are great incentives for industrialized countries to take part in

[175] Throughout the Kyoto negotiation, the United States and many other OECD clamoured for a flexible agreement that would allow countries to lessen domestic pressures in GHG emission reduction. Against EU and G77's proposal on flat rate emission reductions for all, the United States and most OECD countries argued for international flexibility. According to the US Chief Negotiator 'the American target is contingent on the acceptance of a comprehensive package that includes the trading of emissions amongst countries and regions'. Thus, while the EU and G77 called for *binding* commitments, the United States clamoured for *binding but flexible* commitments. The United States had insisted that emission reductions could be achieved most cheaply in the economies perceived to be less efficient, that is, the Economies in Transition (EITs) and developing countries. For an excellent and detailed account of the origin of the flexible mechanisms, see J. Depledge, 'Tracing the Origins of the Kyoto Protocol: An Article-by-Article Textual History' UNFCCC, unfccc.int/resource/docs/tp/tp0200.pdf, accessed 19 January 2015; See also T. Jackson, 'The Language of Flexibility and the Flexibility of Language', *International Journal of Environment and Pollution*, 10(3) (1998); M. Grubb, *The Kyoto Protocol, A Guide and Assessment* (London: RIIA/Earthscan, 1999), 29.

GHG mitigation. Markets also allow developing countries to gain in terms of access to newer technologies, more job opportunities and solutions to long-term environmental problems.

Consequently, it is important not to throw away markets but rather ask how new policy measures can be designed to achieve the two-pronged goals of maintaining the market outlook of the international climate change regime, while laying down stricter regulations that respond to the injustices of the carbon market. Thinkers in the field have failed to address the possibility of such a win-win scenario.[176] A win-win scenario can be achieved by restructuring the international climate change regime to include flexible, but binding, obligations on market participants to respect international human rights. Such a mechanism would impose penalties on countries that fail to respect human rights in project execution and implementation. Such a mechanism would represent a balanced mix of mitigation, respect for human rights, equity and development.

Human rights mainstreaming arguably presents an opportunity to achieve a win-win scenario.[177] It provides an opportunity to achieve an instrument that is both incentive-based and command-based. The international climate regime may be strengthened, to be rights-based, by integrating processes, standards and institutions that address market inequalities. By putting individuals and stakeholders at the centre through participation, accountability and access to justice, human rights mainstreaming provides the opportunities to address human rights issues associated with projects and to change the negative perceptions of the market systems. Without this, it might amount to attempting to solve the problem of climate change by creating many other problems.[178]

[176] See S. Jodoin, 'Human Rights and the Environment: Key Challenges and Best Practices' (2011), 10–11, Center for International Sustainable Development Law Brief presented to the United Nations Office of the High Commissioner for Human Rights, 10–11, www.cisdl.org/publications/collaborative005.html, accessed 03 August 2014. Jodoin notes that scholars have not been creative enough to explore the possibilities of a win-win scenario of this nature.

[177] See M. Dowell-Jones, 'International Finance and Human Rights: Scope for a Mutually Beneficial Relationship', *Global Policy*, 4(3) (2012), 467.

[178] See E. Meijer and J. Werksman, 'Keeping It Clean: Safeguarding the Environmental Integrity of the Clean Development Mechanism' in D. Freestone and C. Streck (eds.), *Legal Aspects of Implementing the Kyoto Protocol* (Oxford University Press, 2005), 191, 201; Cameron, 'Human Rights and Climate Change: Moving from an Intrinsic to an Instrumental Approach', (n 138), 673, 699; The World Bank, *Development and Climate Change: World Development Report 2010* (The World Bank, 2010); Oxfam International, *Climate Wrongs and Human Rights: Putting People at the Heart of Climate-Change Policy*, Oxfam Briefing Paper 117 (Oxfam International, 2008); P. Baer, T. Athanasiou,

3.4.4 *Conflicts of human rights*

The human rights mainstream paradigm has also been criticized for neglecting the issue of how applying human rights principles in one context might lead to conflict in another context. As Steiner notes, 'only a theory of rights that avoids conflict is rational'.[179] Advocates of human rights mainstreaming have not paid sufficient attention to this question, leading to the charge that the human rights mainstream paradigm as a whole is irrational and ambiguous.[180]

Human rights conflicts can manifest in two ways.[181] Firstly, the human rights of a person might conflict with the rights of another person.[182] Conflicts here are manifested when individuals are faced with the dilemma of having the right to perform incompatible actions. The question becomes, whose rights should prevail? For example, my right to health might conflict with the property rights of my neighbour. Preventing my neighbour from building a hydroelectric power plant on his/her land in order to avoid pollution that could arise from the project results is an affront to my neighbour's right to property. It results in a dilemma as to if, and how, both interests could be balanced in policymaking.[183]

Secondly, human rights may conflict with another human right of the same person.[184] For example, my right to health might conflict with my right to liberty. A classic illustration is the right of a patient brought to the hospital by an ambulance with no time to sign consent forms for treatment. In an attempt by authorities to protect, respect and fulfil the patient's right to health and right to life, the patient's right to liberty and right to privacy might be infringed. The dilemma here is which of

and S. Kartha, *The Right to Development in a Climate Constrained World: The Greenhouse Development Rights Framework* (Heinrich Böll Foundation, 2007), 132–60.

179 H. Steiner, *An Essay on Rights* (Oxford: Blackwell, 1997) 59.

180 See Koskenniemi, 'Human Rights Mainstreaming as a Strategy', (n 22), 47; see also Alston, 'Resisting the Merger and Acquisition of Human Rights', (n 22), 815.

181 For an excellent discussion of this, see Freeman, *Human Rights*, (n 40), 82–84; see also J. Thomson, *The Realm of Rights* (Cambridge: Harvard University Press, 1990); J. Feinberg, 'The Nature and Value of Rights', *Journal of Value inquiry*, 4 (1970), 249, reprinted in J. Feinberg, *Rights, Justice, and the Bounds of Liberty* (Princeton, NJ: Princeton University Press, 1980), 143–158.

182 According to Jeremy Waldron, conflicts of rights are conflicts of their corresponding duties. See J. Waldron, 'Rights in Conflict', in *Liberal Rights* (Cambridge: Cambridge University Press, 1993).

183 See P. Montague, 'When Rights Conflict', *Legal Theory*, 7 (2001), 257.

184 See Freeman, *Human Rights*, (n 40), 82–84.

these different rights, possessed by the same person, should be respected or upheld? Which should be accorded precedence? Or should we simply go with the simplistic notions that the right to life is more sacrosanct than the right to privacy, even though the patient might personally think otherwise?

It would be difficult, if not impossible, to evolve ways of resolving all the different instances of human rights conflict. I agree with Freeman's observation that '. . .it may be that no theory of human rights can resolve all conflicts of rights. Life presents us with tragic dilemmas and human rights theory cannot change this'.[185] Conflict between different human rights is sometimes unavoidable but can be mitigated by careful balancing between the different sets of concerns.[186]

Since conflicts are somehow intricately intertwined with the idea of human rights, the way out is to fashion out how such conflicts can be balanced whenever they arise. For example, Freeman discussed the idea of assigning human rights different priorities and differential importance.[187] This idea is arguably not without faults as it neglects the inter-connectedness and the universal nature of rights. The idea of assigning rights an order of priority runs contrary to the interdependent nature of rights.[188]

Montague discusses specification as a way of avoiding such conflicts. He advocates for an approach that prioritizes rights by laying down conditions that are necessary for individuals to possess rights in particular cases. Specification implies that in putative conflict situations, at least one party to the 'conflict' does not actually have a right. It rejects a general assumption that everyone is entitled to the enjoyment of rights at all times and lays down specific requirements that must be met at any given time for human rights to be triggered. It allows rights-principles to contain grounds for concluding that one party to the conflict may not exercise a given right at the time. Rather, individuals have rights 'to. . .only if. . .'.[189] Thus, specification seeks to guarantee that situations

185 *Ibid.*, 84.

186 See A. Eide, 'The Sub-commission on Prevention of Discrimination and Protection of Minorities' in P. Alston (ed.), *United Nations and Human Rights: A Critical Appraisal* (Oxford: Oxford University Press, 1992), 211.

187 See Freeman, *Human Rights*, (n 40), 82.

188 See De Schutter, *International Human Rights Law*, (n 112), 553; See also Steiner, Alston and Goodman, *International Human Rights in Context*, (n 112), 280; N. Aryeh, 'Social and Economic Rights', (n 112), 1; Plant, 'Social and Economic Rights Revisited', (n 112), 13.

189 See Montague, 'When Rights Conflict', (n 183), 274–278.

will not arise in which individuals have rights to perform incompatible actions.

This idea of specification is arguably too ambiguous and indeterminate. How do we ascertain the conditions necessary for the enjoyment of specific rights? To do this might involve examining all human rights and adding pre-conditions to each. Such an exercise would only be speculative as it would be difficult to establish the potential conflict situations that could be encountered in the interpretation of rights. Specification also bears close resemblance to well-rejected assumptions that human rights could be forfeited.[190] By establishing pre-conditions for the enjoyment of rights, there is an inherent assumption that when such pre-conditions necessary for possessing rights are not met, the particular right is forfeited. Specifying conditions for enjoying rights in conflict situations makes human rights enjoyment indeterminate; it also obscures the idea of rights in general.

Another proposal is the utilitarian approach, which suggests that conflicts of rights should be resolved in whichever way the protection of rights is maximized.[191] This approach flows from Bentham's theory that every human decision is motivated by some calculation of pleasure and pain and that every political decision should be made on the same calculation to maximize the net produce of pleasure over pain. This approach, however, reduces human rights to a felicific calculus that is impracticable and practically impossible to follow.[192] Agreeably, adding and subtracting the benefits of one right against the other to determine what would produce the greatest net balance of happiness is a calculation that would be very difficult to make.

Jones however fine-tunes this approach with what he calls a rule-utilitarianism approach, which suggests that we ought to live by those rules that best promote the common good.[193] This is by taking decisions that favour the society as a whole. Freeman also suggests that a better approach is to respect all rights, but when some sacrifice is unavoidable, it is pertinent to distribute the sacrifice fairly among the society. As he

[190] Even though Montague tries to distinguish specification from forfeiture, the distinction drawn seems obscured. Forfeiture approach claims that in certain situations people *forfeit* their rights by what they do – which is to say that by their actions, they violate conditions that are necessary for possessing rights. See Montague, 'When Rights Conflict', (n 183), 257. For a discussion on forfeiture as relied upon by Montague, see S. Uniacke, *Permissible Killing* (Cambridge: Cambridge University Press, 1994), 196.

[191] See Shestack, 'The Philosophical Foundations of Human Rights', (n 26), 44–45; see also Freeman, *Human Rights*, (n 40), 82.

[192] See Shestack, *ibid.* [193] P. Jones, *Rights* (London: Macmillan, 1994), 203.

notes, the society can distribute the sacrifice such that no individual bears all the sacrifice of human rights violation.[194] Freeman's idea could be potentially useful to resolve the dilemma of rights. By distributing the sacrifice, we can avoid inequalities arising from governmental policies that site toxic factories and projects in poor and desperate neighbourhoods. The common good approach sees that it is unjust to protect the rights of some individuals by infringing on the rights of others.

Even though this approach is theoretically sound, it could also prove indeterminate. This approach leaves open a very important question: who is the judge of 'what is the common good?' At the end, we face a greater dilemma of leaving in the hands of the state the discretion to violate the rights of some people for the 'common good'. In a number of the 'not in my backyard' injustices of the past, and the massive governmental land grabs as presented in some of our case studies, governments have argued that these decisions to violate rights were taken for the common good or for public purposes.[195] The idea of violating some rights for the common good would be controversial and may even open a flood gate of many more injustices.

Arguably, the better approach would be one of due process. When there is a dilemma in decision-making on the need to balance human rights, it might be better to initiate a fair, equal and transparent process which allows every member of the society whose rights and interests might be affected to decide on what is the common course of action. A due process approach provides opportunities for a fair and opened process of negotiations and discussions between different right-bearers, right-holders, stakeholders and governments to decide and determine among themselves what is good for society as a whole. Through such an invitation to dialogue, a fair and balanced decision could be reached on how to distribute the pain and pleasure of a particular decision or project equally among society. Using our initial conflict situation as an example, if building a dam conflicts with my neighbour's right to health, a practical approach would be to initiate dialogue among holders of incompatible rights to discuss the conflict, explore options to mitigate or avoid conflict and arrive at a solution which both parties would consider fair and conclusive. This is the due process approach for resolving a conflict of rights. By allowing holders of rights an open and transparent opportunity for dialogue, arguably conflict situations could be avoided.

[194] See Freeman, *Human Rights*, (n 40), 82. [195] See Chapter 2 for these examples.

The human rights mainstream paradigm represents a due process approach that could avoid some of the conflicts generated by different human rights. It allows for a deliberative process through which potential dilemmas in policymaking are confronted head on, examined practically and resolved in the process of planning. It presents a practical chance for policymakers to reduce the gaps between policy intentions and actual results. By initiating an authentic deliberative process for balancing rights and other social considerations, human rights mainstreaming presents an integrated solution that avoids the conflicts between human rights, as well as the human rights versus development arguments. It also provides a long-term, strategic process through which conflict resolution is systematically integrated into all systems, structures and practices reducing the chances of conflicts resurfacing.

3.5 Conclusion

The *leitmotiv* of human rights mainstreaming is to prompt changes in institutional practices and to promote progress towards the goals of protecting human rights guarantees in climate change governance. Essentially, human rights mainstreaming is about collaboration and partnership-integration of human rights and climate change obligations, to achieve consistent, coherent and systemic implementation.

Human rights mainstreaming is not about forcing human rights norms into the international climate change regime. As discussed in this chapter, the principle of systemic interpretation in international law, the inter-doctrinal legal transplant theory, the PDT and the full belly thesis all provide support for the fact that a robust, holistic and integrated understanding of legal problems, such as the problems of climate change, would provide harmonized solutions, that is, solutions that do not create new problems in an attempt to solve one. The human rights mainstream paradigm enables the development of a human rights culture in climate change governance, such that human rights protection would not only come to the table in the face of protests and petitions; rather, human rights considerations would form part of the rules of the game in the design, approval and execution of carbon projects.

To adopt and implement the human rights-based approach to carbon finance, it is important to consider questions on how human rights standards can be practically mainstreamed in the international climate regime architecture. Theoretical issues aside, one knotty debate in the human rights mainstream literature is the question of approach: if we were to agree that human rights mainstreaming is important and necessary, how should it be done? What is the best framework or approach for achieving the goal of human rights mainstreaming: should it be through a substantive human rights approach or through a procedural approach? This question is the focus of the next chapter.

4

The human rights mainstream paradigm and the question of approach

4.1 Introduction

If human rights are to be integrated and applied in the design, approval, finance and implementation of carbon projects, the starting point is to assimilate human rights language, norms and safeguards into the international legal regime on climate change. In order to empower supervisory bodies to consider human rights issues while approving and financing carbon projects, there is the need to establish legal basis and authority for this level of project review.

One of the most contentious questions in human rights mainstream literature is the question of approach. Despite wide acknowledgement of the need to integrate human rights standards into the global regime and governance structures on climate change, scholars have yet to agree on the appropriate legal and institutional framework for achieving the linkages and legal partnerships between human rights and climate change under international law. As McCrudden rightly suggested, there is a need for considerably more discussion as to the most effective practical means of achieving human rights mainstreaming to avoid counter-productive outcomes.[1]

This chapter unpacks legal debates on the question of approaches for mainstreaming human rights in the international climate change architecture. It evaluates two main strategies that have been proposed in the literature: the substantive rights approach and the procedural rights approach (human rights-based approach). It provides justifications for the choice of the procedural rights approach, as illustrated by the UN

[1] See J.C. McCrudden, 'Mainstreaming Human Rights' in Colin Harvey (ed.), *Human Rights in the Community: Rights as Agents for Change* (Oxford: Hart Publishing, 2005). See also J. Ruhl, 'The International Right to Equitable Climate Change Adaptation—Substance or Procedure?', *Yale Law Journal*, 37(2) (2012); S. Jodoin, 'Lost in Translation: Human Rights in the Climate Change Negotiations', (Centre for International Sustainable Development Legal Working Paper, January 2010), www.cisdl.org/pdf/working_papers_climate/Jodoin.pdf, accessed 2 June 2015.

Human Rights-Based Approach (HRBA), as a holistic policy framework for mainstreaming human rights norms into the international climate change regime. It demonstrates that a process-based approach, as typified by the HRBA, provides preventive, transformative and long-term solutions for dealing with some of the tensions and conflicts generated by climate change mitigation. It avoids human rights versus environment arguments; prompts changes in institutional practices; and also provides a preventive/long-term process through which human rights are systematically integrated into climate change policies, regimes and governance structures.

4.2 Linking human rights and climate change: suggested approaches

Debates on the best approach to recognize the linkages between human rights and climate change in international law can be traced back to older debates on the linkages between human rights and the environment.[2] Scholars have, over the last two decades, suggested different approaches for recognizing the linkages between human rights and the environment.[3] The first approach understands environmental protection as a pre-condition to the enjoyment of internationally guaranteed human rights, especially the rights to life and health. This indirect approach draws closely from the Stockholm Declaration, which discusses environmental protection as a prerequisite for enjoying existing human rights. The second approach calls for a direct recognition of a substantive human right to environment under international law.[4]

[2] A. Boyle, 'The Role of International Human Rights Law in the Protection of the Environment', in A. Boyle and M. Anderson (eds.), *Human Right Approaches to Environmental Protection* (Oxford: Oxford University Press, 1996), 43.

[3] J. Knox, 'Linking Human Rights and Climate Change at the United Nations' *Harvard Environmental Law Review*, 33 (2009), 477–498; L. Rajamani, 'The Increasing Currency and Relevance of Rights-Based Perspectives in the International Negotiations on Climate Change', *Journal of Environmental Law*, 22(3) (2010); A. Parsons, 'Human Rights and Climate Change: Shifting the Burden to the State?', *Sustainable Development Law and Policy*, 9(2) (2009), 22–23; M. Stallworthy, 'Environmental Justice Imperatives for an Era of Climate Change', *Journal of Law and Society*, 36(1) (2009), 55–74; J. Hohmann, 'Igloo as Icon: A Human Rights Approach to Climate Change for the Inuit?', *Transnational Law and Contemporary Problems*, 18(2) (2009), 296–316; E. Posner and C. Sunstein, 'Climate Change Justice', *Georgetown Law Journal*, 96 (2008), 1565–1597.

[4] D. Shelton, 'Developing Substantive Environmental Rights', *Journal of Human Rights and the Environment*, (2010), 89–120; S. Turner, *A Substantive Environmental Right: An Examination of the Legal Obligations of Decision-Makers towards the Environment*

The third approach calls for re-interpretation of existing procedural human rights to cater for the environment.[5] This expansionist approach tries to formulate a link between human rights and environmental protection in procedural terms. This approach is well illustrated by Principle 10 of the Rio Declaration on Environment and Development and the Aarhus Convention, which both provide that access to information (ATI), public participation and access to effective judicial and administrative proceedings, including redress and remedy, should be guaranteed because environmental issues are best handled with the participation of all concerned citizens. This book develops on and argues for this third approach. This approach builds on procedural human rights such as ATI, participation and access to justice as the necessary foundation for integrating climate change actions and projects with human rights standards. This approach also forms the cornerstone of the UN HRBA – a policy frame that has been promoted by the United Nations since 1997 and that builds on five human rights: ATI, participation, accountability, equality and non-discrimination, and access to justice.

Though not legally binding, the HRBA has been adopted by a number of UN agencies as a policy frame for mainstreaming human rights norms into areas of their mandate. In 2012, the United Nations Development Group (UNDG) launched the UN Human Rights Mainstreaming Programme, which called on all agencies of the United Nations to adopt the HRBA as a policy frame for mainstreaming human rights principles into their areas of work. Despite the fact that the UNEP and the UNFCCC are part of the UNDG, not much has been done to mainstream human rights principles into the international climate change regime using the HRBA frame. One identified reason is the absence of a theoretical framework and robust guidelines on how to apply the HRBA in the area of climate change.

(The Netherlands; Frederick MA: Kluwer Law International, 2009); M. Paellemarts, 'The Human Right to a Healthy Environment as a Substantive Right', in M. Dejeant-Pons and M. Paellemarts (eds.), *Human Rights and the Environment* (Council of Europe, 2002), 11–15.

5 L. Rajamani, 'The Increasing Currency and Relevance of Rights-Based Perspectives', (n 3); M. Fitzmaurice, *Contemporary Issues in International Environmental Law* (Northampton, MA; Cheltenham, UK: Edward Elgar, 2009), 170; A. Boyle, 'Human Rights or Environmental Rights? A Reassessment', *Fordham Environmental Law Review*, XVIII (2007), 471 at 497; S. Kravchenko, 'Right to Carbon or Right to Life: Human Rights Approaches to Climate Change', *Vermont Journal of Environmental Law*, 9 (2008), 513–547.

Considering the confusion that surrounds the different approaches that link human rights discussions to climate change, it is important to clarify how the human rights-based approach differs from other approaches. Similarly, as we seek innovative solutions on how best to mainstream human rights norms in the international climate change regime, particularly in the light of the two dimensional threats posed by climate change to human rights, it is pertinent to examine how each of the previously suggested approaches have been implemented and the successes, pitfalls and the lessons we can draw as to why one approach might best guide us with respect to climate change. Previous prescriptions and approaches can guide us towards addressing the indirect threats posed by carbon projects to human rights if appropriate lessons are drawn from their respective strengths and weaknesses. The question therefore is, 'what is the value of one approach over the other?'

The different approaches are discussed under two broad headings: the substantive rights approach – comprising both the direct new human rights approach and the re-interpretation approach – and the procedural rights approach, consisting mainly of the human rights-based approach.

4.2.1 Substantive human rights approach

Since the publication of the path-breaking treatise edited by Boyle and Anderson, scholars have increasingly argued in favour of an approach that protects the environment using substantive human rights instruments.[6] The crux of this approach is that environmental protection can be achieved when it is recognized as a freestanding human right and defined in core international and national human rights instruments.[7] Arguably, environmental protection can be recognized in the body of international human rights law either by establishing a new human right

6 See Boyle and Anderson (eds.), *Human Right Approaches to Environmental Protection*, (n 2).

7 For excellent and detailed discussions on the substantive recognition of a right to environment under international and domestic law, see generally International Bar Association (IBA), Climate Change Justice and Human Rights Task Force Report, *Achieving Justice and Human Rights in an Era of Climate Disruption* (International Bar Association, 2014) 124–125 [*IBA Report*], where the IBA Task force called for the express recognition of a freestanding human right to a healthy environment by human rights bodies and states; also S. Turner, *A Substantive Environmental Right: An Examination of the Legal Obligations of Decision-Makers towards the Environment* (Leiden: Kluwer Law International, 2009) at 1–10; James R. May, 'Constitutionalizing Fundamental Environmental Rights Worldwide' *Pace Environmental Law Review*, 23 (1), (2006), 113 at 118–125 (discussing the importance of establishing constitutional environmental rights).

to environment under international law, or re-interpreting existing human rights such as the right to life or the right to health to include environmental protection as an element. Boyle calls this the 'greening' of already existing rights and remedies in order to cater to environmental protection.[8] This indirect approach calls for states, courts and international organizations to derive environmental protection from existing human rights instruments.

4.2.1.1 Re-interpreting existing rights: the indirect approach

This approach focuses on an expansive re-interpretation of existing human rights to cover environmental protection. As the United Nations Independent Expert on human rights and environment rightly noted, one firmly established aspect of the relationship between human rights and the environment is that 'environmental degradation can and does adversely affect the enjoyment of a broad range of human rights'.[9] The indirect approach therefore focuses on an expansive re-interpretation of existing human rights to cover environmental protection. As Anderson argued, 'existing rights must be re-interpreted with imagination and vigour in the context of environmental concerns which were not prevalent at the time existing rights were first formulated'.[10]

The extent to which environmental human rights can be derived from other human rights has been extensively analysed. As rightly noted by the United Nations Independent Expert on human rights and environment, the obligations of states to adopt legal and institutional frameworks that protect against, and respond to, environmental harm that may interfere with the enjoyment of human rights, can be derived from a number of existing human rights such as the rights to life and health.[11] Existing human rights such as the right to life, health and right to information can

[8] A. Boyle, 'Human Rights or Environmental Rights? A Reassessment', (n 5). See also IBA Report at 119, stating that the process of 'greening' refers to implementing existing human rights obligations in the context of environmental and climate change harms.

[9] See United Nations Human Rights Council, *Report of the Independent Expert on the issue of human rights obligations relating to the enjoyment of a safe, clean, healthy and sustainable environment, John H. Knox* – Preliminary report, A/HRC/22/43 (24 December 2012), para. 34.

[10] M. Anderson, 'Human Rights Approaches to Environmental Protection: An Overview', in Boyle and Anderson (eds.), *Human Rights Approaches to Environmental Protection*, (n 2), 7–8.

[11] United Nations Human Rights Council, Report of the Independent Expert on the issue of human rights obligations relating to the enjoyment of a safe, clean, healthy and sustainable environment, John H. Knox, A/HRC/25/53 (30 December 2013), at para. 47.

be consistently invoked to address human rights risks in environmental issues.[12] As Konne rightly argues, failure to effectively monitor and enforce environmental protection is a violation of the right to health under regional and international human rights instruments.[13]

Lessons on the workability of this approach must be drawn from a review of cases in which litigants sought to pursue environmental claims by placing them under existing human rights.[14] In a number of cases, this approach has yielded favourable responses and jurisprudence. For example, in a unique ruling in *Urgenda Foundation v. Kingdom of Netherlands*, the Hague District Court held that failure of the Netherlands government to combat the dangers caused by climate change violates several human rights and legal obligations.[15] The court ruled that government must take more action to reduce greenhouse gas (GHG) emissions in the Netherlands and the imminent danger caused by climate change. The court further ruled that government must ensure that Dutch emissions in the year 2020 will be at least 25 per cent lower than those in 1990. This case provides path-breaking recognition of the linkages between the direct impacts of climate change on the human rights of the public, and the legal obligation of the Dutch government to prevent these negative impacts.

Similarly, in *Gabcikovo Nagymoros Case*, Judge Weeramantry of the International Court of Justice (ICJ) recognized the protection of the environment as a '*sine qua non* for numerous human rights such as right to health and the right to life itself'.[16] The African Commission on Human and Peoples' Rights (African Commission) adopted this approach in the case of *Endorois Welfare Council v. Kenya.*[17] In that

[12] A. Harding, 'Practical Human Rights, NGOs and the Environment in Malaysia', in Boyle and Anderson (eds.), *Human Rights Approaches to Environmental Protection*, (n 2), 227.

[13] B. Konne, 'Inadequate Monitoring and Enforcement in the Nigerian Oil Industry: The Case of Shell and Ogoniland' *Cornell Intl Law Journal*, 47, (2014), 181 at 183 (arguing that failure of the Nigerian Government to prevent pollution in the Niger Delta is a violation of its obligations under the International Convention on Economic, Social, and Cultural Rights and the African Charter on Human and People's Rights).

[14] See H. Post, 'Hatton and Others: Further Clarification of the "Indirect" Individual Right to a Healthy Environment', *Non State Actors and International Law*, 2 (2002), 259.

[15] *Urgenda Foundation v. The State of the Netherlands (Ministry of Infrastructure and the Environment)* [2015] Case C/09/456689/HA ZA 13-1396.

[16] Gabcikovo Nagymoros (Hungary/Slovakia) (1998) 37 ILM 162, 206 (Separate Opinion of Judge Weeramantry).

[17] The African Commission on Human and Peoples' Rights is a quasi-judicial body established pursuant to the African Charter on Human and Peoples' Rights, to interpret the Charter, consider individual complaints of violations of the Charter and to promote

case, the Endorois community, an indigenous community, challenged the Government of Kenya for displacing them from their ancestral lands, and failing to adequately compensate them for the loss of their property, disruption of the community's pastoral enterprise and violations of the right to practice their religion and culture, as well as their overall process of development.[18] The African Commission found that failure of the Government of Kenya to conduct prior social and environmental impact assessment, or consult the Endorois people on the use of the ancestral lands, 'is tantamount to a violation of Article 14, the right to property', and the right to development under the African Charter.[19] The Commission noted that the Endorois people have, as a result, suffered loss and environmental degradation of their land and a violation of their rights to freely dispose of their wealth and natural resources.[20] The decision in this case emphasizes how failure to effectively enforce and ensure environmental protection is a violation of the rights to property and development.

Similarly, in *SERAC v. Nigeria*, the African Commission recognized the linkages between environmental protection and a number of human rights. According to the commission, the failure of the Nigerian government to mandate oil companies to produce basic health and environmental impact studies relating to oil production, had violated many rights of the Niger Delta people, including their right to life, health, food, and property.[21] Similarly, in *Gbemre v. Shell Petroleum and Development Company Ltd*, the Nigerian Court declared gas flaring as illegal and as a breach of the fundamental human rights to life and dignity.[22]

The case of *Mossville Environmental Action Now v. United States* – an environmental pollution petition brought before the Inter-American Commission against the US government – is another example.[23] In this

human and peoples right in the African continent. See African Commission on Human and Peoples' Rights, *About*, www.achpr.org/about/, accessed 7 October 2015.

[18] *Center for Minority Rights Development (Kenya) and Minority Rights Group International on behalf of the Endorois Welfare Council v Kenya* (2009), African Comm on Human and Peoples' Rights, No 276/2003, *27th Activity Report of the ACHPR* [*Endorois Welfare Council v. Kenya*].

[19] *Ibid.*, at para, 228, 298. [20] *Ibid.*, para. 245, 266–267.

[21] *The Social and Economic Rights Action Center (SERAC) and Center for Economic and Social Rights v. Nigeria* (2002), African Comm. on Human and Peoples' Rights, No 155/9, 15th Activity Report of the ACHPR at 3–4, 9–10.

[22] *Gbemre v. Shell Petroleum and Development Company Nigeria Ltd*, [2005] AHRLR 151 at 2–3, 5 (HC Nigeria).

[23] Organization of American States, Inter-Am. Commission on HR, *Mossville Environmental Action Now v. United States*, Report no 43/10, Petition no 242/05, Admissibility, OEA/Ser LV/II 138 (17 March 2010), 5.

case, the US government had opposed the admissibility of the petition on the grounds that 'there is no such right as the right to a healthy environment either directly, or as a component of the rights to life, health, privacy and inviolability of the home, or equal protection and freedom from discrimination'.[24] The court, however, held that while the US government may not recognize the right to environment as alleged in the petition, the Commission finds the petition admissible based on a broad interpretation of international and Inter-American human rights systems.

The European Court of Human Rights has also recognized this linkage approach in a number of cases.[25] For example, in the *Lopez Estra case*, the court held that 'severe environmental pollution may affect individuals' well-being and prevent them from enjoying their homes in such a way as to affect their private and family life adversely.[26] The Inter-American Commission on Human Rights also recognized the linkages in *Yanomami Indians v. Brazil*, where it held that Brazil had violated the Yanomami Indians' right to life, liberty and personal security by not taking measures to prevent the environmental damage that resulted in the loss of life and cultural identity among the Yanomami.[27]

However, this approach could, in some cases, be faced with the problem of uncertainty and indeterminacy. In the absence of an express recognition of a right to environment in any of the core international human rights treaties or an express recognition in the specific treaty relied upon before an international tribunal, the success or otherwise of this approach depends on the liberality of the treaty body or court concerned. As such, some courts have insisted that laws should be interpreted as they are, as opposed to a whimsical expansion of a rights instrument to include a right to environment. For example, in *Hatton v. United Kingdom*, a Grand Chamber of the European Court of Human Rights overruled an earlier decision by one of its smaller sections on a complaint about noise pollution arising from night flights at London's Heathrow Airport. In that case, eight applicants, all British citizens,

[24] Response of the Government of the United States, www.ehumanrights.org/docs/US-Response-to-Mossville-Petition-09–06.pdf, accessed 12 May 2015.

[25] See also Guerra *v.* Italy, 4 BHRC 63 (1999); *Hatton v. United Kingdom* (2002) 34 E.H.R.R. 1; Kyrtatos *v.* Greece (2003) 36 EHRR 242 [52]; *Taskin v. Turkey* (2006) 42 EHHR 50; *Moreno Gomez v. Spain* (2005) 41 EHRR 40; *Fadeyeva v. Russia* (2007) 45 EHRR 10; *Tatar v. Romania* (App no 67021/01) 27 January 2009; *Leon and Agnieszka Kania v. Poland* (App no 12605/03) 21 July 2009; *Olujic v. Croatia* (App no 22330/05) 5 February 2009.

[26] ECHR (16798/1994).

[27] Inter-Am. C.H.R. 7615, OEA/Ser.L.V/II/66 doc. 10 rev. 1 (1985).

argued that night-time noise from the airport, especially in the early morning, interfered with their right to private and family lives, right to peaceful enjoyment of their homes and their environmental rights. The Chamber held that failure of the government to reduce night flights from Heathrow Airport in the interests of the economic well-being of the country did not breach the human rights of those living beneath the flight path. It held that considering the small number of people afflicted by sleep disturbance, the interference could not be regarded as disproportionate and that a fair balance had been struck, taking into account the economic well-being of the country and the economic significance of the airport.[28] According to the Chamber:

> Environmental protection should be taken into consideration by Governments in acting within their margin of appreciation and by the Court in its review of that margin, but it would not be appropriate for the Court to adopt a special approach in this respect by reference to a special status of environmental human rights.[29]

The Chambers refused to ascribe any special recognition or status to environmental human rights. Similarly, US courts have consistently held that there is no such right as a right to environment under the US constitution or international law.[30] Previous attempts to seek redress for international environmental pollution under the US Aliens Torts Act (ATCA) have recorded little success.[31] One of the challenges is that US courts have not recognized environmental pollution to constitute a violation of the Law of Nations. The US Supreme Court has held in *Sosa v. Alvarez-Machain* that the ATCA only provides a cause of action for violations of international norms that are 'specific, universal, and

[28] 2003 (Application no. 36022/97) 003-VIII Eur. Ct. H.R. 189, 37 Eur. H.R. [29] *Ibid.*

[30] See the decision in *Tanner v. Armco Steel* (1972) 340 F. Supp 532, 537 where the court held that 'no legally enforceable right to a healthful environment...is guaranteed by the Fourteenth Amendment or any other provision of the Federal Constitution'. See also *De Shaney v. Winnebago County Dept. of. Soc. Serv.*, (1989) 489 U.S 1189, 195–196, decision affirmed in 556 F.2d 576 (4th Cir. 1977); *Upper W. Fork River Watershed Association v. Corps of Engineers* (1977) 414 F. Supp. 908 (N.D.W.Va); *Lindsey v. Normet* (1972) U.S 405 U.S 56. In these cases, the courts refused to recognize a right to environment.

[31] Section 1350 of Alien Torts Act 28 USC § 1350 (ATCA) allows non-citizens of the United States to sue in US federal courts for violations of the law of nations committed outside the United States. This statute has been employed by victims of human rights abuses to seek redress in US courts. For example, in *Filartiga v. Pena-Irala*, the US court allowed a case of torture brought by two Paraguayan citizens resident in the United States, against a Paraguayan former police chief who was also living in the United States, the court held that torture constituted a violation of the Law of Nations.

obligatory'.[32] As such, in *Amlon Metals Inc. v. FMC Corp*, the US court held that the Stockholm Declaration on Human Environment does not form part of the Law of Nations.[33] Similarly, in *Beanal v. Freeport McMoran, Inc.*, the US court held that the polluter pays principle (PPP), the precautionary principle and the proximity principles of international environmental law have not been sufficiently defined to constitute part of the 'Law of Nations'.[34] In *Aguinda v. Texaco*, the Court held that allegations of oil spillage do not fit neatly enough into existing categories of international law to constitute a violation of the Law of Nations under ATCA. The court therefore dismissed the case.[35]

These cases show the shortcomings of seeking environmental protection through an indirect human rights approach or litigation. Specifically, they are instructive on the potentials and pitfalls of trying to establish or derive the linkages between human rights and climate change through litigation.[36] There have been more arguments now than ever that climate change is a human rights issue.[37] The United Nations has, in a number of resolutions, identified that 'climate change poses an immediate and far-reaching threat to people and communities around the world and has implications for the full enjoyment of human rights'.[38] Though not legally binding, these resolutions emphasize that human rights can

[32] 542 US 692 (2004). See also *Flores v. Southern Peru Copper Corporation* 343 F.3d 140, 143–144 (2d Cir. 2003), *Sarei v. Rio Tinto* 550 F.3d 822 (9th Cir. 2008).

[33] 775 F. Supp. 668, 674 (S.D.N.Y. 1991). [34] 969 F.Supp. 362 (E.D.La. 1997).

[35] 142 F. Supp. 2d 534 (S.D.N.Y. 2001).

[36] See S. Hsu, 'A Realistic Evaluation of Climate Change Litigation Through the Lens of a Hypothetical Lawsuit', *University of Colorado Law Review*, 79(3) (2008), 701–766; D. Adam, 'Climate Change in Court', *Nature Climate Change*, 1(3) (2011), 127–130; M. Averill, 'Linking Climate Litigation and Human Rights', *Review of European Community and International Law*, 18 (2009), 139; E. Posner, 'Climate Change and International Human Rights Litigation: A Critical Appraisal', *University of Pennsylvania Law Review*, 155 (2007), 1925; H. Osofsky, 'The Inuit Petition as a Bridge: Beyond Dialectics of Climate Change and Indigenous Peoples' Rights', *American Indian Law Review*, 31 (2007), 675; Post, 'Hatton and Others: Further Clarification', (n 14); H. Osofsky, 'The Geography of Climate Change Litigation: Implications for Transnational Regulatory Governance', *Washington University Law Quarterly*, 83 (2005), 1789.

[37] See M. Robinson, 'Climate Change is an Issue of Human Rights', (2012), www.independent.co.uk/voices/commentators/mary-robinson-climate-change-is-an-issue-of-human-rights-1059360.html. accessed 7 October 2015.

[38] UN Human Rights Council Res. 7/23, in UN Human Rights Council, Report of the Human Rights Council on Its Seventh Session, at 65, UN Doc A/HRC/7/78 (14 July 2008); UN Human Rights Council Res. 10/4, in UN Human Rights Council, Report of the Human Rights Council on its Tenth Session 13, at 15, UN Doc A/HRC/10/L.11 (31 March 2009); UN Human Rights Council Res. 18/22, in UN Human Rights

underpin and influence developments with respect to climate change mitigation and adaptation. In essence, there is growing consensus that climate change is a human rights issue and that failure to mitigate climate change could result in a violation of several existing human rights. Despite the consensus, however, the search for a more robust recognition of these linkages in the international climate change regime and more specifically laying down mandatory threshold and guidelines to prevent the indirect impacts of climate change actions, projects and measures on human rights, continues.[39]

Some scholars have therefore suggested the potentials of deriving these linkages through litigation.[40] More than ever, there have been calls that human rights affected by climate change could be protected through litigation. This goes back to the question whether such indirect linkage approach and enforcement through litigation could provide robust solutions to both the direct and indirect impacts of climate change on human rights. If lessons from previous experiences of deriving environmental rights through litigation are anything to go by, the strengths and pitfalls of the climate change litigation or indirect approach become very evident.

Council, Report of the Human Rights Council on Its eighteenth Session, at A/HRC/RES/18/22 (30 September 2011).

39 Interestingly, this approach succeeded in *Massachusetts v. EPA* (2007) 127 S. Ct. 1438, where the US Supreme Court held that green house gas emissions from motor vehicles, including carbon dioxide, count as 'pollutants' under the Clean Air Act. For a discussion of the ability of litigation to encourage federal responses to climate change, see Osofsky, 'The Geography of Climate Change Litigation', (n 36), 1789, 1855; see also C. Carlarne, 'Climate Change Policies an Ocean Apart: EU & US Climate Change Policies Compared', *Pennsylvania State Environmental Review*, 14 (2006), 435, 453–457; H. McGee, Jr., 'Litigating Global Warming: Substantive Law in Search of a Forum', *Fordham Environmental Law Review*, 16 (2005), 371, 372–377; D. Goldberg and M. Wagner, 'Petitioning for Adverse Impacts of Global Warming in the Inter-American Human Rights System', in V. Grover (ed.), *Climate Change: Five Years After Kyoto* (Enfield, NH: Science Publishers, 2004), 191.

40 See Posner, 'Climate Change and International Human Rights Litigation', (n 36), 1925; D. Markell and J.B. Ruhl, 'An Empirical Assessment of Climate Change in the Courts: A New Jurisprudence or Business as Usual?', *Florida Law Review*, 64(1) (2012), 15–86; M. Gerrard, 'What Litigation of a Climate Nuisance Suit Might Look Like', *Yale Law Journal Online*, 121 (2011), 135–177; W. Burns and H. Osofsky (eds.), *Adjudicating Climate Change: State, National, and International Approaches* (Cambridge: Cambridge University Press, 2009); L. Case, 'Climate Change: A New Realm of Tort Litigation, and How to Recover When the Litigation Heats Up', *Santa Clara Law Review*, 51 (2011), 265–298; J. Cofre, et al., 'Climate Change Litigation', in P. Watchman and Q. Paul (eds.), *Climate Change: A Guide to Carbon Law and Practice* (London: Globe Business Publishing Ltd, 2008).

While litigating climate harm under core human rights ingenious approach, as it could raise awareness on the importance of recognizing the direct and indirect effects of climate change as human rights violations, the greatest danger in this 'pigeonhole' approach is that claimants would have to be as imaginative as possible in fitting violations, arising directly and indirectly from climate change, into any of the recognized human rights in order to bring their claims under the jurisdiction of an existing international human rights body. Such efforts have not been the most straightforward, particularly when brought before a conservative judicial forum. This approach could fail where the court is unwilling to expand existing rights to cover climate change. For example, in December 2005, the Inuit Circumpolar Conference (ICC) filed a petition with the Inter-American Commission on Human Rights against the United States for its acts and omissions with respect to climate change. It alleged that the United States, as the world's largest emitter of GHGs, was violating the human rights of the Inuit residents of the Arctic through its climate change policies, including its decision not to ratify the Kyoto Protocol. The petition alleged that the rights to property, physical well-being and cultural life of the Inuits had been adversely affected by the acts and omissions of the United States. In its response, the Commission held that 'information provided does not enable us to determine whether the alleged facts would tend to characterize a violation of right protected by the American Declaration'.[41]

Considering that claims brought before any international human rights review committee or court must disclose a *prima facie* case for the complaint to scale the initial hurdle of admissibility before a consideration of the merits, the chances of admissibility for claims linking climate change actions and omissions to human rights may be speculative and largely derivative.[42] Arguing that carbon projects violate

[41] J. George, 'ICC Climate Change Petition Rejected' (Nunatsiaq News, 15 December 2006), www.nunatsiaq.com/archives/61215/news/nunavut/61215_02.html; see also A.C. Revkin, 'World Briefing Americas: Inuit Climate Change Petition Rejected' (NY Times, 16 December 2006), A9; D. Zabarenko, 'Climate Ills, Rights Linked; Global Warming Endangering Inuit "sentinels", Nobel Nominee Argues' (Toronto Star, 5 March 2007) at A7. Even though the petition was rejected, the Commission later invited and heard testimony on the relationship between human rights and climate change from representatives for the Inuit in 2007.

[42] For example, since the right to property is not recognized in the ICCPR and the ICESCR, the UNHRC has refused to comment on or declare a violation of the right to property in cases where displacements from homes were part of the claims brought before it. See, for example, *Oló Bahamonde v. Equatorial Guinea Communication No. 468/1991, UN Doc CCPR/C/49/D/468/1991 (1993); Ackla v. Togo Diergaardt of Rehoboth Baster Community et al. v. Namibia*, in these cases even though the claims brought before the UNHRC

human rights might work in some cases where it is easy to establish direct links to recognized human rights, while it might not in other cases where such projects have not resulted in a violation of any recognized right, for example if it has not brought about any loss of life, or has not violated other rights such as personal liberty, political participation and the freedom of association.[43] In such cases, it will be difficult to bring claims under any of these rights. Communities may suffer significant human rights impact from a project but remain unable to seek redress if, for example, it does not lead to death (thus the right to life is still intact) or does not impair their movements in any way (the right to free movement is still intact).[44] As Francioni rightly notes with respect to environmental protection generally, it does not make much sense to engage human rights language to combat environmental degradation only when such degradation affects the rights to life, property and the privacy of certain directly affected individuals. Francioni calls for a 'more imaginative and courageous jurisprudence' which adapts the language and technique of human rights discourse to environmental risks as a whole.[45]

Francioni's logic arguably demonstrates the dangers in a reductionist use of human rights to protect rights affected by carbon projects. The indeterminacy of this approach is in the fact that there might be cases where dangers posed or caused by a project cannot be linked to any existing category of rights. Also, whether a claim would succeed before the complaint mechanism of a human rights treaty would depend on how imaginative an applicant might be in linking the violations by a project under any of the rights covered by that treaty. Success would therefore vary on a case-by-case basis. Success might also depend on how liberal the court or review body is in agreeing to expand the application of such rights. The indeterminate nature of this approach means that the likelihood or chances of obtaining remedies will depend to a large extent on

showed *prima facie* evidence of unjustifiable interference with property rights, the UNHRC only commented on the right against discrimination. See R. Jacobs, 'Treading Deep Waters: Substantive Law Issues in Tuvalu's Threat to Sue the United States in International Court of Justice', *Pacific Rim Law and Policy Journal*, 14 (2005), 103.

43 See J. Marshall, 'The Human Right to a Clean Environment, Phantom or Reality? The European Court of Human Rights and English Courts Perspectives on Balancing Rights in Environmental Cases', *Nordic Journal of International Law*, 76(2) (2007), 103.

44 D. Shelton, 'Human Rights, Environmental Rights, and the Right to Environment', *Stanford Journal of International Law*, 28 (1991), 134.

45 F. Francioni, 'International Human Rights in an Environmental Horizon', *European Journal of International Law*, 21(1) (2010), 55; see also Hsu, 'A Realistic Evaluation of Climate Change Litigation', 701–766.

the understanding or standard applied by the adjudicating international panel or tribunal.

This aside, climate change litigation as an approach would arguably not provide an opportunity to address the root causes of the indirect impacts of climate change projects on the public, neither would it provide an opportunity to prevent these concerns in the first place, without a need for recourse to remedial measures. There is a need for an approach that is preventive and anticipatory in nature, such that these concerns are dealt with as part of climate change governance, rather than a reductionist approach that tries to bring in human rights when there is already a violation or problem.[46]

A direct recognition of a substantive climate change right under international law has been identified as a possible way to prevent such situations.[47] Some have argued that the linkages could be easily established as part of a substantive human right on the environment or climate change under international law.

4.2.1.2 The new human rights approach

There have been unprecedented demands within the last decade on the need to link environmental protection and human rights through the creation of a substantive human right. While some scholars argue that an international recognition of a substantive right to environment would provide human rights coverage for climate change cases, since climate change is an element of the environment,[48] some have also argued for a direct recognition of a fundamental human right against the direct and

[46] R. Gainza-Carmenates, *et al.*, 'Stakeholder-based Scenarios for Post-2012 Climate Policy: A Participatory Approach', *Carbon and Climate Law Review*, 3 (2009), 248; K. Engel, 'Harmonizing Regulatory and Litigation Approaches to Climate Change Mitigation: Incorporating Tradable Emissions Offsets Into Common Law Remedies', *University of Pennsylvania Law Review*, 155 (2007), 1563; A. Spain, 'Beyond Adjudication: Resolving International Resource Disputes in an Era of Climate Change', *Stanford Environmental Law Journal*, 30(2) (2011), 343–390; M. Gerrard, 'What the Law and Lawyers Can and Cannot Do About Global Warming', *Southeastern Environmental Law Journal*, 16 (2007), 33.

[47] C. Gearty, 'Do Human Rights Help or Hinder Environmental Protection?', *Journal of Human Rights and the Environment*, 1 (2010), 7–22. As Gearty rightly notes, 'environmental protection needs a more direct guarantee in human rights law than can be provided by the right to privacy or by civil and political rights generally'.

[48] Knox, 'Linking Human Rights and Climate Change at the United Nations', (n 3), 477–498; J. Von Doussa, 'Human Rights and Climate Change: A Tragedy in the Making', *The University of New South Wales Law Journal*, 31(3) (2008), 953–964; see also A. Postiglione, 'Human Rights and the Environment', *The International Journal of Human Rights*, 14(4) (2010), 524–541; B. Lewis, 'Climate Change and Human Rights:

indirect effects of climate change, also framed as the right to be cold.[49] The substantive right approach to environmental protection is arguably the most debated approach to mainstreaming human rights norms into environmental protection.[50]

Calls for a substantive human right to environment under international law have been extensively examined by scholars such as Shelton,[51] Turner,[52] and Atapattu,[53] who have developed strong arguments on the potentials and challenges of the new rights approach. Proponents of a substantive human right to a healthy environment call for its adoption under international law as the best approach to protect the environment from pollution caused by climate change; transboundary movement of hazardous wastes; depletion of the ozone layer; and other environmental concerns.[54] The crux of the argument here is that environmental protection is fundamental enough to be accorded human rights status under international law.[55]

Perspectives of Environmental and Indigenous Rights', *Journal of the Australasian Law Teachers Association*, 1(1) (2008), 153–161.

[49] See S. Watt-Cloutier, *The Right to Be Cold* (Winnipeg: Penguin Canada, 2015), also S. Watt-Cloutier, 'A New Legal Frontier in the Fight against Global Warming', *Fordham Environmental Law Review*, 16 (2005), 292. See also Watt-Cloutier, *The Right to be Cold* (Evidence provided for Humanity on Trial 2009), humanityontrial.wordpress.com/2009/09/28/sheila-watt-clouiter-on-the-right-to-be-cold/, accessed 12 January 2015. Watt-Cloutier explored the parallels between safeguarding the Arctic and the survival of Inuit culture from climate change and argued that the issue of climate change had to do with the 'right to be cold' in Inuit communities. Watt–Cloutier argues that climate change is a human rights issue and one to which all of us on the planet are inextricably linked. See also J. Harrington, 'Climate Change, Human Rights, and the Right to Be Cold', *Fordham Environmental Law Review*, 18 (2006–2007), 513.

[50] United Nations, 'Outcome Document of the High Level Expert Meeting on the New Future of Human Rights and Environment: An Agenda for Moving Forward' (Nairobi, 30 November–1 December 2009), www.unep.org/environmentalgovernance/LinkClick.aspx?fileticket=cTNdmZ_wGK0%3d&tabid=2046&language=en-US, accessed 12 March 2014.

[51] Shelton, 'Developing Substantive Environmental Rights', (n 4).

[52] S. Turner, 'Factors in the Development of a Global Substantive Environmental Right' *Oñati Socio-Legal Series*, 3(5), (2013), 893.

[53] S. Atapattu, 'The Right to a Healthy Life or the Right to Die Polluted?: The Emergence of a Human Right to a Healthy Environment under International Law', *Tulane Environmental Law Journal*, 16 (2002) at 72–73. See also S. Atapattu, *Human Rights Approaches to Climate Change Challenges and Opportunities* (Oxford: Routledge, 2016).

[54] Turner, 'Factors in the Development of a Global Substantive Environmental Right', (n 52) at 903–905, also IBA Report, at 124-125.

[55] See J. Lee, 'The Underlying Legal Theory to Support a Well-Defined Human Right to a Healthy Environment as a Principle of Customary International Law', *Columbia Journal of Environmental Law*, 25 (2000), 305.

The right to environment has been defined and recognized in a number of countries. Currently, at least 177 countries of the world's 193 UN member nations have granted legal recognitions to a right to environment, either through their constitutions, environmental legislation, or court decisions.[56] Regionally, Article 24 of the AfCHPR also provides that 'all peoples shall have the right to a general satisfactory environment favourable to their development'. Article 11 of the 1988 San Salvador Protocol to the 1969 Inter-American Convention on Human Rights also provides that everyone shall have the right to live in a healthy environment and to have access to basic public services.[57]

Even though there is no recognition or mention of a right to environment in any of the core international human rights treaties, it has gained increased recognition in a number of international instruments, for example the Aarhus Convention.[58] This increased recognition provides

[56] The question on the extent to which this spate of constitutionalization of environmental rights have resulted in concrete implementation or environmental protection is, however, a different and difficult one. Many of the referenced constitutional rights are placed in the non-justiciable sections of the Constitution, making them unenforceable in courts. Apart from countries such as Venezuela, Ecuador, Bolivia and Kenya, which have constitutions that establish justiciable environmental rights, many of the other referenced constitutions establish non-enforceable environmental rights. This continues to raise questions on how environmental rights can be better recognized, protected and fulfilled at the national level. See D. Boyd, *The Environmental Rights Revolution: A Global Study of Constitutions, Human Rights, and the Environment* (Vancouver: University of British Columbia Press, 2012).

[57] Art. 24, African Charter on Human and Peoples' Rights (Adopted 27 June 1981, entered into force 21 October 1981) OAU Doc. CAB/LEG/67/3 Rev. 5, reprinted in 21 ILM 58 (1982) (providing a 'right to a general satisfactory environment favourable to. . .development'); Additional Protocol to the American Convention on Human Rights in the Area of Economic, Social and Cultural Rights, Art. 11. (Adopted 17 November 1988, entered into force 16 November 1999) O.A.S.T.S. No. 69, 28 ILM 161 ('Everyone shall have the right to live in a healthy environment and to have access to basic public services').

[58] See the preamble to the Aarhus Convention ('every person has the right to live in an environment adequate to his or her health and well-being'). See also Article 1 of the Aarhus Convention. There are also other non-binding declarations and instruments that recognize the linkages between human rights and the environment. See United Nations' Conference on the Human Environment (Stockholm Declaration) UN Doc A/CONF.48/PC.13 (1971), www.unep.org/Documents.Multilingual/Default.asp?DocumentID=97&ArticleID=1503, accessed 10 March 2014. ('Man has the fundamental right to freedom, equality and adequate conditions of life, in an environment of a quality that permits a life of dignity and well being. . .'); World Charter for Nature, G A Res. 37/7, Annex, 123, UN Doc A./RES/37/7/Annex (28 October 1982) (Providing the right for all persons to access 'redress when their environment has suffered damage'); Hague Declaration on the Environment (Adopted 11 March 1989), 28 ILM 1308 ('[R]emedies to be sought involve, the right to live in dignity in a viable global environment. . .'); United Nations Conference on Environment and Development 3–14 June 1992, Rio Declaration

an opportunity to invoke human rights language for environmental protection. Despite these recognitions, however, the absence of a direct recognition of the right to environment in any of the core international human rights instruments continues to generate scholarly interest on the status of a right to environment under international law and the available mechanisms for its enforcement.[59] Even though the Aarhus Convention is by design an open global Convention, as it provides opportunities for ratification by countries outside the Economic Commission for Europe, the ratification and membership of the Convention remains concentrated in Europe.[60] This underscores the importance of reflecting procedural human rights provisions of the Aarhus Convention in the international climate change regime to address the specific linkages between human rights and climate change, most especially indirect impacts of response measures and projects.

on Environment and Development, Principle 1, UN Doc A/CONF. 151/5/Rev.1 (13 June 1992), reprinted in 31 ILM 874 ('Human beings are entitled to a healthy and productive life in harmony with nature').

[59] See M. Salomon, 'The Future of Human Rights', *Global Policy*, 3(4) (2012); L. Collins, 'Are We There Yet? The Right to Environment in International and European Law', *McGill International Journal on Sustainable Development Law and Policy*, 3(2) (2007), 119–153; M. Salomon, *Global Responsibility for Human Rights: World Poverty and the Development of International Law* (Oxford: Oxford University Press, 2007); see also D. Cassel, Commentary, 'Do We Have a Human Right to a Healthy Environment under International Law?' (2009), www.internationallawobserver.eu/2009/10/12/right-to-a-healthy-environment/, accessed 12 January 2015 (where the author critiques the existence of a right to environment under international law. He argues that there are still many complex underlying legal challenges that have to be fully comprehended before we can lay claim to the existence of this right under international law). Despite these debates, a right to environment has, arguably, gained sufficient recognition in international law and has evidently become applicable in more countries, than ever before. However, as it will be demonstrated in this chapter, there is still a need for a direct recognition and clarification of the specific linkages between human rights and climate change in the international climate change regime. Carbon projects produce indirect human rights consequences that must be specifically clarified and addressed as part of the international climate change regime. Arguably these concerns would be better addressed and enforced through an anticipatory and process-based framework that sets a mandatory threshold for the design and implementation of mitigation projects.

[60] As of 2 April 2013, there are forty-seven parties to the Convention. Currently, no African country has ratified the Aarhus Convention. All the developing countries discussed in this book (Honduras, Panama and Nigeria) are also not parties to the Aarhus Convention. Similarly, CDM giants such as India, China, Indonesia, and Brazil are not parties to the Aarhus Convention; this limits the prospects of invoking the Aarhus provisions and processes to address some of the procedural and human rights questions in the implementation of carbon projects in developing countries.

4.2.1.3 Climate change and human rights

The impetus to combat global climate change has brought about specific debates and proposals for substantive human rights to 'climate change mitigation' or a substantive 'human right to be cold'.[61] Some scholars have argued for the recognition of the fact that failure to mitigate climate change in one country is a threat to human rights in another country.[62] Sheila Watt-Cloutier, one of the most consistent advocates of this right, argues that failure to mitigate climate change violates the right to be cold in Inuit communities. She notes that GHG emissions are a major cause of Arctic warming and loss of ice, which is destroying the environment that has both formed and nurtured the traditional Inuit way of life.[63] She notes that the right to be cold would impose obligations on industrialized countries, particularly Canada and the United States, where a majority of the Inuits live, to address the impacts of human-induced climate change on livelihood, particularly in the Arctic communities.[64] Hall and Weiss also develop

[61] Watt-Cloutier, *The Right to be Cold*, (n 49); Harrington, 'Climate Change, Human Rights, and the Right to Be Cold', (n 49), 513; Ruhl, 'The International Right to Equitable Climate Change Adaptation', (n 1), 1–2; see also Africa Initiative Congress on Climate Change, 'Mitigating Climate Change is a Human Right', (2009), www.cigionline.org/blogs/2009/11/mitigating-climate-change-basic-human-right, accessed 21 January 2015; see also S. Watt-Cloutier, *Petition to the Inter American Commission on Human Rights Seeking Relief From Violations Resulting From Global Warming Caused by Acts and Omissions of the United States*, www.earthjustice.org/library/legal-docs/petition-to-the-inter-american-commission-on-human-rights-on-behalf-of-the-inuit-circumpolar-conference.pdf [hereinafter *ICC Petition*].

[62] See I. Burton, S. Huq, B. Lim, O. Pilifosova, and S. Schipper, 'From Impacts Assessment to Adaptation Priorities: The Shaping of Adaptation Policy', *Climate Policy*, 2 (2002), 145–159; see also D. Hunter, 'Human Rights Implications for Climate Change Negotiations', *Oregon Review of International Law*, 11 (2009), 331, 332; J. Knox, 'Climate Change and Human Rights Law', *Virginia Journal of International Law*, 50 (2009), 163, 168.

[63] See *ICC Petition*, (n 61).

[64] See The Utqiaġvik Declaration, Inuit Circumpolar Conference (2006); see also Arctic Climate Impact Assessment (ACIA), Impacts of a Warming Arctic 2004 Overview, 8: Sheila Watt Cloutier, Remarks upon receiving the Canadian Environment Awards Citation of a Lifetime Achievement, Vancouver, BC, 5 June 2006. The Arctic Climate Impact Assessment (ACIA) supports these claims. It warns that 'reductions in sea ice will drastically shrink marine habitat for polar bears, ice-inhabiting seals and some seabirds, pushing some species toward extinction. It notes that species that the Inuits rely on for feeding are disappearing and that hunting routes near shorelines have disappeared due to erosion brought on by the thawing of permafrost. Villages have experienced increased flooding in winter due to lessened or disappearing pack ice that normally protects shorelines from surging water'. See ACIA, *Impacts of a Warming Arctic: Arctic Climate Impact Assessment* (Cambridge University Press, 2004) 10.

a comprehensive argument for viewing climate change adaptation at the international level as a substantive human right.[65]

The scope and content of substantive climate change rights are yet to be fully explored and developed.[66] Similarly, most of the discussions with respect to substantive climate change rights continue to neglect the secondary human rights impacts of measures aimed at addressing climate change, particularly in developing countries.[67] For example, the right to be cold as it is currently framed does not speak to the reverse impacts of climate change mitigation actions or carbon projects on human rights. Will the right to be cold or right to climate change mitigation cover concerns generated by carbon projects, or would violations of this right include financing carbon projects that violate human rights? They fail to provide an elaborate understanding of how this right might influence the impact of climate change policies, programmes, projects and interventions on human rights. They also fail to evolve an anticipatory, preventive and harmonized approach that could help protect and fulfil other human rights impacted by

[65] M. Hall and D. Weiss, 'Avoiding Adaptation Apartheid: Climate Change Adaptation and Human Rights Law', *Yale Journal of International Law*, 37(2) (2012), 309. The authors frame equitable climate change adaptation as a human right. They argue that adaptation-related human rights claims – together with analyses of diagonally conceived human rights – can help afford global citizens a more robust international legal framework within which to address climate change. For a critique of this paper see Ruhl, 'The International Right to Equitable Climate Change Adaptation', (n 1), where Ruhl asked whether there is anything new about climate change adaptation, even when viewed through the human rights lens, that calls for the evolution of substantive international law doctrine, or even further, for the creation of a distinct realm of international law or whether climate change adaptation simply presents more of the same human rights issues that have been in play for decades, albeit with more quantity and intensity – more frequent and intense floods, droughts, heat waves, and population dislocations? He asks, if there is something new and exceptional about climate change adaptation for international human rights law, what is it and what new substantive doctrine is required?

[66] See J. Carlson, 'Reflections on a Problem of Climate Justice: Climate Change and the Rights of States in a Minimalist International Legal Order', *Transnational Law & Contemporary Problems*, 18(1) (2009), 45–68; E. Burleson, 'Climate Change Consensus: Emerging International Law', *William and Mary Environmental Law and Policy Review*, 34(2)(2010), 543–588; B. Ackerly and M. Vandenbergh, 'Climate Change Justice: The Challenge for Global Governance', *Georgetown International Law Review*, 20 (2008), 553.

[67] See S. McInerney-Lankford, M. Darrow and L. Rajamani, *Human Rights and Climate Change: A Review of the International Legal Dimensions* (World Bank, 2011); Burleson, 'Climate Change Consensus', 543–588; Ackerly and Vandenbergh, 'Climate Change Justice', 553

climate change.[68] These questions would have to be addressed to provide a more robust examination of the cross-cutting linkages between the direct and indirect impacts of climate change on human rights.[69] There is a need to examine how human rights principles can guide efforts to combat climate change such that the indirect impacts of carbon projects are anticipated and addressed within the international climate change regime. Arguably, a better approach is a process-based approach as it provides opportunities to coherently integrate human rights processes and guidelines as part of the international climate change regime, such that the secondary impacts of climate change actions and projects are anticipated and prevented in the first place.[70]

Similarly, calls for a new climate change right will undoubtedly reawaken debates on whether this is not another attempt to proliferate rights and whether such a right will not be redundant as such claims could easily fit within environmental rights.[71] Arguably, a better approach is an approach that examines how existing human rights instruments can be employed to protect rights affected by carbon projects.[72] There is a need to focus on an approach that locates and

[68] See S. McInerney-Lankford, 'Human Rights and Climate Change: Perspectives on the Value Added of Human Rights Law, (2012)', stating that there is a need for an approach that draws upon human rights obligations to inform the design and implementation of goals set forth in climate agreements. Lankford notes that exploiting the legally binding structure of human rights law obligations to respect, protect and fulfil can result in enhanced accountability in climate policy.

[69] See McInerney-Lankford, 'Human Rights and Climate Change: Perspectives on the Value Added', *ibid.*

[70] Hohmann, 'Igloo as Icon', (n 3), 295, where the author explores what a human rights approach to climate change can offer redress to Inuit communities whose right to housing are increasingly threatened by the direct impacts of climate change.

[71] Ruhl raised the question whether there is anything new about a substantive approach to climate change or whether it is just part of efforts to proliferate human rights. Ruhl also doubts whether the right of equitable climate adaptation would bring anything new to international law or international institutions in terms of implementation and combating climate change. Ruhl asks that if there is something new and exceptional about climate change adaptation for international human rights law, what is it and what new substantive doctrine is required? Ruhl argues that a better approach is to leverage procedural rights to achieve the goals of protecting human rights affected by climate change. See Ruhl, 'The International Right to Equitable Climate Change Adaptation', (n 1); see also McInerney-Lankford, 'Human Rights and Climate Change: Perspectives on the Value Added', (n 68).

[72] See Hohmann, 'Igloo as Icon', (n 3), 295, where the author explores what a human rights approach to climate change can offer Inuit communities whose right to housing is being threatened by the direct impacts of climate change. See also A. Johl and S. Duyck, 'Promoting Human Rights in the Future Climate Regime', *Ethics, Policy and Environment*, 15(3) (2012), 300; Wewerinke and Doebbler, 'Exploring the Legal Basis of

mainstreams internationally agreed obligations and existing human rights into the international climate change regime; for example, an approach that considers how the robust procedural environmental human rights in the Aarhus Convention could be leveraged to inform the design and implementation of carbon projects. Existing human rights provide robust foundations on which we can build in an attempt to highlight the cross-cutting issues in climate change mitigation and international human rights law. Furthermore, it is arguably a more practicable approach as it allows us to effectively link climate change responses with existing human rights obligations, helping to ensure that human rights are not traded off in emission reduction schemes and projects. This approach avoids questions and debates on whether a new human right to climate change could be recognized under international law.

4.2.2 *Procedural rights approach: the human rights-based approach*

4.2.2.1 Justifications

Debates surrounding the scope of a substantive climate change right have put more emphasis on the need to develop more innovative proposals for recognizing the cross-cutting issues in human rights and climate change. The participatory or due process approach is arguably a less contentious approach that offers significant promise for the quest to mainstream human rights into the international climate regime. This approach emphasizes mainstreaming existing human rights norms, standards and principles into legal regimes, policies and action plans on climate change mitigation, so as to ensure that the interests of the public are protected at all times. The procedural rights approach calls for an interpretation and mainstreaming of the rights of members of the public to information, participation in environmental decision-making and access to review procedures.[73] As Anderson noted while commenting on the importance of procedural rights to environmental protection, 'the more

a Human Rights Approach to Climate Change', *Chinese Journal of International Law*, 10 (2011), 141–160; R. Bronen, 'Climate-Induced Community Relocations: Creating an Adaptive Governance Framework Based in Human Rights Doctrine', *New York University Review of Law & Social Change*, 35 (2011), 357–407.

73 See Common Wealth, 'Human Rights and Climate Change: An Approach That Puts People in the Forefront of the Debate' Commonwealth Secretariat Discussion Paper 5 (July 2009); see also J. Ebbesson, 'Information, Participation and Access to Justice: the Model of the Aarhus Convention' (2009) United Nations Expert Meeting, Background Paper No. 5; Hohmann, 'Igloo as Icon', (n 3), 295.

flexible, honest and context-sensitive approach is to endow people with robust procedural rights that will foster open and thoroughgoing debate on environmental matters'.[74] Rajamani also rightly warns that human rights cannot 'simply be layered on, as yet another preambular recital; they would need to be integrated and mainstreamed'. She argues that human rights approaches are important to enable existing human rights meet the needs and constraints of the international climate regime.[75]

Rajamani's cautionary views are accurate and apt. To think that mere recognition of a substantive right to environment or a right to be cold, or the addition of new paragraphs of human rights commitments to the international climate regime will be enough to address the indirect concerns associated with climate change, arguably, misses the point. Human rights mainstreaming must spearhead a reform of the international climate change regime to reflect human rights considerations at all levels of decision-making. This point brings us back to the question of why human rights are seen as relevant to climate change. Should human rights come to the table only when there is an act or omission, or should human rights be respected and protected at all stages of decision-making with respect to climate change? Should we continue to think of human rights as relevant only to address the consequences of climate change or should we recognize the more pressing need to restructure the international climate change regime such that human rights become part of the rules of the game when it comes to the design, approval, finance and implementation of carbon projects. Arguably, the latter is the more sustainable approach that emphasizes the need to reflect human rights safeguards in climate actions and projects. A procedural rights approach would prompt changes in governance structures and institutional practices such that rights-holders are incorporated in the decision-making processes prior to the approval and registration of a project. This could help prevent violations and equally reduce the need for litigation or resistance of carbon projects by the public.

A procedural rights approach anticipates sources of human rights violations and mainstreams opportunities to address them at the stage of project planning and execution. This approach could also help to reduce litigation arising from climate change mitigation efforts as it allows citizens to take part in the process for planning and designing

[74] Anderson, 'Human Rights Approaches to Environmental Protection', (n 10), 10–11; see also Collins, 'Are We There Yet?', (n 59), 130–131.

[75] Rajamani, 'The Increasing Currency and Relevance of Rights-Based Perspectives', (n 3), 406; see also Engel, 'Harmonizing Regulatory and Litigation Approaches', (n 46), 1563.

responses and projects.[76] Through participatory rights, the design of climate actions could be strengthened so that their execution and approval would be dependent on the consent and cooperation of stakeholders, individuals and communities. This way human rights infringements resulting from carbon projects may be avoided. Procedural rights provide preventive/long-term processes through which human rights are systematically integrated into climate change governance structures so as to avoid the source of the problems in the first place.[77]

This approach is in tandem with the precautionary principle, which requires states to anticipate and avoid environmental damage before it occurs, especially where failure to do so would result not only in environmental degradation but in human rights violations as well.[78] Article 3(3) of the UNFCCC urges states to take 'precautionary measures to anticipate prevent or minimize the causes of climate change and mitigate its adverse effects'.[79] According to the principle, where there are threats of serious or irreversible damage, governments should take all effective measures to prevent the adverse effects of climate change and that the lack of full scientific certainty should not be used as a reason for postponing such preventive measures. This approach calls on governments to anticipate and respond to the adverse effects of climate change. This would arguably include not only the direct adverse effects of climate change on human rights but also the adverse effects of climate change measures on human rights. By providing opportunities for the public to raise genuine environmental and human rights concerns about a project such concerns could be effectively addressed and prevented at the earliest opportunity.

Stakeholders who would most likely be affected by a project are the most credible sources of information when it comes to its environmental and human rights threats. By providing them opportunities to bring

[76] See S. Kas, 'Integrated Justice: Human Rights, Climate Change, and Poverty', *Transnational Law & Contemporary Problems*, 18(1) (2009), 115–138; J. Carlson, 'Reflections on a Problem of Climate Justice: Climate Change and the Rights of States in a Minimalist International Legal Order', *Transnational Law & Contemporary Problems*, 18(1) (2009), 45–68.

[77] See Ruhl, 'The International Right to Equitable Climate Change Adaptation', (n 1); Johl and Duyck, 'Promoting Human Rights', (n 72), 300.

[78] See Principle 15 of the Rio Declaration, which states that 'where there are threats of serious or irreversible damage, lack of full scientific certainty shall not be used as a reason for postponing cost effective measures to prevent environmental degradation'.

[79] The United Nations Framework Convention on Climate Change, unfccc.int/resource/docs/convkp/kpeng.html, accessed 10 January 2015.

forward the existence of such threats governments can better anticipate and respond to these threats and take necessary precautions to avoid them. The precautionary principle includes elements of risk assessment, risk management and risk communication.[80] Environmental risks can be avoided when information on such risks is gathered transparently, disseminated efficiently and dealt with collectively with the participation of those citizens concerned. As noted in Principle 10 of the Rio Declaration:

> Environmental issues are best handled with the participation of all concerned citizens, at the relevant level. At the national level, each individual shall have appropriate access to information concerning the environment that is held by public authorities, including information on hazardous materials and activities in their communities, and the opportunity to participate in decision-making processes. States shall facilitate and encourage public awareness and participation by making information widely available. Effective access to judicial and administrative proceedings, including redress and remedy, shall be provided.

Though not legally binding, Principle 10 recognizes public access to information, public participation in decision-making proceedings and access to review procedures and remedies (access to justice) as prerequisites of a procedural rights approach. These three elements have also been recognized in many international environmental treaties as pillars of a procedural approach.[81] For example, the Aarhus Convention is a legally binding Convention that provides a suite of procedural safeguards such as the right to information, to participate in decision-making and to seek remedies for violations.[82] Despite some of the implementation challenges posed by the procedural rights guaranteed by the Aarhus Convention, it remains a good example of how environmental agreements can mainstream human rights safeguards that anticipate and

[80] See Communication from the Commission on the Precautionary Principle, COM (2000) 1, 02/02/ 2000. The Council adopted a resolution endorsing the broad lines of the Commission's communication in December 2000.

[81] See D. Shelton, 'Human Rights and Environment Issues in Multilateral Treaties Adopted between 1991 and 2001' (2002) Paper presented at the Joint UNEP-OHCHR Expert Seminar on Human Rights and the Environment (Geneva 14–16 January 2002).

[82] The Aarhus Convention is legally binding on all states that have ratified it. Currently, there are 47 parties to the Convention primarily European and Central Asian countries and the European Union. Article 19 (3) of the Aarhus Convention opens the door to accession by any other state, outside the Economic Commission for Europe (ECE) provided they are members of the UN and that the Meeting of the parties of the Aarhus Convention approves the accession. The extent to which countries outside the ECE, particularly countries in Africa, Asia and in the Caribbean will ratify and become parties to the Aarhus Convention, however, remains unclear.

prevent violations, ensure due process, deliberative democracy, accountability and access to justice into decision-making processes. This process-based approach provides opportunities to ensure that members of the public are given adequate opportunities to prevent projects that could violate human rights.[83]

To effectively mainstream human rights into the international climate change regime, specifically rules governing carbon projects, the better approach is, arguably, an approach that identifies the underlying, structural and institutional causes of rights violations and societal exclusions in project approval processes and then provides tailored procedures and governance structures to address and prevent such causes. This arguably is what a procedural human rights approach does. Table 4.1 summarizes how a procedural rights approach differs from a substantive rights approach to human rights mainstreaming.

Table 4.1 *Procedural vs substantive rights approaches*

Substantive rights approach	Procedural rights approach
Focus on immediate causes of human rights violations	Focus on underlying, structural and institutional causes of rights violations
Individuals are objects and recipients of human rights	Individuals are subjects and are empowered to protect and claim their rights
Emphasis on protecting human rights	Emphasis on realizing human rights
Focus on inputs and outcome	Focus on process and outcomes
Are focused and sector based	Emphasis on holistic, integrated and multidimensional approach to issues
Corrective, remedial and reactive approach to human rights	Preventive, precautionary and anticipatory approach
Offer universal protection of fundamental rights and freedoms	Target selection and focus on marginalized members of the society
Focus on top-down approaches for rights protection	Focus on bottom-up and locally owned approaches to rights protection

[83] See McCrudden, 'Mainstreaming and Human Rights', 7, where McCrudden argues that mainstreaming approaches are intended to be anticipatory, rather than essentially remedial, to be extensively participatory in the definition of the issue and how it might be addressed, and to be integrated into the activities of those primarily involved in policy-making. It aims to complement existing approaches to compliance, rather than replace them.

4.2.2.2 The United Nations human rights-based approach

Ever since the UN Secretary General launched the UN Programme for Reform in 1997, calling on all entities of the United Nations to mainstream human rights norms into their respective mandates and activities,[84] the idea of integrating human rights principles into development efforts began to gain significant recognition and popularity from UN agencies, the European Union, the World Bank and national governments as a planning and programming framework.[85] From the early 2000s, human rights mainstreaming began to develop increasingly but somehow haphazardly by different international organizations, leading to a lack of coordination or common understanding on its meaning, scope and content.

To resolve the ambiguities and confusion, UN agencies came together in 2003 with a human rights-based approach framework. The *Common Understanding on HRBAs*, which spells out the meaning, nature, design and essential attributes of a HRBA framework, was adopted.[86]

[84] See UN/UNDOC/GEN/N97/A/51/950/1997. Also, the 1998 Report of the Secretary General to the UN Economic and Social Council recommended (i) adoption of a 'human rights-based approach' to activities carried out within the respective mandates of components of the United Nations system; (ii) development of programmes/projects addressing specific human rights issues; (iii) reorientation of existing programmes as a means of focusing adequate attention on human rights concerns. See UNESCO, 'Guidelines for Human Rights-Based Programming Approach' Executive Directive 1998–004; see also J. Theis, *Evolution and Future of Rights-based Approaches* (Sweden: Save the Children, 2003), 10.

[85] See T. Greiber, M. Janki, M. Orellana, A. Savaresi-Hartmann and D. Shelton, *Conservation with Justice: A Rights-Based Approach* (International Union for Conservation of Nature, 2009); A. Frankovits, 'Human Rights Based Approach and the United Nations System' (2006) UNESCO, unesdoc.unesco.org/images/0014/001469/146999e.pdf, accessed 14 October, 2014; C. Nyamu-Musembi and A. Cornwall, *What is the 'Rights-Based Approach' all about? Perspectives from International Development Agencies* (UK: Institute of Development Studies Working Paper, 2004), 234; K. Svadlenak-Gomez, *Human Rights and Conservation: Integrating Human Rights in Conservation Programming* (Bronx, NY: World Conservation Society and Washington, DC: Translinks, USAID, 2007); I. VeneKlasen, *Rights-Based Approaches and Beyond: Challenges of Linking Rights and Participation* (UK: Institute of Development Studies Working Paper, 2004), 235; J. Theis, *Promoting Rights-based Approaches: Experiences and Ideas from Asia and the Pacific* (Child Rights Information Network Working Documents, 2004); B.I. Hamm, 'A Human Rights Approach to Development', *Human Rights Quarterly*, 23 (2001), 1005–1031; G. Castillo and M. Brouwer, 'Reflections on Integrating a Rights Based Approach in Environment and Development' in Conservation and Human Rights (IUCN Commission on Environment, Economic and Social Policy (CEESP), 2007), cmsdata.iucn.org/downloads/pm15.pdf, accessed 15 May 2015.

[86] See United Nations, 'The Human Rights Based Approach to Development Cooperation: Towards a Common Understanding among UN Agencies' (2003), www.undg.org/archive_docs/6959-The_Human_Rights_Based_Approach_to_Development_Cooperation_Towards_a_Common_Understanding_among_UN.pdf, accessed 12 October 2014.

It identifies the HRBA as a policy frame for mainstreaming human rights norms, standards and principles into legislation, policies and planning so as to ensure that citizens' interests are protected at all times. The *Common Understanding* spells out the basics of the HRBA and encourages agencies covered by UN operations to develop specialized standards tailored to cover their activities.[87]

The UN HRBA is a policy framework founded on a procedural rights approach.[88] Since 2003, the HRBA has been promoted by the United Nations as a conceptual framework that is normatively based on international human rights standards and operationally directed to promoting and protecting human rights.[89] The HRBA seeks to identify and address root causes of inequalities, social exclusions and human rights violations that impede the efficacies of development programmes and policies. HRBAs proceed on the notion that traditional approaches that relegate human rights to the background in planning processes are counterproductive, ineffective, unsustainable and often contentious.

Under the HRBA, human right does not come to the table when there is a protest or problem; human rights are treated as parts of the rule of the game of development and planning. HRBAs help to promote the sustainability of development policy or project by empowering people themselves – especially the most marginalized – to participate in policy

[87] The *Common Understanding* is based on three essential principles: that all programmes of development co-operation, policies and technical assistance should further the realization of human rights as laid down in the Universal Declaration of Human Rights and other international human rights instruments; that human rights standards contained in, and principles derived from, the Universal Declaration of Human Rights and other international human rights instruments guide all development cooperation and programming in all sectors and in all phases of the programming process; and that programmes contribute to the development of the capacities of duty-bearers to meet their obligations and of 'rights-holders' to claim their rights. See United Nations, 'The Human Rights Based Approach to Development Cooperation', (n 86).

[88] See D. Shelton, 'Whiplash and Backlash—Reflections on a Human Rights Approach to Environmental Protection', *Santa Clara Journal of International Law*, 13(1) (2015), 11–29, 21, stating that a 'rights-based approach' to environmental protection avoids many of the problems found in private litigation, as well as the limitations of environmental regulation and market-based incentives. See also D. Shelton, 'Equitable Utilization of the Atmosphere: A Rights-Based Approach to Climate Change', in S. Humphreys (ed.), *Human Rights and Climate Change* (Cambridge University Press, 2010), 91–126; Castillo and Brouwer, 'Reflections on Integrating a Rights Based Approach', (n 85).

[89] Office of the United Nations High Commissioner for Human Rights, *Frequently asked questions on a human rights-based approach to development cooperation*, www.ohchr.org /Documents/Publications/FAQen.pdf, accessed 12 April 2015.

formulation and hold accountable those who have a duty to act. The HRBA places emphasis on addressing and mitigating human rights impacts of development projects.[90] It draws attention to the potential impacts of actions, measures and projects on the most vulnerable groups, such as women, children, disabled persons and indigenous peoples.[91] It provides normative procedural frameworks for addressing systemic and structural injustices, social exclusions and human rights repressions in the development of climate change solutions.

Through the HRBA, procedural human rights, such as participation, accountability, equality and non-discrimination, and access to justice, are harmonized and integrated into policies and project activities, thereby giving citizens a basis to demand enforcement. [92] The HRBA therefore represents a shift from a needs-based approach to an approach that requires governments and project proponents to consider the impact of a particular project on the enjoyment of existing human rights.[93]

The HRBA has gained increased recognition in literature as the new wonder drug of equal opportunities, and as a policy framework that allows

[90] C.J. Dias, 'Understanding the UN Common Understanding on a Human Rights-Based Approach to Development Programming', in C.R. Kumar and S.K. Srivastava (eds.), *Human Rights and Development: Law, Policy and Governance*, (Hong Kong: Lexis Nexis, 2006), 5–7; see also UN High Commissioner for Human Rights, *Claiming the Millennium Development Goals: A Human Rights Approach*, UN/ HR/PUB/08/3 (2008), www.ohchr.org/Documents/Publications/Claiming_MDGs_en.pdf, accessed 12 October 2014.

[91] *See* UN HRBA Portal, The Human Rights Based Approach to Development Cooperation: Towards a Common Understanding among UN Agencies (2003), hrbaportal.org/the-human-rights-based-approach-to-development-cooperation-towards-a-common-understanding-among-un-agencies; see also Dias, 'Understanding the UN', *ibid.*

[92] According to the United Nations, other elements of good programming practices that are also essential under an HRBA include: (i) People are recognized as key actors in their own development, rather than passive recipients of commodities and services; (ii) Participation is both a means and a goal; (iii) Strategies are empowering, not disempowering; (iv) Both outcomes and processes are monitored and evaluated; (v) analysis includes all stakeholders; (vi) Programmes focus on marginalized, disadvantaged, and excluded groups; (vii) The development process is locally owned; (viii) Programmes aim to reduce disparity; (ix) Both top-down and bottom-up approaches are used in synergy; (x) situation analysis is used to identity immediate, underlying, and basic causes of development problems; (xi) Measurable goals and targets are important in programming; (xii) strategic partnerships are developed and sustained; (xiii) programmes support accountability to all stakeholders. See United Nations, 'The Human Rights Based Approach to Development Cooperation', (n 86).

[93] A. Cheria, E. Sriprapha Petcharamesree, *A Human Rights Approach to Development: Resource Book* (India: Books for Change, 2004) 2–4.

for the merger and acquisition of human rights and development in international law.[94] Firstly, the HRBA is rightly viewed as a potentially less contentious and less problematic approach to mainstream human rights norms into the international climate change regime.[95] Since it builds on existing human rights, and already recognized standards, principles and methods of human rights, it arguably avoids the proliferation, definition and legitimacy questions that have trailed substantive rights proposals. By focusing on due process and an opportunity for stakeholders to iron out grievances, the HRBA also avoids some of the problems of human rights conflicts that are difficult to avoid under substantive approaches.[96]

The legal basis of the HRBA is arguably less problematic because most of the parties to core human rights treaties are also parties to the international climate change regime.[97] As such, these countries have obligations under international law to respect human rights and to take action in terms of climate change. The HRBA provides a holistic framework though which parties can meet their obligations under both instruments. It allows states to respect, protect and fulfil human rights in the process of meeting their climate change obligations. As such, the HRBA does not seek to impose any new obligation other than those which parties have already agreed to under international human rights law. The most significant advantage of the paradigm shift is to anchor the debate about equity and justice in international principles already

[94] See T. Greiber (ed.) *Conservation with Justice* (Gland: IUCN, 2009) 6, where the IUCN describes the rights-based approach as an approach that could bring about conservation with justice. The IUCN defines 'conservation with justice' to mean a situation where all state and non-state actors planning or engaged in policies, projects, programmes, and activities with potential impact on nature conservation secure to all potentially affected persons the substantive and procedural rights that are guaranteed by national and international law.

[95] See McInerney-Lankford, 'Human Rights and Climate Change: Perspectives on the Value Added', (n 68); Shelton, 'Equitable Utilization of the Atmosphere', (n 88), 91–126; Castillo and Brouwer, 'Reflections on Integrating a Rights Based Approach', (n 85).

[96] See E. Petersmann, 'Time for a United Nations "Global Compact" for Integrating Human Rights into the Law of Worldwide organizations: Lessons from European Integration', *European Journal of International Law*, 13 (2002), 626; see also E. Petersmann, 'The "Human Rights Approach" Advocated by the UN High Commissioner for Human Rights and by the ILO: Is it Relevant for WTO Law and Policy?', *Journal of International Economic Law* (2004), 7 (3), 605–628.

[97] There are 166 parties to the ICCPR, 160 to the ICESCR. The majority of these parties have signed the UN Framework Convention on Climate Change and the Kyoto Protocol. See Rajamani, 'The Increasing Currency and Relevance of Rights-Based Perspectives', (n 3), 391.

endorsed by the international community and that have some legitimacy in international law.[98]

In the climate change context for instance, applying the HRBA will mean integrating procedural human rights safeguards such as access to justice, public participation in decision-making, equality and non-discrimination, ATI and accountability in the international climate change regime such that climate change actions are designed and implemented with the full participation and involvement of the public.[99]

The HRBA also empowers the public to play more active roles in the design, approval and execution of climate change mitigation actions and projects.[100] It supports rights-holders to demand their rights and to be involved in climate change mitigation efforts. This addresses the root causes of human rights violations and well-entrenched power imbalances that tend to exclude members of the public from playing active roles in climate change mitigation.[101] Through the HRBA, human rights could be harmonized and integrated into the international climate change regime so as to give the public a legal basis to freely air their views about a project and to demand a review when such views are not taken into consideration in project design and implementation.[102] As Mary Robinson rightly notes:

> In each situation we confront, a rights-based approach requires us to ask: what is the content of the right? Who are the human rights claim-holders? Who are the corresponding duty-bearers? Are claim-holders and duty-bearers able to claim their rights and fulfill their responsibilities? If not, how can we help them to do so? This is the heart of a human rights based approach.[103]

[98] See R. Offenheiser and S. Holcombe, 'Challenges and Opportunities in Implementing a Rights-Based Approach to Development: An Oxfam America Perspective', *Non Profit and Voluntary Sector Quarterly*, 32 (2003), 300; see also Lankford, Darrow and Rajamani, *Human Rights and Climate Change: A Review*, (n 67), 29–30.

[99] See Hohmann, 'Igloo as Icon', (n 3), 296–316; see also Lewis, 'Climate Change and Human Rights', (n 48), 158.

[100] J. Theis, *Promoting Rights Based Approaches: Experiences and Ideas from Asia and the Pacific*, (Thailand: Keen Publishing Co Ltd., 2004) 2–10; R. Eyben, 'The Rise of Rights: Rights-Based Approaches to International Development', *Institute of Development Studies Policy Briefing*, 17 (2005).

[101] F. Nguyen, 'Emerging Features of a Rights-Based Development Policy of UN, Development Cooperation & NGO Agencies' (2002) Discussion paper, OHCHR Asia-Pacific Regional Office, www.un.or.th/ohchr/issues/rba/Pdf/discussrba.pdf, accessed 12 October 2014.

[102] See United Nations, 'The Human Rights Based Approach to Development Cooperation', (n 86).

[103] M. Robinson, 'Comments at 2nd Interagency Workshop on Implementing a Rights-Based Approach in the Context of UN Reform', (2003), www.humanrights.se/upload/

The HRBA places emphasis on identifying human rights obligations that are relevant to a project; using human rights standards to guide project planning and implementation; engaging with the rights implications as a matter of obligation; and supporting efforts to address the underlying causes of rights violations. This is achieved by addressing inequitable power relations that prevent the public from playing active roles in project design and implementation; building the capacity of both rights-holders and duty-bearers to claim their rights and meet their respective responsibilities; taking all available measures to respect rights in all cases and supporting their protection and further fulfilment wherever possible, particularly for the most vulnerable; and supporting efforts to provide access to justice and redress for violations.[104] For example, with respect to carbon projects, such as the CDM, the HRBA would enable us to identify the public as rights-holders and to emphasize their entitlement to play active roles in project design and execution. It identifies project proponents, host countries, the CDM EB and the Conference of the Parties to the UNFCCC as the corresponding duty-bearers and emphasizes their obligations to work towards strengthening the capacities of rights-holders to make their claims.[105]

The HRBA also stresses the indivisibility of rights.[106] The HRBA stresses respect for all human rights whether civil and political or economic, social and cultural rights. Under the HRBA all rights are considered one and the same and as important. As such, projects that violate any category of human rights would be considered as incompatible with international human rights obligations. Members of the public are provided ample opportunities to demonstrate how a particular mitigation project could affect their human rights. This supports the development of the capacities of duty-bearers to meet their obligations and of 'rights-holders' to claim their rights.[107] The HRBA bridges theoretical gaps

files/2/R%C3%A4ttighetsperspektivet/Common%20Understanding%20FN%202003.pdf, accessed 12 January 2015.

104 See IUCN, 'Resolution 4.056, Rights-Based Approaches to Conservation', (IUCN, 2008), community.iucn.org/rba1/RBA%20Wiki/Resolution%204.056%20Rightsbased%20approaches%20to%20conservation.aspx, accessed 12 October 2014.

105 See also United Nations, *Claiming the Millennium Development Goals: A Human Rights Approach*, (2008) UN/ HR/PUB/08/3.

106 UNDP, 'Integrating Human Rights with Sustainable Development: A UNDP Policy Document', (UNDP, 1998); H. Quane, 'Further Dimension to the Interdependence and Indivisibility of Human Rights? Recent Developments Concerning the Rights of Indigenous People', *Harvard Human Rights Journal*, 25 (1) (2012).

107 United Nations, 'The Human Rights Based Approach to Development Cooperation', (n 86).

between political, civil, social and economic rights by understanding how they are interconnected in practice.[108]

The aim of the HRBA is to integrate and harmonize all human rights norms and obligations into the processes of planning and executing carbon projects. HRBA therefore de-emphasizes the unnecessary and artificial bifurcations of rights that have often resulted in the infringements of certain rights in order to uphold the other. When stakeholders are given a chance to play active roles in decision-making, conflicts between human rights are better addressed in a balanced way.

Thus, the HRBA is not necessarily irreconcilable with substantive human rights. The aim of the HRBA is to establish and institutionalize due procedures and processes that can contribute directly to the coherent realization of all human rights. The HRBA provides a due process approach that focuses on identifying and using indicators for ensuring the incorporation of human rights into the design, implementation and monitoring of climate change mitigation efforts and projects. As such, the HRBA goes hand in hand with the substantive approach as it seeks to provide an institutional framework through which all human rights can be protected and enforced. The HRBA could be a useful approach to eliminate human rights issues facing climate change mitigation efforts. By mainstreaming human rights principles such as participation, access to justice and ATI into international climate governance architecture, the HRBA approach could offer a preventive approach to human rights violations and risks occasioned by such projects.

The HRBA also provides a legal framework for the public to hold non-state actors, such as project proponents or multinational corporations that finance carbon projects, accountable for human rights violations. The increasing obligations of business enterprises to address systemic human rights risks in climate and energy investments have been well documented. What is however required is more and better clarity on what responsible corporations should be doing to mitigate such risks.[109] The HRBA provides the public with the opportunity to demand transparency and accountability from

[108] Offenheiser and Holcombe, 'Challenges and Opportunities', (n 98), 286; also Quane, 'Further Dimension to the Interdependence and Indivisibility of Human Rights', (n 106).

[109] See J. Ruggie, 'United Nations Guiding Principles on Business and Human Rights: Implementing the United Nations "Protect, Respect and Remedy" Framework' (2011) UN Document A/HRC/17/31 (21 March 2011), paras. 8–12; see also Economist Intelligence Unit, *The Cost of Inaction: Recognising the Value at Risk from Climate Change* (Economist Intelligence Unit, 2015).

corporations that sponsor and benefit from carbon projects, which violate human rights. As Jochnick argues, the 'narrow focus of human rights law on state responsibility is not only out of step with current power relations but tends to obscure them'.[110] Focusing on the state in a globalizing world may shield other actors (the private sector and international institutions) from responsibility and leave the poor with both hands tied behind their backs. He noted that by demanding explanations and accountability, human rights expose the hidden priorities and structures behind violations. For example, the roles played by a German development bank in the Barro Blanco project in Panama attracted so much concern and put enormous pressure on the CDM EB to condemn and delist the bank from funding CDM projects.[111] There is a need to recognize the increasing roles played by such organizations by empowering supervisory bodies, such as the CDM EB, to investigate and block projects that entrench human rights violations. Through the HRBA, procedural safeguards that allow the public to obtain answers from financing intermediaries would be incorporated into the international legal regime on climate change.

Practically, what this approach means is that the international climate change regime would be revitalized and reformed to establish project approval guidelines that would include elements of participation, accountability, equality and non-discrimination, ATI and access to justice. Project registration and approval guidelines would provide a threshold that require governments and project proponents to demonstrate that these elements of the HRBA have been complied with and guaranteed in project planning and execution. Any project that does not satisfy the elements would either be referred back or refused approval by supervisory bodies of carbon projects, for example CDM EB and designated operational entities (DOEs). It would also include establishing complaint mechanisms and procedures for stakeholders or private individuals whose human rights have been infringed to seek redress, to block the approval of such projects or to seek the review of already approved projects.

[110] C. Jochnick, *The Human Rights Challenge to Global Poverty* (Brooklyn: Center for Economic and Social Rights, 1999), www.cesr.org/publications.htm, accessed 23 December 2014.

[111] See S. Jodoin, 'Rights-Based Frameworks for Climate Finance' (Center for International Sustainable Development Law, Legal Working Paper Series 2011), 6–7.

4.3 Epistemic debates on the human rights-based approach to carbon finance

Despite the considerable promise of the HRBA as a holistic and less contested policy framework for mainstreaming human rights safeguards into the international climate change regime, the HRBA is not without contentions and debates. A number of scholars have expressed pessimism on the practicality of applying the HRBA framework within the climate change context. Epistemic concerns include questions on whether outsiders to human rights have the capacity to fully mainstream human rights; whether there is an operational framework that will guide climate change administrators through such a radical reform process; or whether the existing governance structures on climate change have the resources to cope with the nature of responsibilities associated with a HRBA. These questions have not been considered in sufficient detail in emerging literature, raising logistical concerns and doubts on the practicability of a human rights-based approach to carbon finance. This section evaluates these debates with the aim of providing conceptual clarification on the practical aspects of implementing a HRBA to carbon finance.

4.3.1 Can outsiders interpret and mainstream human rights?

The HRBA framework advocates a legal partnership between human rights and the international climate change regime, as well as the reflection of human rights norms in the international climate change regime. This could include textual updates of international legal instruments on climate change to contain human rights provisions, and enlargement of climate change governance structures to provide for human rights assessment and review of carbon projects. These are fundamental transformations that could expand the scope of activities of climate change institutions and governance structure into uncharted areas such as interpreting human rights and making decisions on the human rights impacts of a project. The question therefore is whether climate governance entities, such as the UNFCCC Secretariat and CDM EB, and DOEs, auditors, development agencies or international financial institutions, which comprise mainly environmental and development practitioners, have the capacity to implement such a complex reform.

Put simply, do environmentalists, scientists and outsiders to human rights have the capacity to mainstream human rights?[112] Capacity has been defined as the ability of individuals, institutions and societies to perform functions, solve problems, and set and achieve objectives in a sustainable manner.[113] Adopting this definition, the question is whether these bodies have the abilities and resources to perform human rights-related functions of reviewing carbon projects for human rights compliance before verification, registration or approval for funding.

McCrudden has extensively discussed this question. According to him, these debates arise from epistemic communities and human rights advocates who see human rights interpretation as a task and function for human rights experts only.[114] He defines an epistemic community as a network of professionals with recognized expertise in a particular domain and an authoritative claim to knowledge within that domain. They have a shared set of normative beliefs, shared causal beliefs, shared notions of validity and a common policy enterprise.[115] The crux of the argument is that the interpretation of human rights instruments should be concentrated in bodies whose primary function is human rights interpretation; otherwise human rights will become domesticated, stripped of their radical promise.[116] McCrudden has discussed the dangers and demerits of placing the function of interpreting human rights in the hands of professional administrators.[117]

[112] J. Tallant, 'Forging Stronger Cooperation between Human Rights and Climate Change Communities', (February 2012) United Nations Seminar to Address the Adverse Impacts of Climate Change on the Full Enjoyment of Human Rights 23–24 February 2012 – Palais des Nations, Salle XII – Geneva, Switzerland, www.ohchr.org/Documents/Issues/ClimateChange/Seminar2012/DanielTaillant_24Feb2012.pdf, accessed 12 April 2015.

[113] C. Lopes and T. Theisohn, *Ownership, Leadership, and Transformation: Can We Do Better For Capacity Development?*, (United Nations Development Program (New York: UNDP, 2003).

[114] McCrudden, 'Mainstreaming Human Rights', (n 1); see also P. Haas, 'Introduction: Epistemic Communities and International Policy Coordination', *International Organization*, 46 (1992), 1–35.

[115] McCrudden, *Ibid.*; see also Tallant, 'Forging Stronger Cooperation between Human Rights and Climate Change Communities', (n 112).

[116] McCrudden, 'Mainstreaming Human Rights', (n 1). See also J.C. McCrudden, 'Mainstreaming Equality in Northern Ireland 1998–2004: A Review of Issues Concerning the Operation of the Equality Duty in Section 75 of the Northern Ireland Act 1998' in E. McLaughlin and N. Faris (eds.), Section 75 *Equality Duty: An Operational Review* (North Ireland Office, 2004).

[117] *Ibid.*

In essence, some consider as dangerous, proposals to place the task of interpreting and applying human rights norms in the hands of professional bodies or entities such as the CDM EB and DOEs, which are composed mainly of professional administrators, scientists and analysts who do not belong to the epistemic community of human rights law and who arguably know little or nothing about international human rights. This is based on the notion that since such administrators are not particularly trained or educated in human rights law, their attempt to mainstream human rights might be counterproductive and could lead to the erosion of human rights principles.

Koskenniemi takes these arguments further when he noted that entrusting human rights interpretation to outsiders would further blur the already thinning line between human rights experts and outsiders.[118] He considers that this may lead to a situation whereby any group may present itself as a human rights group by articulating its agenda as a human rights agenda. This may lead to a further devaluation of the human rights currency and the dilution of human rights concepts. Alston also raised this view in his rebuttal of proposals by Petersmann that the World Trade Organization (WTO) administrators should be granted the authority to mainstream human rights into the WTO.[119] According to Alston, the result of following such an approach would be to hijack international human rights law in a way that would fundamentally redefine its contours and make it subject to libertarian principles.[120] This, he noted, could lead to the 'economization' of human rights principles.[121] Petersmann however argued that such an approach can be enhanced by a radical restructuring of UN organizations to foster global integration.[122] This would be by proper staffing of institutions with

[118] See M. Koskenniemi, 'Human Rights Mainstreaming as a Strategy for Institutional Power', *Humanity* 1(1) (2010), 47.

[119] See E. Petersmann, 'From "Negative" to "Positive" Integration in the WTO: Time for "Mainstreaming Human Rights" into WTO Law', *Common Market Law Review*, 37 (2000), 1363–1382; see Petersmann, 'Time for a United Nations "Global Compact"', (n 96), 621.

[120] See P. Alston, 'Resisting the Merger and Acquisition of Human Rights By Trade Law: A Reply To Petersmann', *European Journal of International Law*, 13(8) (2002), 815.

[121] *Ibid.*

[122] For the Alston/ Petersmann debate, see Petersmann, 'Time for a United Nations "Global Compact"', (n 96), 621; Alston, 'Resisting the Merger and Acquisition of Human Rights', 815;E. Petersmann, 'Taking Human Dignity, Poverty and Empowerment of Individuals More Seriously: Rejoinder to Alston', *European Journal of International Law*, 13 (2002), 845–851.

human rights experts who can assist with the human rights mainstreaming process.

Alston's views that outsiders to human rights may find it difficult to properly interpret and mainstream human rights as intended by the HRBA are, arguably, apposite. They, however, would not constitute enough reasons for intellectual surrender. As Tallant rightly notes, this epistemic distinction is fuelled by the tendency of actors to remain within the formal confines of their areas of mandate, that is of human rights institutions/or within the negotiation circles of the UNFCCC process.[123] This is due to the absence of a formal agenda or obligations to collaborate between actors; the lack of fluid programmatic activity between intergovernmental agencies, for example the United Nations Environment Programme (UNEP) and the Office of the High Commissioner for Human Rights (OHCHR), or between civil society groups where human rights and environmental groups are also divided by agenda differences; there is also the personal unwillingness of actors to collaborate across sectors/agendas, spurred by the absence of a formal mandate to do so. As such human rights have, over the years, had little traction visibility or political support in climate change discussions, while climate change is not a priority issue in the everyday affairs of human rights institutions. This explains why human rights and climate change communities presently do not overlap or communicate. Tallant's views that there is need for a more conducive policy, economic and institutional space that would bring about a collaboration and constructive engagement between the two epistemic communities is a way forward in this regard.

There is a need for institutional integration that will ensure that programmes and projects to address climate vulnerability are designed, financed and implemented together by human rights and climate change communities.[124] The chances to achieve a harmonized international legal order whereby UN organizations, international financial institutions and funding bodies mainstream human rights into their activities could be boosted by a structural restructuring which would include appointing human rights experts into such organizations. For example, for the World Bank or UNFCCC to properly integrate human rights issues into their scope and mandate there would be a need to appoint experts and administrators with training and experience in human rights. This could either be by establishing a separate Human Rights Unit internally

[123] Tallant, 'Forging Stronger Cooperation between Human Rights and Climate Change Communities', (n 112).

[124] *Ibid.*

or by expanding the membership of existing committees and project review boards to include human rights experts. This would ensure that human rights instruments and norms are sufficiently understood, internalized, implemented and enforced.[125]

Human Rights Education (HRE) could also be beneficial to administrators, policymakers and regulators in the areas of climate change. Though not legally binding, the United Nations Declaration on Human Rights Education and Training encourages all international and regional organizations to ensure human rights training for all officers within their areas of mandate. The Declaration emphasizes that human rights education and training is essential for the 'promotion of universal respect for and observance of all human rights and fundamental freedoms for all'.[126] Article 11 provides that the United Nations and international and regional organizations should provide human rights education and training for their civilian personnel and military and police personnel serving under their mandates.[127] This Declaration underscores the need for development agencies, auditors, international financial institutions and stakeholders in the design, approval, finance and implementation of carbon projects to provide adequate HRE for officers and staff in order to prepare them for identifying and mitigating human rights risks in carbon projects.

HRE is essential so that field personnel and administrators can be acquainted with human rights standards, including the human rights regime and their monitoring and implementation mechanisms.[128] For example, administrators that have not acquired direct and extensive training on human rights could be enlisted for short courses and advanced training on human rights norms. This would provide some basic knowledge necessary to understand and handle mundane human rights issues arising within the secretariat. For example, the international certificate course on 'Mainstreaming Human Rights' could provide some level of training and general education on human rights and on how to apply

[125] See Knox, 'Linking Human Rights and Climate Change at the United Nations', (n 3), 477–498.

[126] United Nations Declaration on Human Rights Education and Training, UN Human Rights Council Res. 16/1, in UN Human Rights Council, Report of the Human Rights Council on its Sixteenth Session UN Doc A/HRC/RES/16/1 (19 December 2011).

[127] *Ibid.*

[128] Hamm illustrates this with the example of an expert from the German Ministry of Economic Cooperation and Development (BMZ), who notes that development experts in his ministry lack knowledge of human rights and thus they are not in a position to implement the HRBA. See Hamm, 'A Human Rights Approach to Development', (n 85), 1005–1031.

human rights norms.[129] Such training might provide general understanding of human rights concepts and human rights mainstreaming, which could prove sufficient to handle human rights issues such as reporting, verifications and project assessments. It could also reduce the wide gap in knowledge between the human rights epistemic communities and the perceived outsiders.[130]

4.3.2 *Question of resources*

The nature of transformation in governance structures advocated under the HRBA comes with high resource and cost implications; for example, cost of staffing, training and programme funding.[131] Unlike traditional human rights enforcement approaches, the HRBA requires the expansion of existing structures and institutions and the building of new ones to review and monitor carbon projects for human rights compliance. As McCrudden rightly noted, due to limited resources and competing budget priorities, the rights-based approach, which requires a radical transformation, may run into implementation problems.[132]

To reduce the cost of a human rights-based approach, there is a need for a framework that builds on existing capacities and resources.[133] This requires building on internal programmes and capacities and strengthening them to take on a more focused role. Linking new programmes to already existing ones would save some cost and provide ready infrastructures to work with. For example, to mainstream human rights norms into the existing international climate change regime, there would be a need to leverage on the resources of both the UNFCCC and UN human rights bodies. One of the missions of the Office of the High Commissioner for Human Rights (OHCHR) is to offer the best expertise and support to the

[129] Human Rights Education, Course 1T09: Mainstreaming Human Rights, www.hrea.org/index.php?doc_id=676, accessed 12 March 2015.

[130] UNDP, *Capacity Development (Practice Note)* (UNDP, 2005), 4.

[131] See IUCN, *Conservation with Justice: A Rights-Based Approach* (IUCN, 2009); Theis, 'Promoting Rights-based Approaches', (n 85); Hamm, 'A Human Rights Approach to Development', (n 85), 1005–1031, see also S. Fredman, *Human Rights Transformed: Positive Rights and Positive Duties* (Oxford: Oxford University Press, 2008), where Fredman discusses the considerable cost implications of realizing and enforcing human rights.

[132] McCrudden, 'Mainstreaming Human Rights', (n 1).

[133] United Nations, 'The Human Rights Based Approach to Development Cooperation', (n 86).

different human rights monitoring mechanisms in the UN system.[134] The OHCHR supports the work of treaty bodies and assists them in harmonizing their working methods and reporting requirements. It aims to improve the integration of human rights standards and principles within the UN entities.[135] A HRBA to climate change would benefit from leveraging some of the resources, facilities and best practices available from the OHCHR to bolster implementation. The emphasis here is on strong inter-agency linkages and partnerships to reduce the cost of mainstreaming and to coordinate efforts. Human rights experts from the OHCHR could be consulted to better assist the UNFCCC in designing its human rights mainstreaming programmes and governance structures.

The cost of mainstreaming could also be reduced by appointing a lead committee or creating a specialized office on human rights and climate change within the climate change governance structure to spearhead the mainstreaming effort. This Office would bring together key committees and avoid duplication of roles. It would identify areas of priorities and distribute resources in accordance with the priorities identified. It would also identify institutions or offices that could be compressed and condensed to make room for only necessary offices. Offices that duplicate roles could be merged or scrapped.

4.3.3 Lack of an operational framework or precedence

Another question is the lack of an operational framework or precedence that could provide guidance on how to go about implementing a human rights-based approach to carbon finance.[136] Despite the rapid ascendancy of the HRBA, its application in the area of climate change has not been given sufficient practical consideration and application. As of today, little exists in the literature on the idea of applying the HRBA to carbon finance.[137]

[134] See UNHCR, *Frequently Asked Questions on the Human Rights Based Approach*, (2005), www.ohchr.org/en/hrbodies/Pages/HumanRightsBodies.aspx, accessed 12 December 2014.

[135] *Ibid.*

[136] See C. Sunderland, T. Greiber and G. Oviedo (eds.), *Rights-Based Approaches: Exploring Issues and Opportunities for Conservation* (Center for International Forestry Research (CIFOR) and IUCN, 2009).

[137] Current discussions on climate change and human rights have focused mainly on the impact of climate change itself on the enjoyment of human rights and not on how human rights norms can be utilized to respond to the impacts of climate change mitigation and adaptation projects. See Lankford, Darrow and Rajamani, *Human Rights and Climate Change: A Review*, (n 67), 29–30; S. Caney, 'Climate Change, Human Rights and Moral

As such, the practical implications of applying the HRBA framework in the implementation of carbon projects have not been properly communicated. Questions on the legal and institutional framework required for implementing a rights-based approach in the context of climate actions and projects have been rarely posed, let alone answered in existing literature.[138] This is the gap this book sets out to fill.

For a normative framework at its early stage of development, the lack of an operational guideline or precedence is understandable and perhaps inevitable. It is, arguably, not enough reason to question the potentials of adopting a human rights-based approach to the international legal regime on climate change. This question on 'how' is extensively discussed in the subsequent chapters of this book that develop and propose legal and institutional guidelines or frameworks for mainstreaming and applying a human rights-based approach to carbon finance in the extant and emerging international climate change regime. Pertinent legal reforms, in addition to the 'web' of institutionalization necessary to translate the idea of mainstreaming human rights norms into the international climate change regime from theory to practice, are explored in subsequent parts of this book.

There could be no better time to think of such a framework. As the global community continues to map out the future of the international climate change regime, and its associated carbon finance architecture, it is important that human rights issues facing the implementation of carbon projects are brought to the fore and addressed to restore the integrity of market-based mechanisms and the international climate change regime generally.

4.4 Conclusion

Of the two main approaches proposed for exploring and recognizing the intersections between human rights and climate change, the procedural rights approach arguably stands out as a promising approach through which human rights norms could be integrated and mainstreamed into the international climate change regime to prevent tradeoffs. Through

Thresholds', in Humphreys (ed.), *Human Rights and Climate Change* (Cambridge University Press, 2010), 69–90; E. Page, *Climate Change, Justice and Future Generations* (Cheltenham: Edward Elgar, 2006), 132–160; S. Vanderheiden, *Atmospheric Justice Political Theory of Climate Change* (Oxford University Press, 2008).

138 United Nations Environment Program (UNEP), *Climate Change and Human Rights* (UNEP, 2015) 14, noting that very little has been done to elaborate on how countries should fulfill the goal of respecting human rights in climate actions.

procedural rights, the international climate change regime could be strengthened to anticipate, avoid and prevent some of the human rights questions arising in emission reduction schemes and projects, much unlike traditional approaches that are reactive in nature. By ensuring that the execution and approval of carbon projects are dependent on the consent and cooperation of stakeholders, individuals and communities, human rights risks in carbon projects may be avoided.

The HRBA is a transformative policy framework built on procedural human rights such as ATI, participation, equality, accountability and access to justice. Its adoption by international organizations such as the United Nations and governments at regional and national levels, coupled with its rapid ascendancy in scholarly works, demonstrates its potential and value as a practical framework for mainstreaming human rights safeguards into the international climate change regime, specifically the procedures and structures for approving carbon projects.

However, in order to properly understand the scope and practical implications of applying the HRBA in carbon finance context, there is a need for a detailed examination of the normative contents of the HRBA. The next chapter, begins the transition from theory to practice. It examines the normative contents, elements and practical implications of implementing the United Nations human rights-based approach to carbon finance.

PART III

The human rights-based approach to carbon finance

Nature, elements and content

5

Normative contents/elements of the human rights-based approach

5.1 Introduction

This chapter explores the normative content and scope of the human rights-based approach (HRBA) to carbon finance. Five key elements are integral to the HRBA to carbon finance; they are participation, access to information (ATI), non-discrimination and equity, accountability and access to justice.[1] The HRBA advocates the integration and mainstreaming of these inter-connected procedural human rights norms in the design, approval, finance and implementation of carbon projects, to prevent and address human rights risks. Adopting the HRBA to carbon

[1] The *UN Common Understanding* identified the five elements of the HRBA to include access to information (ATI); participation and inclusion; accountability; equality and non-discrimination; and access to justice. These categorizations have been widely accepted by scholars and practitioners as constituting the elements of the HRBA. See United Nations (UN), 'The Human Rights Based Approach to Development Cooperation: Towards a Common Understanding among UN Agencies' (2003), www.undg.org/archive_docs/6959-The_Human_Rights_Based_Approach_to_Development_Cooperation_Towards_a_Common_Understanding_among_UN.pdf, accessed 12 October 2014; see also C. Nyamu-Musembi and A. Cornwall, *What Is the 'Rights-Based Approach' All about? Perspectives from International Development Agencies* (UK: Institute of Development Studies Working Paper, 2004), 234; J. Ackerman, 'Human Rights and Social Accountability' (World Bank, 2005), 86; J. Theis, *Evolution and Future of Rights-based Approaches* (Sweden: Save the Children, 2003); F. Nguyen, 'Emerging Features of a Rights-Based Development Policy of UN, Development Cooperation & NGO Agencies' (2002) Discussion paper, OHCHR Asia-Pacific Regional Office, www.un.or.th/ohchr/issues/rba/Pdf/discussrba.pdf, accessed 12 October 2014; B.I. Hamm, 'A Human Rights Approach to Development', *Human Rights Quarterly*, 23 (2001), 1005–1031; G. Castillo and M. Brouwer, 'Reflections on Integrating a Rights Based Approach in Environment and Development' in Conservation and Human Rights (IUCN Commission on Environment, Economic and Social Policy (CEESP), 2007), cmsdata.iucn.org/downloads/pm15.pdf, accessed 15 May 2015; United Nations Population Fund (UNFPA), 'Core Concepts of a Human Rights Based Approach' (2010), www.unfpa.org/webdav/site/global/shared/documents/publications/2010/hrba/introduction.pdf, accessed 15 October 2014; Rights and Humanity Interaction, 'Definitions of Rights Based Approaches to Development: By Perspective' (2004), www.interaction.org/rba/documents.html#intro, accessed 8 December 2014.

finance would mean that any project proposal that does not guarantee or satisfy these requirements would not be approved, no matter the volume or value of carbon credits that such project could generate. This chapter explores the core contents of each of these elements. It analyses the normative elements and key requirements of the HRBA.[2]

5.2 Elements of the HRBA

Earlier parts of this book have discussed core international human rights instruments that recognize elements of the HRBA as fundamental human rights. This section expatiates on the normative contents of each of the five elements of the HRBA and the key requirements for their implementation.

5.2.1 Access to information

5.2.1.1 Contents of the right to information

In this book, ATI has been defined as the right to 'seek, receive and impart' information.[3] The question then is what are the elements of the right to seek, receive and impart information? Different studies have extensively elaborated the content of ATI as a human right.[4] What is arguably the most exhaustive description of this right is contained in

[2] According to the United Nations, the following elements are necessary, specific and unique to a HRBA: (a) assessment and analysis of the human rights claims of rights-holders and the corresponding human rights obligations of duty-bearers as well as the immediate, underlying and structural causes of the non-realization of rights; (b) programmes to assess the capacity of rights-holders to claim their rights and of duty-bearers to fulfil their obligations. They then develop strategies to build these capacities; (c) programmes to monitor and evaluate both outcomes and processes guided by human rights standards and principles; (d) programming should be informed by the recommendations of international human rights bodies and mechanisms. See United Nations, 'The Human Rights Based Approach to Development Cooperation', *ibid.*

[3] See Chapter 2.

[4] See P. Birkinshaw, 'Freedom of Information and Openness: Fundamental Human Rights', *Administrative Law Review*, 58 (2006), 177 at 210; J. Echeveria and J. Kaplan, *Poisonous Procedural Reforms: In Defense of Environmental Right to Know* (Georgetown University, 2002); S. Jagwanth, 'The Right to Information as a Leverage Right' in R. Calland and A. Tilley (eds.) *The Right to Know, The Right to Live: Access to Information and Socio-Economic Justice* (Capetown: Open Democracy Advice Centre, 2002), 3–16; M. Fitzmaurice, 'Some Reflections on Public Participation in Environmental Matters as a Human Right in International Law', *Non-State Actors and International Law*, 2 (2002), 1–22; S. Kuhn, 'Expanding Public Participation is Essential to Environmental Justice and the Democratic Decision-making Process', *Ecology Law Quarterly*, 25(4) (1999), 647;

'The Public's Right to Know: Principles on Freedom of Expression Legislation', a report packaged by the Article 19 group and endorsed by the UN Special Rapporteur.[5] This document clarifies the content of the right to information and sets out the important elements that underpin it.[6] Though not legally binding, these principles have been recognized as representing the normative contents of the right to information under international law and 'the fundamental basis and criteria to secure effective access to information'.[7] This chapter adopts these principles as a framework for analysis.

Summarily, contents of ATI, as a human right, include maximum disclosure of information, limited scope of exceptions, obligation to publish, process to facilitate timely access, affordable cost of access and public education and awareness. A good example of an international public disclosure programme that reflects these elements is the UNECE Kiev Protocol on Pollutant Release and Transfer Registers (PRTR Protocol).[8] This protocol was established to implement ATI rights provided in the Aarhus Convention.[9] The protocol mandates the frequent release of pertinent information on hazardous chemicals based on the right of the people to know, whether or not the information is requested. It is the first legally binding international instrument that mandates a disclosure of information on pollutant release and transfer registers.

S. Paul, *Holding the State to Account: Citizen Monitoring in Action* (Bangalore, India: Books for Change, 2002).

[5] Article 19 is a London-based international human rights non-governmental organization that focuses on promoting the enforcement and realization of the right to information and expression. According to the group, they take their name from Article 19 of the UDHR which states that 'Everyone has the right to freedom of opinion and expression; this right includes freedom to hold opinions without interference and to seek, receive, and impart information and ideas through any media and regardless of frontiers.' The group monitors, researches, publishes, advocates, campaigns, sets standards and litigates on behalf of freedom of expression wherever it is threatened. See Article 19, www.article19.org/pages/en/who-we-are.html, accessed 12 May 2015.

[6] Article 19, 'The Public's Right to Know', *ibid.*

[7] See Report of the Special Rapporteur, *Promotion and Protection of the Right to Freedom of Opinion and Expression*, UN Doc E/CN.4/2000/63, 18 January 2000, para. 42; see also IACHR Press Release No. 2/98, 6 March 1998, paras. 14–15.

[8] The Protocol became binding on its Parties on 8 October 2009; see United Nations Economic Commission for Europe (UNECE), www.unece.org/env/pp/prtr.html, accessed 15 December 2014.

[9] See UNECE, Aarhus Convention on Access to Information, Public Participation in Decision-Making and Access to Justice in Environmental Matters (Adopted 25 June 1998, entered into force 30 October 2001) No. 37770, treaties.un.org/doc/Publication/MTDSG/Volume%20II/Chapter%20XXVII/XXVII-13.en.pdf, accessed 19 March 2015 (Aarhus Convention), Articles 4, 5 (9).

Its international importance as an 'open' global protocol is underscored by the fact that it is opened for ratification by non-ECE members and members of the United Nations. The PRTR Protocol offers a good example of the elements of ATI and how ATI can be implemented within the context of an international environmental agreement such as the Kyoto Protocol.[10]

Articles 4 (g–j) and 11 of the PRTR Protocol define the core elements of information disclosure. Both articles require each party to establish an information register that is 'publicly accessible' through the Internet; 'free' of charge; 'searchable' according to separate parameters; 'user-friendly' in its structure and provides links to other relevant registers; presents standardized and 'timely data' on a structured, computerized database; and provides 'detailed information' on releases and transfers of pollutants covered by the Protocol, such as greenhouse gases, acid rain pollutants, ozone-depleting substances, heavy metals, and certain carcinogens, such as dioxins. The Protocol also includes 'limited confidentiality' provisions and allows for 'public participation' in its development and modification of registers. It provides detailed guidelines on how best to design and structure a public disclosure programme in order to implement ATI.

Each of the elements established in the PRTR Protocol could serve as parameters or templates on how to enhance ATI with respect to carbon projects. Information on carbon projects must be fully disclosed, made publicly accessible, searchable and proactively disclosed to ensure transparency and to ensure that the public are aware of the relevant details of such projects. Each of these elements of information disclosure will be briefly discussed.

Maximum disclosure This requires that all information held by public bodies and project proponents must be subject to disclosure. Information on carbon projects must be completely disclosed, including the potential social, financial, political, cultural and environmental impacts of the project. For example, Article 4 of the PRTR Protocol requires that registers should be designed to provide detailed information on all releases and transfers of pollutants covered by the Protocol. Providing detailed information in the context of carbon projects will mean that all

[10] There are currently thirty countries, mostly European, that have become Parties to the Protocol on Pollutant Release and Transfer Registers (PRTRs). See UNECE, www.unece.org/env/pp/prtr.html, accessed 15 December 2014.

project information is released to the public as it becomes available. To further widen ATI, 'information' must be broadly defined to include all information pertaining to a mitigation project except when confidentiality applies. This will include reports on environmental impact assessments and human rights impact assessments, emission reduction potentials of a project, project completion timescale, the social and economic benefits of the project, geographical information on communities that would be affected by the project and other relevant information that would assist the public to make informed decisions on the potentials and pitfalls of a project. This information must be subject to complete disclosure and exceptions should only apply in very limited circumstances. This requirement places an onus on the public authority to justify any denials of ATI requests by citizens.

Failure to disclose project information must be understood as a serious barrier to the fulfilment of human rights. Consequently, it is not enough for governments to claim confidentiality or to declare that disclosing project information would jeopardize public interest.[11] Denial of ATI must be clearly demonstrated and justified by public authorities; this would prevent arbitrary denial of legitimate information requests. Refusal should not be based on the aim to protect governments from embarrassment or the exposure of wrongdoing; a complete list of the legitimate aims, which may justify non-disclosure, should be provided and exceptions should be narrowly drawn so as to avoid including material which does not harm the legitimate interest.[12]

Limited scope of exceptions As noted earlier, requests for project information should be denied only when they fall under legitimate exceptions. This second criterion requires that such exceptions imposed on the release of project information should be limited. Exceptions to disclosure should not unnecessarily overshadow the principle of disclosure. Requests for information should be met unless the public body shows that the information falls within a narrow category of exceptions.

[11] See K. Dingwerth and M. Eichinger, 'Tamed Transparency: How Information Disclosure Under the Global Reporting Initiative Fails to Empower', *Global Environmental Politics*, 10(3) (2010), 74–96; A. Florini, 'Introduction: The Battle Over Transparency' in A. Florini (ed.), *The Right to Know: Transparency for an Open World* (New York: Columbia University Press, 2007), 1–2.

[12] Article 19, 'The Public's Right to Know', (n 5).

This has been a major threat to information access in many developing countries.[13]

The question of what constitutes legitimate scope of exceptions seems to have been laid to rest by a comprehensive tripod test laid down by the UN Special Rapporteur.[14] Though it does not have any legally binding status, it provides normative clarifications and guidelines on what constitutes legitimate exceptions.[15] The first test is that the information must relate to a legitimate aim listed in the law. Denial of information must be primarily for the purpose of protecting and keeping in line with an existing law or regulation. When information is withheld as part of compliance with an existing law or regulation, such denial would fall

[13] For example, the Nigerian FOI Act is replete with more exception sections and clauses that tend to deny access to information than those that grant it, a situation which allows government agencies to deny information access. For example, while only two sections (sections 1 and 3) deal with access to information, ten sections (sections 7, 11, 12, 14, 15, 16, 17, 18, 19 and 28) contain exceptions that, arguably, deny the public access to information. Also, the time limit provided in the law for granting or refusing requests as provided for in Section 4 of the Act is very short. In the case of *Nigerian Contract Monitoring Coalition v. The Power Holding Company of Nigeria (PHCN)* (2012) FHC/ABJ – a suit initiated to challenge the refusal of PHCN to release details of a World Bank-funded PHCN contract for the supply and installation of High Voltage Distribution systems in its facilities in Abuja, Lagos and Ibadan. Information requested by the coalition included the report of the needs assessment that preceded the contract, documentation on the design and specification requirements, bidding documents issued to all bidders, a list of all the contractors that submitted bids, copies of the letters of award, as well as the names and addresses of all distribution companies on whose behalf the procurement was undertaken. The request was turned down by both PHCN Abuja and Electricity Distribution Company Plc. The applicant filed a motion *ex parte* seeking the court's order granting it leave to apply for the prerogative order of mandamus to compel PHCN, its general manager and the Attorney General of the Federation to provide it with details of the contract. In a significant ruling on access to information in Nigeria, Justice Ademola granted the coalition's application condemning the decision of PHCN to ignore the request for such vital public information. The Court noted that open and transparent governance is not enhanced when citizens feel that they need to resort to the long and arduous path of litigation before they are able to obtain information from public institutions. Apart from the length of time it would take for litigations and appeals, there is also the considerable expense of the entire legal process that is beyond the reach of many Nigerians. This case demonstrates the culture of impunity in many developing countries in which access to public information is denied to the public on the grounds of confidentiality. It emphasizes the need for mandatory international standards if access to information on mitigation projects is to be protected from governmental secrecy.

[14] See Report of the Special Rapporteur, *Promotion and Protection of the Right to Freedom of Opinion and Expression*, (n 7).

[15] See Organization of American States (OAS) Special Rapporteur on Freedom of Expression in his 1999 Report, Volume III of the Report of the Inter-American Commission on Human Rights to the OAS.

under the legitimate exceptions. Secondly, disclosure must threaten substantial harm to that aim. Information denial is justified if disclosure threatens to cause substantial harm to a legitimate aim protected by law. It is not enough to quote or cite laws that could be affected by a disclosure; public officials must demonstrate how the requested information might lead to a substantial violation of an existing law or treaty obligation. This exception is triggered, for instance, when releasing public information might result in the violation of another public law. The law would not command the doing of the impossible; neither should the law contradict itself at any given time. Thirdly, the harm to the aim must be greater than the public interest in disclosure.[16] It must be demonstrated that denying access to such information would be more beneficial to the public than releasing the information. This test includes the weighing and balancing of the loss or harm that the release of publicly held information might bring. When the harm outweighs the gain, then the third exception is triggered.

A good application of this test is found in Article 12 of the PRTR Protocol which provides that disclosure registers must have limited confidentiality provisions and that the grounds for confidentiality shall be interpreted in a restrictive way, taking into account the public interest served by disclosure and whether the information relates to releases into the environment. Article 12(1) provides what constitutes legitimate exception to disclosure and when confidentiality should apply. Confidentiality will arise when such information would adversely affect international relations, national defence or public security; the course of justice, the ability of a person to receive a fair trial or the ability of a public authority to conduct an enquiry of a criminal or disciplinary nature; the confidentiality of commercial and industrial information, where such confidentiality is protected by law in order to protect a legitimate economic interest; intellectual property rights; or the confidentiality of personal data and/or files relating to a natural person if that person has not consented to the disclosure of the information to the public, where such confidentiality is provided for in national law.

These requirements describe the nature of legitimate exceptions to disclosure. Even though governments can develop other exceptions for context-specific purposes, the aim must always be to ensure that

[16] 'The Public's Right to Know, Principles on Freedom of Information Legislation' Article XIX. IAHCR, Annual Report 1999, Volume III, Report of the Office of the Special Rapporteur for Freedom of Expression, OEA/Ser.L/V/II.106.

exceptions to disclosure do not unnecessarily stifle or overshadow the principle of disclosure. Public authorities must also exercise caution in applying the discretion to withhold information. Improper uses of such discretion have been the subject of much litigation.[17] Applying the legitimate exceptions test to carbon projects would mean that project information must be proactively disclosed, except when such disclosure would jeopardize security or public interests. The discretion to withhold project information should involve two steps. First, public authorities must determine whether the confidentiality exemptions apply. If they do, the reviewer must go on to ask whether, having regard to all relevant interests, including the public interest in disclosure, disclosure should be made or whether information should be withheld.[18] All project information outside these categories must be proactively released.

Obligation to publish To protect and guarantee the right to information, public bodies must do more than accede to requests for information. They must also proactively disclose relevant and material information, even where there is no request.[19] For example, the PRTR Protocol provides that information must be made 'publically available' regardless of specific requests. Project developers and public bodies must actively publish and disseminate key categories of information that are of significant public interest as soon as such information is available. What this means is that the right to information should not only be reactive but also be proactive. The right to information includes the right to such information that must be published and disseminated *suo motu* (proactively) by public authorities on their own volition. Public bodies are to publish and disseminate widely, documents that are of significant public value and interest, for example, information about how the public body functions and the content of any decision or policy affecting the public.[20]

[17] For excellent examples of how the doctrine of legitimate exception has been interpreted domestically, see *Maranda v. Richer* (2003) SCC 67 [2003] 3 S.C.R. 193; *Lavallee, Rackel & Heintz v. Canada (Attorney General)* (2002) SCC 61 [2002] 3 S.C.R. 209, at paras. 36–37; *Pritchard v. Ontario (Human Rights Commission)* (2004) SCC 31 [2004] 1 S.C.R. 809; *Lank v. Canada (Minister of Justice)* (2006) SCC 39 [2006] 2 S.C.R. 319; *Canada (Privacy Commissioner) v. Blood Tribe Department of Health* (2008) SCC 44 [2008] 2 S.C.R. 574. The only exceptions recognized to the privilege are the narrowly guarded public; *Smith v. Jones* [1999] 1 S.C.R. 455; *R. v. Brown* (2002) SCC 32 [2002] 2 S.C.R. 185.

[18] See *Maranda v. Richer* (2003) SCC 67 [2003] 3 S.C.R. 193.

[19] Article 19, 'The Public's Right to Know', (n 5).

[20] See Report of the Special Rapporteur, *Promotion and Protection of the Right to Freedom of Opinion and Expression*, (n 7).

This places a duty on governments to run disclosure programmes that allow individuals to gain regular access to publicly held information. This could be by keeping a public database or by releasing periodic documents on governmental affairs. The aim is to reduce the culture of secrecy in governance and to foster transparency.

Examples of information that should be released through public disclosure programmes include operational information about a project, budgeting and costs, information on project complaints, procedures for public participation and input, and the content of decisions taken on issues affecting the public.[21] Implementing this element of the right to information requires governments to establish comprehensive public disclosure programmes on projects and policies. An example is the disclosure policy contained in the US Safe Drinking Water Act, which requires public water companies to inform their customers when they violate contamination standards, fail to follow a treatment requirement or mandatory testing procedure, or neglect to perform required monitoring.[22] Under the Act, when a violation 'has the potential to have serious adverse effects on human health as a result of short-term exposure', companies must follow detailed procedures for alerting citizens about the dangers and the remedial steps being taken.[23] The Act stipulates a dual notification system for customers. It provides that the water company should report a contamination first locally and directly and then nationally and electronically in newspapers and media houses.[24] This proactive public disclosure programme brings relevant information to stakeholders. It is pertinent for the international climate change regime to mandate project proponents to establish such public disclosure programmes for mitigation projects. The requirement would be to publish and inform citizens about known hazards about a project as soon as such information is available even when no request has been made. This requirement to disclose constitutes a fundamental element of the right to know which takes the right beyond information requests. It places an obligation on custodians of publicly held information that is vital and important to the people to publish it *suo motu*.

The nature of project information to be disclosed to the public will include information on the proposed activity such as a description of physical and technical characteristics of the proposed activity, including an estimate of the expected residues and emissions, a description of the

[21] See Article 19, 'The Public's Right to Know'. [22] 42 USC § 7412 (7) (H) (I) (III).
[23] 42 USC § 300g-3(c)(20 (C). [24] *Ibid.*

significant effects of the proposed activity on the environment, a description of the measures envisaged to prevent and/or reduce the effects, including emissions, a non-technical summary of the foregoing information and an outline of the main alternatives studied by the applicant, the main reports and advice issued to the public authority, and the time when the public concerned shall be informed of the proposed activity.[25]

Information disclosure should not stop at the project design stage. For example, the PRTR Protocol provides that the information on the register should be continuously and immediately available through electronic means.[26] Continuous information is needed to provide important updates that might change the sustainability of a project over time. The effects of projects change from time to time, so what was environmentally viable at the stage of design might become a threat post-registration. As such, the public should be carried along and updated with new project information from stage to stage as it becomes available. It should include reasonable time frames for the different phases, allowing sufficient time for informing the public as new information becomes available.[27] Failure by a project proponent to proactively disclose information concerning the project should constitute a ground for project review, and/or withdrawal of project approval.

Process to facilitate timely access Project information must be disclosed upon request, in a timely manner, and as soon as such information becomes available. Public authorities must facilitate ATI and should not delay or frustrate requests for information. Timely release of information, at all stages of project planning, is essential for the full enjoyment of ATI. Article 6 of the Aarhus Convention provides relevant guidance on what timely information means. Article 6(2) provides that the public concerned shall be informed, either by public notice or individually as appropriate, 'early' in an environmental decision-making procedure and in an adequate, timely and effective manner. Article 6(3) also provides that the public participation procedures shall include reasonable time frames for the different phases, allowing sufficient time for informing the public and for the public to prepare and participate effectively during the environmental decision-making. This emphasizes the need to release

25 Article 6(6) of the Aarhus Convention. 26 See Article 5(4), *ibid.*
27 See Article 6(3) of the Aarhus Convention, *ibid.*

project information as soon as possible in order to allow the public adequate time to understand and process the information.

The second requirement is that public bodies must facilitate access. The process of obtaining information should not be so tedious and complex, neither should it be designed to frustrate and discourage individuals from seeking publicly held information. As Article 5(4) of the PRTR Protocol suggests, information disclosure registers must be designed for 'maximum ease' of public access through electronic means, such as the Internet'. The protocol states that registers must be user-friendly in structure and should provide links to other relevant registers; they should also present standardized and timely data on a structured, computerized database. This requires that appropriate systems should be put in place to ensure that information is easily accessible. Governments can facilitate access by utilizing electronic options for gathering and disseminating information. This could be by creating a publicly accessible database or information-sharing portal through which interested members of the public can access information without protocol or delay. The Protocol also provides that where the information is not easily publicly accessible by direct electronic means, each party shall 'facilitate electronic access' to its register in publicly accessible locations, for example in public libraries, offices of local authorities or other appropriate places. In all, the Protocol places an overall duty on public authorities to facilitate access either through electronic means or through other public places.

With respect to carbon projects, the international climate change regime must establish information disclosure requirements that require designated national authorities (DNAs) and national governments to facilitate access by making it easy for people to obtain project information at their disposal either through electronic means or by publishing such information in newspapers or other public medium. Information must also be made public as soon as it is available and in a timely manner. For information to be considered timely, it must be provided at the design stage and before the projects are approved. Project information must also be released continuously after approval and during execution. Similarly, the DNA or other national bodies responsible for handling requests should process all requests for information quickly and fairly. Unnecessary and duplicative procedures and administrative bottlenecks that complicate the process of seeking and obtaining project information must be removed to make access easier and less demanding. ATI includes putting in place institutions and procedures that facilitate and simplify the process of obtaining project information.

Facilitating access also includes delivering project information in a language and format understood by locals. It includes interpreting project documents into multiple local languages, as required, and disseminating information through a wide variety of channels, including radio, video, graphics, documentaries, photos, and town criers, in order to ensure that information is extensively circulated to all relevant stakeholders, most especially those in remote and rural communities.

Costs The cost of accessing project information should not be so high as to deter requests. ATI includes seeking and obtaining information at a negligible cost. To promote ATI, project information requests should, to the extent possible, not be subject to fees, or if they must, such fees must be minimal and highly reduced.[28] By establishing effective and yet cost-free disclosure programmes, the public will be encouraged to seek and obtain publicly held information. For example, Articles 11(3) and (4) of the PRTR Protocol provide that each party shall ensure that ATI contained in its register is free of charge and that competent authority may make a charge for reproducing and mailing the information, but such charge shall not exceed a reasonable amount.[29]

A disincentive to public information requests in many developing countries is the administrative cost of 'buying' project information.[30] Individuals and communities with genuine interests are unable to access project information because they cannot afford to pay the costs. To remove cost barriers to information access, and to further protect ATI rights with respect to carbon projects, the international climate change regime must establish a public disclosure requirement, which like the PRTR Protocol will mandate project proponents and national authorities to disclose project information free of charge or at a very negligible cost. To reduce the costs of implementation on DNAs and national governments, such information could easily be disclosed and

[28] See Report of the Special Rapporteur, *Promotion and Protection of the Right to Freedom of Opinion and Expression*, (n 7).

[29] See also the Council of Europe's Convention on Civil Liability for Damage Resulting from Activities Dangerous to the Environment. Article 14(1) provides that parties may make a charge for supplying environmental information, but such a charge may not exceed a reasonable cost. CETS No.: 150 opened for signature on 21 June 1993.

[30] See 'Nigeria's Access to Information Law Is Not Working', www.icirnigeria.org/index.php?id=28, accessed 12 January 2015.

updated on a government website from which anyone could access and download relevant information free of charge. The context in which each DNA will implement such a requirement will vary based on local realities. For example, in developing countries such as Nigeria, where Internet access might be a challenge especially in rural areas, such project information could easily be made readily available through the DNA office or local government offices, while project proponents could also be required to publicize project information in local newspapers. The most important aspect of implementation is for international supervisory bodies such as the CDM EB to require host countries and project proponents to demonstrate before project approval and registration the extent to which project information has been made freely or easily accessible.

Public education and awareness To protect and fulfil ATI rights, the public must be well educated and informed about the nature and scope of their rights.[31] Governments and public authorities must promote ATI rights by designing programmes aimed at educating and informing the public about their rights to access project information, the scope of information available and the manner in which the right can be exercised. This is different from the requirement to publish project information, as it lays down a requirement to promote public awareness on the existence of the right to information and how it can be enforced.

ATI would be meaningless and impossible if right-holders are not aware of their rights or are not aware of ongoing carbon projects and how to access more information about them. Public awareness on ATI rights can be achieved by disseminating information on key institutions and processes through which project information could be obtained. This element would impose an obligation on countries to provide relevant information on national institutions that are in charge of public information and the process for accessing such information. To fulfil this element, governments and DNAs must be encouraged to design public awareness programmes on the existence of this right and on the institutions through which project information can be accessed. Detail of public awareness programmes should also be included as part of the mitigation master plans and project guidelines designed and released by DNAs and national governments.

[31] Article 19, 'The Public's Right to Know', (n 5). See Report of the Special Rapporteur, *Promotion and Protection of the Right to Freedom of Opinion and Expression*, (n 7).

The foregoing elements are crucial for the effective protection, realization and fulfilment of the right to seek, receive and impart information, as guaranteed under international human rights law. ATI as an element of the HRBA requires the mainstreaming of the right to information to project planning and execution. To implement ATI, the international climate change regime would require parties to establish public disclosure programmes that allow easy and cost-effective dissemination of project information. Public disclosure programmes must be designed in line with the requirements of this right and provide a detailed, accessible and cost-free procedure through which members of the public can access such information.

National authorities must also be required to demonstrate that the above ATI requirements have been put in place pre-registration and are sustained post-registration. This would place mandatory public disclosure requirements on project proponents and national governments in host countries to publish on a regular basis all information related to a project, including new risks and challenges, even when the public do not specifically make requests for such information. Important categories of information relevant to carbon projects include project description documents, which contain details about the project activity, the proposed monitoring methodology and baseline, the crediting period of the project, the project participants and the method by which the participants will communicate with the CDM EB. Other relevant information includes environmental impact assessment results, outcomes of human rights impact assessments, operational information, duration of projects, budgeting and costs, analysis of the stakeholders' comments, information on complaints procedures and procedures for public input, and the content of decisions taken on issues affecting the public.

One practical challenge to administering public disclosure programmes is the cost and resources involved. To reduce the costs and resources required to establish and maintain public disclosure programmes, it is pertinent for national authorities to progressively leverage already existing structures and resources of different government programmes and institutions. For example, many countries already have freedom of information laws that establish national institutions to manage public access to information. There is a need to examine the flaws of extant national programmes and institutions, in order to determine how to improve them to better deliver accessible, detailed and cost-free procedures through which members of the public can access project information. This will include strengthening or reforming inefficient institutions in line with the discussed ATI standards to make them better prepared to administer public disclosure programmes. DNAs would also serve as the focal point to coordinate information disclosure for carbon projects from different overlapping national institutions, for example the different ministries and departments involved in a carbon project, that are custodians of pertinent project information.

5.2.2 Participation

Apart from lack of timely and adequate information on carbon projects, another main criticism of the international climate change regime is that key stakeholders are not given sufficient chance to participate in project planning. As illustrated by some of the CDM projects discussed earlier on, there have been projects designed, planned and approved by DNAs without adequate input or contribution by citizens and stakeholders. In this book, we have defined participation as the right to 'take part' and 'to influence' decision-making processes on carbon projects. As discussed earlier, participatory rights have been recognized in a number of core international human rights instruments.[32] Environmental treaties such as the Aarhus Convention also provide robust illustrations of the scope and content of an effective participatory process. Article 7 of the Aarhus Convention provides that each party shall make appropriate practical and/or other provisions for the public to participate during the preparation of plans and programmes relating to the environment within a transparent and fair framework, having provided the necessary information to the public. It also provides that each party shall endeavour to provide opportunities for public participation in the preparation of policies relating to the environment.

Using the Aarhus Convention as a template, it is clear that in order to fulfil and protect the right to participation, a free, transparent and fair framework that allows the public to take part and influence the decision-making process must be put in place. This goes beyond mere consultation.[33] Generally, participation as an element of the HRBA stipulates that members of the public are entitled to active, free and meaningful participation in any decision-making process on any mitigation project that could affect the enjoyment of their civil, economic,

[32] See Chapter 2.

[33] The question of what constitutes 'stakeholders' and 'public' is addressed later in this chapter. The World Bank, *The World Bank Participation Sourcebook* (The World Bank, 1996), 3.

social, cultural and political rights. This means such projects or activities would not be carried out without allowing the public to have a say in an authentic deliberative process.

To guarantee and fulfil the rights of stakeholders to take part in, and influence decision-making processes, on carbon projects, several relevant questions must be addressed. For example, what constitutes participation, what are the elements of participation, what is an authentic deliberative process and when can it be said that a person's right to participate has been fulfilled. These are questions that do not lend themselves to straightforward answers. Some studies have pointed to this difficulty in conceptualizing participation as the chief reason why participation is difficult to actualize in real life. This section of the book will explore the contents of the right to participation by unpacking the meaning, nature and scope of the two elements in the definition of participation, which is the right to 'take part' and 'to influence' decision-making processes. After analysing these two concepts, this section will also address the question as to who should take part in decision-making processes.

5.2.2.1 Elements of participation

The right to participate in decision-making processes arguably connotes the right to take part in, and to influence, decision-making processes. This definition includes two vital elements: 'taking part' and 'influencing'.

5.2.2.2 Taking part

Taking part simply means having the chance to freely express one's views without pressure and manipulation in an authentic deliberative process. The idea of taking part is rooted in the theory of deliberative democracy, which calls for a free and transparent deliberative process.[34] The fundamental tenet of deliberative democracy is that democracy should give all citizens fair and equal opportunities to

[34] The term 'deliberative democracy' was originally coined by Joseph Bessette in his 1980 work 'Deliberative Democracy: The Majority Principle in Republican Government' in *How Democratic Is the Constitution?* (Washington, DC: AEI Press, 2011), 102–116. Barton has identified three inter-related concepts of participatory development: participatory democracy, deliberative democracy and civic republicanism. These three concepts have been constantly formulated as key participatory responses. Of these three, it is the concept of deliberative democracy that is the most relevant to this book and it will be the focal point of our discussions. See B. Barton, 'Underlying Concepts and Theoretical Issues in Public Participation in Resources Development' in D. Zillman, N. Donald, A. Lucas and G. Pring (eds.), *Human Rights in Natural Resource Development* (Oxford University Press, 2002), 82.

take part in decision-making.[35] It offers a space and an opportunity whereby stakeholders can contribute to the process of policy transformations. As Dahl argues, the people are the best judges of what is good for them and are therefore the safest guardians of their rights.[36] The concept of deliberative democracy emphasizes that for a governmental decision to be legitimate it must be preceded by authentic and ideal deliberations.[37] For example, Articles 6 and 8 of the Aarhus Convention define the elements of an effective public participation. Article 8 notes that steps to ensure effective public participation includes that time frames that are sufficient for effective participation should be fixed; draft rules should be published or otherwise made publicly available; the public should be given the opportunity to comment, directly or through representative consultative bodies; and the result of the public participation shall be taken into account as far as possible.

The next question is, what constitutes an effective or authentic deliberation? The concept of Free Prior Informed Consent (FPIC) has emerged in a number of international human rights and environmental instruments as one of the most consistent elaborations of an authentic participatory process.[38] FPIC arguably represents a broader elucidation of the right to self determination which is clearly provided in Article 1 of both the International Covenant on Civil and Political Rights (ICCPR) and the International Covenant on Economic, Social and Cultural Rights (ICESCR), which states that '[a]ll peoples have the right of self-determination. By virtue of that right they *freely* determine their political status and *freely* pursue their economic, social and

35 J. Elster, *Deliberative Democracy* (Cambridge Studies in the Theory of Democracy, 1998).

36 R. Dahl, *Democracy and Its Critics* (Yale University Press, 1989). See also J. Waldron, 'A Rights Based Critique of Constitutional Rights', *Oxford Journal of Legal Studies*, 13(1) (1993), 18–51.

37 J. Fiskin, 'When the People Speak Deliberative Democracy: The Majority Principle in Republican Government' in *How Democratic is the Constitution?* (Washington, DC: AEI Press, 2011), 102–116. See also Cohen's typology on an ideal deliberation, J. Cohen, 'Deliberative Democracy and Democratic Legitimacy' from A. Hamlin and P. Pettit (eds.), *The Good Polity* (Oxford: Blackwell, 1989), 17–34.

38 T. Ward, 'The Right to Free, Prior, and Informed Consent: Indigenous Peoples' Participation Rights within International Law', *Northwestern University Journal of International Human Rights*, 10 (2011), 54; B. Clavero, 'The Indigenous Rights of Participation and International Development Policies', *Arizona Journal of International and Comparative Law*, 22 (2005), 41, 42; K. Masaki, 'Recognition or Misrecognition? Pitfalls of Indigenous Peoples Free, Prior, and Informed Consent (FPIC)', UN Communication on Human Rights (2009).

cultural development'.[39] The FPIC concept provides a normative guideline through which the right to participate can be implemented. It is a template for ensuring that the public meaningfully participate in decisions directly impacting their lands, territories, and resources and for ensuring that the right to self-determination is respected and protected by states.[40]

FPIC underlines the idea that consent must be freely given, obtained prior to implementation of activities and be founded upon an understanding of the full range of issues implicated by the activity or decision in question; hence the formulation: free, prior and informed consent.[41] It has been described as the right of indigenous peoples, as land and resource owners, to say 'no' to proposed development projects at any point during negotiations with governments and/or extractive industries.[42]

The FPIC concept has also been recognized and elaborated in the UN REDD Programme Guidelines.[43] Though not legally binding, this document defines 'Free' as implying no coercion, intimidation or manipulation. 'Free' refers to a process that is self-directed by the community from whom consent is being sought, unencumbered by coercion, expectations or timelines that are externally imposed.[44] It requires that representatives

39 Emphasis added. See also S. Anaya, 'Indigenous Peoples' Participatory Rights in Relation to Decisions about Natural Resource Extraction: The More Fundamental Issue of What Rights Indigenous Peoples Have in Lands and Resources', *Arizona Journal of International and Comparative Law*, 22 (2005), 7.

40 Special Rapporteur on the Rights of Indigenous Peoples, *Promotion and Protection of All Human Rights, Civil, Political, Economic, Social and Cultural Rights, Including the Right to Development: Report of the Special Rapporteur on the Situation of Human Rights and Fundamental Freedoms of Indigenous People*, UN Doc A/HRC/12/34 (15 July 2009).

41 See UN, UN-REDD Programme Guidelines on Free Prior Informed Consent (UNEP, 2012) 7. See also M. Colchester and F. MacKay, *In Search of Middle Ground: Indigenous Peoples, Collective Representation and the Right to Free, Prior and Informed Consent* (Moreton-in-Marsh: Forest Peoples Programme, 2004), 8–14; A. Perrault, K. Herbertson and O. Lynch, 'Partnerships for Success in Protected Areas: The Public Interest and Local Community Rights to Prior Informed Consent (PIC)', *Georgetown International Environmental Law Review*, 19(3) (2007), 477.

42 The UN Workshop on Indigenous Peoples, Private Sector Natural Resource, Energy and Mining Companies and Human Rights, held in Geneva from 5 to 7 December 2001. (E/CN.4/Sub.2/AC.4/ 2002/3, para. 52). The duty to obtain their free, prior and informed consent is a corollary of a myriad of universally accepted human rights, including the right to self-determination, right to participation, right to property, right to cultural integrity and right to equality, that are contained in numerous international human rights instruments.

43 See UN, *UN-REDD Programme Guidelines on Free Prior Informed Consent*, (n 41).

44 It is free when information is transparently and objectively offered at stakeholders' request; process is free from coercion, bias, conditions, bribery or rewards; meetings

are free to express themselves in the way that is customary and comfortable for them. 'Prior' implies that consent has been sought sufficiently in advance of any authorization or commencement of activities and respect time requirements of indigenous consultation/consensus processes. It is a requirement that information be provided before activities can be initiated, at the beginning or initiation of an activity, process or phase of implementation, including conceptualization, design, proposal, information, execution, and following evaluation.[45] Information should be given well in advance to provide time for stakeholders to understand, access, and analyse information on the proposed activity.

'Informed' refers to the full disclosure of the intent and scope of the projects and policies. This element focuses on the type of information that should be provided prior to seeking consent and also as part of the on-going consent process. Information should be accessible, clear, consistent, accurate, and transparent and should be delivered in appropriate language and format.[46] 'Consent' refers to the decision made by indigenous peoples and other local communities reached through their customary decision-making process. Consent must be sought and granted or withheld according to the unique formal or informal political-administrative dynamic of each community. Such consent must be freely and collectively given. Decision-making must be by consensus and must be transparently made. There should be the absence of coercion, bribery or manipulation in the decision-making process.

The FPIC principle was first introduced in 1989 by the International Labour Organisation (ILO), to protect the rights of indigenous peoples in developing economies that were subject to involuntary resettlement. ILO's *Convention on Indigenous and Tribal Peoples in Independent Countries – 169/1989* provides that before indigenous peoples can be relocated from their lands, they are entitled to free and informed

and decisions take place at locations and times and in languages and formats determined by the stakeholders; and all community members are free to participate regardless of gender, age or standing. See Report of the International Workshop on Methodologies Regarding Free Prior and Informed Consent E/C.19/2005/3, endorsed by the *United Nations Permanent Forum on Indigenous Issues* (UNPFII) at its Fourth Session in 2005.

45 UN, *UN-REDD Programme Guidelines on Free Prior Informed Consent*, (n 41); see also United Nations Permanent Forum on Indigenous Issues, 'Report of the International Workshop on Methodologies Regarding Free Prior and Informed Consent', (19 January, 2005), undesadspd.org/IndigenousPeoples/MeetingsandWorkshops/InternationalWorkshoponFPIC.aspx, accessed 12 December 2015.

46 Report of the International Workshop on Methodologies Regarding Free Prior and Informed Consent, *ibid.*

consent.[47] The Convention aims at ensuring that every effort is made by the states to fully consult with indigenous communities in the context of development, land and resources. Articles 6, 7, 16, 17 and 22 of the Convention provide that government shall consult the people concerned, through appropriate procedures in particular through their representative institutions and establish means by which these peoples can freely participate in the decision-making process. In general, the Convention specifies that consultation should take place when: considering legislative or administrative measures that are likely to affect indigenous and tribal peoples;[48] prior to exploration or exploitation of sub-surface resources;[49] when any consideration is being given to indigenous and tribal peoples' capacity to alienate their lands or to transmit them outside their own communities;[50] prior to relocation, which should take place only with FPIC;[51] and on the organization and operation of special vocational training programmes.[52]

Perhaps what is the most comprehensive elaboration of the FPIC is contained in the UN Declaration on the Rights of Indigenous Peoples (UNDRIP). Though not legally binding, UNDRIP articulates FPIC and provides normative guidance on the need for states to obtain the consents of indigenous communities on development activities that might affect them, as opposed to mere consultation. Article 19 of the Declaration provides that:

> States shall consult and cooperate in good faith with the indigenous peoples concerned through their own representative institutions in order to obtain their free, prior and informed consent before adopting and implementing legislative or administrative measures that may affect them. . .[53]

FPIC was also referred to in Articles 1, 12, 20, 27 and 30. Article 32 answers the question on when FPIC is required. It provides that:

> States shall consult and cooperate in good faith with the indigenous peoples concerned through their own representative institutions in order to obtain their free and informed consent *prior* to the approval of any project affecting their lands or territories and other resources,

[47] (ILO No. 169) (1989), 72 ILO Official Bulletin. 59, reprinted in 28 ILM 1382 (Convention on Indigenous and Tribal People).
[48] Article 6.1 (a), *ibid.* [49] Article 15.2, *ibid.* [50] Article 17, *ibid.* [51] Article 16, *ibid.*
[52] Article 22, *ibid.*
[53] United Nations Declaration on the Rights of Indigenous Peoples, UN Doc/ A/61/L.67 (2007).

particularly in connection with the development, utilization or exploitation of mineral, water or other resources. . .[54]

As such, the principle of FPIC must be applied to decisions on proposed activity or projects that would result in the removal of indigenous peoples and/or other relevant rights-holders from traditional or customary lands/territories;[55] and from cultural, intellectual, religious, and spiritual property.[56] FPIC is also required for proposed activity/policy that would include confiscation, occupation or damage of the lands, territories and/or resources of indigenous peoples and/or other relevant rights-holders.[57]

Similarly the UN Committee on the Elimination of Racial Discrimination (CERD) encourages states to 'ensure that members of indigenous peoples have rights in respect of effective participation in public life and that no decisions directly relating to their rights and interests are taken without their informed consent'.[58] In 2001, the UN Committee on Economic, Social and Cultural Rights (CESCR), the supervisory body of the ICESCR, in its report of Columbia in relation to traditional lands specifically urged states 'to consult and seek the consent of indigenous peoples concerned prior to the implementation of timber, soil or subsoil mining projects and on any public policy affecting them'.[59] The CESCR has further expanded on the right of indigenous peoples to FPIC in its General Comment No. 21. It calls on state parties to the Convention to respect the principle of free, prior, and informed consent of indigenous peoples in all matters covered by their specific rights. Though not legally binding, this general comment provides normative interpretations of how a state should implement Article 15 of ICESCR, which outlines the right to participate in cultural life, it concludes that the right to participate cannot be fulfilled if decisions on indigenous lands and resources are taken without the prior and informed consent of the affected peoples.[60]

[54] *Ibid*. Emphasis added. [55] Article 10, *ibid.* [56] Article 12, *ibid.*

[57] See United Nations, *UN-REDD Programme Guidelines on Free Prior Informed Consent* (n 41), 12.

[58] GR XXIII 51 concerning IPs adopted at the Committee's 1235th Meeting, (1997).

[59] Committee on Economic, Social and Cultural Rights [CESCR], *Concluding observations of the Committee on Economic, Social and Cultural Rights: Colombia*, 12 and 33, UN Doc E/C.12/1/Add.74 (6 December 2007); CESCR, Concluding observations of the Committee on Economic, Social and Cultural Rights: Ecuador, 12 and 35, UN Doc E/C.12/1/Add.100 (7 June 2004).

[60] CESCR, *General comment No. 21 Right of everyone to take part in cultural life (Article 15, 1 (a), of the International Covenant on Economic, Social and Cultural Rights)*, UN Doc E/

The FPIC principle has also been recognized in the American Declaration on the Rights of Indigenous Peoples (Draft) of the Organization of American States (OAS).[61] Articles XVII and XXIII of the declaration provides that no decisions directly relating to the rights and interests of indigenous peoples are to be taken without their free and informed consent. Accordingly, states shall obtain FPIC prior to the approval of any project affecting indigenous people's lands, territories and resources, particularly in connection with the development, utilization or exploration of mineral, water or other resources. In the *Awas Tingni case* the IACHR stated that FPIC is generally applicable 'to decisions by the state that will have an impact upon indigenous lands and their communities, such as the granting of concessions to exploit the natural resources of indigenous territories'.[62] As the Inter American Court on Human Rights noted in *Saramaka People v. Suriname*, large scale development projects that would have a major impact within an indigenous people's territory should only proceed with the free, prior, and informed consent of the people, according to their customs and traditions.[63]

Though not legally binding, the Ruggie Guiding Principles on Business and Human Rights also enjoins business enterprises to respect the rights of local communities affected by business activities, in accordance with international human rights standards such as the Indigenous and Tribal Peoples Convention, 1989 (No. 169). In its commentary, it calls on business enterprises to respect the principle of free, prior and informed consent of the indigenous peoples and communities to be affected by their activities and projects.[64] Incorporating FPIC principles into project planning would allow corporations and organizations to better discuss anticipated project risks with the public, such that harmful projects are

C.12/GC/21 (21 December 2009). On the importance of general comments, see M. Flaherty, 'The Concluding Observations of United Nations Human Rights Treaty Bodies', *Human Rights Law Review*, 6 (2006), 27, 33.

[61] Adopted by the Inter-American Commission on Human Rights on 26 February 1997, at its 1333rd session, 95th Regular Session, OEA/Ser/L/V/. II.95 Doc.6 (1997).

[62] *Mayagna (Sumo) Awas Tingni Community v. Nicaragua* Inter-Am. Ct. Hum. Rts. (Ser. C) Case No. 79, para. 103(k) (Judgment of 31 August. 2001), www.corteidh.or.cr/seriecing/serie_c_79_ing.doc, accessed 12 December 2014.

[63] *Saramaka People v. Suriname, Preliminary Objections, Merits, Reparations, and Costs,* Judgment of 28 November 2007. Series C No. 172.

[64] See J. Ruggie, 'United Nations Guiding Principles on Business and Human Rights: Implementing the United Nations "Protect, Respect and Remedy" Framework' (2011) UN Document A/HRC/17/31 (21 March 2011), paras. 8–12.

either redesigned to mitigate the harm or are totally avoided if the harm cannot be efficiently mitigated. This human rights understanding of FPIC would be useful for addressing human rights problems associated with carbon projects.

In 2011, the International Finance Corporation (IFC), the lending arm of the World Bank, incorporated FPIC as a component of its Sustainability Framework (Performance Standard).[65] The IFC Performance Standards build on the Equator Principles, a voluntary set of standards developed by international development banks and private sector financial institutions to reduce the social and environmental impacts in project financing.[66] The IFC Framework rightly notes that FPIC of local communities in development projects promotes sound environmental and social practices, encourages transparency and accountability, and contributes to positive development impacts. In case of projects that may affect indigenous peoples' lands, the IFC requires project proponents to document indigenous peoples' 'informed participation and the successful outcome of the negotiation'.[67]

Though IFC Guidelines are not legally binding human rights instruments, the adoption of FPIC into international lending and investment standards is very important, considering the fact that international lending institutions such as the IFC directly finance many large-scale carbon projects, including CDM projects that have been subjects of human right violations. Also, even though the application of the IFC Guidelines are limited to those projects financed by the IFC, it sets a standard for other financial institutions involved in project financing for carbon projects on why and how to mandate free and prior consultation with local communities as a compulsory requirement for project financing, to avoid unintended consequences of funded projects. The international climate change architecture could also integrate robust FPIC standards and requirements into project approval guidelines, so as to mandate banks and private actors that finance or fund carbon projects to demonstrate full compliance with FPIC standards,

[65] The IFC's Performance Standards define clients' responsibilities for managing their environmental and social risks before projects can be financed by the IFC. See Performance Standard 7 of the International Financial Corporation, www.ifc.org/ifcext/policyreview.nsf/Content/2012-Edition, accessed 12 March 2015.

[66] See the Equator Principles, June 2006, www.equator-principles.com/resources/equator_principles.pdf, accessed 21 January 2015.

[67] See Performance Standard 7 of the International Financial Corporation, www.ifc.org/ifcext/policyreview.nsf/Content/2012-Edition, accessed 15 June 2015.

prior to financing a carbon project. Financial institutions and investors may be required to delay their financing, or decide not to finance a project that does not meet FPIC standards.

As to the responsibility of national authorities, at a minimum, the FPIC concept would require that carbon projects should only be granted host country approval with the free, prior, and informed consent of the people or the public in the community where a project is located. FPIC, as elaborated in both human rights and environmental instruments, requires more than mere consultation; rather it involves allowing stakeholders to raise different proposals, considering them without force or fear and providing stakeholders with opportunities to make free choices from available options. National authorities can enforce FPIC by demanding full compliance with FPIC prior to granting host country approval; and when approved, by establishing procedures to withdraw such approval, commence enforcement actions, impose civil or criminal penalties on a project proponent that fails to continually accommodate stakeholder perspectives throughout the lifecycle of the project.

However, as the FPIC concept continues to evolve, much thought must go into the practical aspects of implementing it. What FPIC would mean varies from one context to another and from one country to another. While in some countries, legal safeguards on participation are well developed and guaranteed; in other countries there are weak structures for making FPIC work. There is also a low level of understanding on how to implement FPIC in autocratic countries with few or no mechanisms to ensure due process.

As such, to guarantee participation in carbon projects, it is not enough to include paragraphs in climate change instruments or project approval guidelines and rules to the effect that states must ensure FPIC in planning a carbon project. It is important to, as a further step, adopt a mandatory project requirement or threshold, which would make effective participation and the FPIC of the public a criterion for project approval and registration.[68] It is also important to provide clear guidelines on the threshold that must be met and demonstrated by project proponents during a participatory process as part of the FPIC requirement.

[68] The International Fund for Agricultural Development (IFAD) adopted this stringent FPIC requirement by making respect for FPIC a criterion for approval of agricultural projects. See IFAD, *Engagement with Indigenous Peoples Policy* (2009), www.ifad.org/english/indigenous/documents/ip_policy_e.pdf, accessed 10 June 2015.

Implementing FPIC Recognizing the right to FPIC and implementing a FPIC process can arguably eliminate human rights, legal, financial and reputational risks of project proponents, host countries, financial institutions, and other key stakeholders involved in the implementation of carbon projects. However, translating FPIC requirements to practice requires a detailed evaluation, by these key stakeholders, of several practical questions. For example, when we say FPIC, Free for whom? Consent by who and how do you ensure FPIC? This section examines these and other questions on the practical and logistical requirements for implementing FPIC in the context of carbon projects. Four essential elements are important for the process to enable the public to take part in decision-making processes, namely broad-based participation, fairness and inclusivity, rationality, and agreement.[69] We will analyse how each of these elements are pivotal to realizing FPIC in the approval and implementation of carbon projects.

Who should participate: broad-based participation Every member of the public affected by a decision irrespective of social status or societal strata must be able to take part in decision-making processes. Broad-based participation calls for the representation of all significant sectors of the society at the decision table.[70] Every member of the public should be able to raise proposals, criticize, and support measures, without any substantive hierarchy.[71] Emphasis here is on the establishment of a system of free deliberation among all stakeholders that would be affected by the outcome of the deliberative process. As such, rules that exclude a section of the society hinder the ability of any and every person affected by a project to take part in decision-making processes. The Aarhus Convention arguably provides best practices in this regard. For example, Article 2(4) of the Aarhus Convention defines 'the public' as 'one or more natural or legal persons and, in accordance with national legislation or practice, their associations, organizations or groups'. It also defines public concerned as the public affected or likely to be affected by,

[69] The analysis here is influenced by some of the elements of deliberative democracy identified by L. Jacobs, F. Cook and M. Carpini, *Talking Together: Public Deliberation and Political Participation in America* (University of Chicago Press, 2009), 9.

[70] See J. Hartz-Karp and M. Briand, 'Institutionalizing Deliberative Democracy', *Journal of Public Affairs*, 9 (2009), 127; J. Poiser, 'A Civic Republican Perspective on the National Environmental Policy Act's Process for Citizen Participation', *Environmental Law*, 26 (1996), 53 at 83.

[71] Cohen, 'Deliberative Democracy and Democratic Legitimacy', (n 37), 17–34.

or having an interest in, the environmental decision-making. This definition of the public includes individuals, NGOs, grassroots organizations, youth, women's groups, and corporations and other business organizations.[72] As Rodenhoff points out, the term 'public' is used in Aarhus to mean 'the sum total of all of society's potential actors'.[73]

This expansive notion of participation in Aarhus allows a larger section of the society to take part in decision-making processes. By carrying along as many members of the public as possible, carbon projects can avoid the negative publicity, protests and pitfalls caused by ignorance or lack of knowledge by the public on the details of a project. Arguably therefore, members of the public that should take part in, and influence, decision-making processes on the design and approval of carbon projects include individuals, landowners, community leaders, corporations and business organizations, and representatives from non-governmental organizations and interests groups. These categories should be recognized as important 'key stakeholders' and the public concerned with respect to any carbon project.

This definition of the public assigns a prominent role to NGOs to take part in decision-making processes. It is important to include NGOs as members of the public that can participate in environmental decision-making processes from the outset. As noted earlier, despite genuine concerns on the difficulties inherent in an overly expansive and untempered participation, especially the tendency of NGOs and interest groups to slow down negotiation processes due to pre-conceived standpoints,[74]

[72] See Arts. 2 and 4 of the Aarhus Convention on Access to Information, Public Participation in Decision-Making and Access to Justice in Environmental Matters (Adopted 25 June 1998, entered into force 30 October, 2001).

[73] V. Rodenhoff, 'The Aarhus Convention and its Implications for the "Institutions" of the European Community', *Review of European Community and International Environmental Law*, 11(3) (2002), 344.

[74] For example, companies usually prefer quick processes with long term commitments, whereas NGOs and local communities often advocate for extensive FPIC processes with flexible outcomes that remain open for adjustments. These differing perspectives on FPIC may result in excessive delays in participatory processes. See M. Lee, C. Armeni, J. De Cendra, S. Chaytor, S. Lock, M. Maslin, C. Redgwell and Y. Rydin, 'Public Participation and Climate Change Infrastructure' *Journal of Environmental Law* (2012), where the authors discuss how untempered public participation might become a 'simple bureaucratic hurdle, frustrating for all concerned, where the authors demonstrate how participation might become a 'simple bureaucratic hurdle', 'frustrating for all concerned'. See also Forest Stewardship Council, 'FSC Guidelines for the Implementation of the Right to Free, Prior and Informed Consent (FPIC)' (Bonn: FSC International Center, 2012) 14–15; M. Lee and C. Abbot, 'The Usual Suspects? Public Participation under the Aarhus Convention', *Modern Law Review*, 66 (2003), 80; Sunstein, 'Deliberative Trouble? Why

broad-based participation by NGOs and interests groups is very useful in monitoring decision-making processes and in raising awareness. An effective participatory process tries to involve important players from the outset. In many countries where carbon projects are resisted or opposed, these oppositions are often spearheaded by NGOs, for example the Environmental Rights Action (ERA) in Nigeria. As such, to gain the support and understanding of all stakeholders for a project, it is important that such stakeholders are part of the planning process from the beginning.

Similarly, NGOs are most times closer to the local communities and as such are able to understand and better frame local concerns and priorities. This is especially true in many developing countries where a large percentage of the society is either uneducated or uninterested. NGOs could help source for local knowledge that could further improve and enrich the decision-making processes. As such, the aim is not for NGOs to act as stumbling blocks or opponents of viable projects, rather NGOs would provide the appropriate check and balances that could help improve the design and implementation of projects to reduce human rights problems.

Considering the importance of participation in protecting human rights, the goal should not be to avoid FPIC or draw down its influence in order to allow projects to move on without hurdles or delay. Rather, the aim should be to set appropriate guidelines and rules that could tackle unnecessary hurdles. While it is true that broad-based participation may allow interest groups with unreasonably high expectations to frustrate the entire deliberative process, the solution is not to gloss over the importance of FPIC. Avoiding or addressing such risks would require proper planning. For example, setting common rules of deliberation and an agenda prior to the deliberative process and keeping strictly to the terms of the set agenda and rules can enhance a participatory process. This will reduce the tendencies for unnecessary deviations and hard stances. For example, a questionnaire could be circulated prior to the discussions to sample general opinion on issues to be discussed. The results of the opinion poll will form the agenda while public authorities such as DNAs would also adopt transparent guidelines on the deliberative process itself.

Groups Go To Extremes', *Yale Law Journal*, 110 (2000), 71; Black, 'Proceduralizing Regulation: Part I', *Oxford Journal of Legal Studies*, 20 (2000), 597, and 'Proceduralizing Regulation: Part II', *Oxford Journal of Legal Studies*, 33 (2001), 33.

Fairness and inclusivity An effective FPIC process must ensure an extensive and inclusive engagement with all stakeholders or local communities that may be affected by a project. Here, focus is on the need to provide fair opportunities for key stakeholders to attend decision-making meetings. An effective decision-making process must not exclude vulnerable members of the society who are interested in a project from taking part in decision-making.[75] Inclusivity means members of the public must have equal opportunities and the chance to have an input during project planning and approval. Inclusivity includes the notion of fairness, which stipulates that all those having an interest in or affected by a decision should have a say in that decision.[76] For example, under the Aarhus Convention, once an interest is established, it does not matter whether the concerned party is a citizen or not. Article 3(9) of the Convention adds that 'the public' shall have the opportunity to participate in decision-making 'without discrimination as to citizenship, nationality or domicile and, in the case of a legal person, without discrimination as to where it has its registered seat or an effective centre of its activities'. Thus, members of the public need not be nationals of the member state in question nor need they be permanently resident in the state concerned. Processes that exclude the public based on citizenship, nationality, domicile or status tend to shut the door of participation against some members of the public that could be affected by a project.

Fairness also includes providing means for stakeholders to take part in decision-making. To enable the public to take part in decision-making processes, barriers to access must be removed. In developing countries, for example, barriers range from simple issues such as distance of the meetings, fear of persecution by agents of the government and also language and technological barriers. For example, holding meetings in the English language in communities where local indigenes can only speak their traditional languages. This denies them the chance to participate. Another example is when the lack of knowledge of technical jargon prevents participants from being able to make real contributions to project plans and designs.[77]

[75] See T. Dietz, 'What is a Good Decision? Criteria for Environmental Decision-Making', *Human Ecology Review*, 10(1) (2003), 33, 35.

[76] T. Webler and S. Tuler, 'Fairness and Competence in Citizen Participation: Theoretical reflections from a Case Study', *Administration and Society*, 32 (2000), 5, 570.

[77] See F. Onojiribholo, 'Kwale Chief Laments Plight of Communities' (*Daily Independent* 16 August 2011). S. Foster, 'Justice from the Ground Up: Distributive Inequities,

To remove barriers to participation during project planning, national authorities must actively facilitate broad-based participation. It is not enough to send out invitations to attend deliberative meetings, DNAs must also facilitate attendance by selecting central and easily accessible venues for meetings and by providing means of ensuring participation, for example providing free transportation for the poor. Inclusion requires pulling down all sorts of artificial barriers to participation.[78] For example, to achieve broad-based participation during the approval process for a project, DNAs would have to play more facilitative roles by reducing any advantage due to class that allows some individuals to influence the outcome of a deliberative process while excluding the poor and the illiterates. DNAs could create a level field by hiring language interpreters for locals, and by reducing technicalities in discussions. Technological advancements could also be utilized by providing opportunities for online participation through online surveys and questionnaires.[79]

Rationality and freedom to make choices There is a need to provide participants the chance to freely consider and explore different options during the deliberative process, and to make an informed choice without fear or force. This involves identifying multiple options that reflect different sets of values held by different sectors of the society and weighing arguments for and against each option through an indefinite period of extensive and continuing discussions.[80] As such, a rational process provides participants with the opportunity to consider and give opinion on the strengths and weaknesses of the various options and project scenarios. The public should be able to consider the strengths and weaknesses of a project proposal and then decide whether the proposal should

Grassroots Resistance and the Transformative Politics of the Environmental Justice Movement', *California Law Review* 85 (1998), 775, 812.

78 As Macias rightly notes, it is hard to speak of fairness when some members of the public cannot afford to attend important decision-making meetings. See R. Macias, *Public Participation in Energy and Natural Resources Development: A Theory and Criteria for Evaluation* (CIRL, 2010), 20.

79 For detailed discussions on the value of e-rulemaking, S. Novek, 'Electronic Revolution in Rule Making', *Emory Law Journal*, 53 (2004), 434; see also C. Coglianese, 'Citizen Participation in Rule Making: Past, Present and Future', *Duke Law Journal*, 55 (2006), 943; A. Fung and E. Olin Wrights (eds.) *Deepening Democracy: Institutional Innovations in Empowered Participatory Governance* (London; New York: Verso, 2003).

80 This point was elaborated extensively by Hartz-Karp and Briand, 'Institutionalizing Deliberative Democracy', (n 70), 127.

stand as it is or whether extensive revisions should be made to its design. Participation should not be framed as a winner takes all or loser loses all scenario. For example, the Aarhus Convention provides that the public should be informed of the nature of possible decisions.[81] It also provides that procedures for public participation shall allow the public to submit, in writing or, as appropriate, at a public hearing or inquiry with the applicant, any comments, information, analyses or opinions that it considers relevant to the proposed activity.[82] Allowing the public to consider a wide range of options and project design scenarios enhances the abilities of the public to make informed and fact-based decisions, thereby enhancing the overall utility and value of the deliberative process.

Generally, DNAs should be required to demonstrate the extent of proposals received and considered before taking a decision to approve a carbon project. It is not enough to tick a box to the effect that a participatory process has been conducted. Compliance with the requirement of participation must be anchored on demonstrating the extent of consultation carried out before reaching a decision to execute a project.

Agreement An effective deliberative process should produce a 'final agreed upon' outcome that reflects the freely expressed consent of the public. This is arguably what distinguishes participation from mere consultation. In the absence of an agreed outcome, the process becomes a one-way formality in which stakeholders are merely informed about an action or project without any real opportunity to influence the final outcome. There must be a measure of agreement that most participants in the deliberative process can accept as a 'decision'.[83] The finality of the process rests in adopting a binding decision that clearly reflects whether consent has been freely given for a project to go ahead or not. For example, Article 6(9) of the Aarhus Convention provides that each party shall ensure that, when the decision has been taken by the public authority, the public is promptly informed of the decision in accordance with the appropriate procedures. Each party shall make accessible to the public the text of the decision along with the reasons and considerations on which the decision is based. This element illustrates the requirement for conclusiveness in a deliberative process. In an authentic deliberative process, not only are matters brought to the fore and discussed, it should

[81] Article 6(2) (b) of the Aarhus Convention, (n 72).
[82] Article 6(7) of the Aarhus Convention, *ibid.*
[83] See Hartz-Karp and Briand, 'Institutionalizing Deliberative Democracy', (n 70), 127.

culminate in a final, agreed upon outcome, which will resolve all conflicts and leave a majority if not all the participants satisfied that they took part in reaching that final outcome.

5.2.2.3 Influencing

Following our definition of participation, that is the right to take part and influence the decision-making process, the second aspect is influencing a decision making process. This is a more straightforward element of the definition. Corell and Betsill rightly stipulate that influence occurs 'when one actor intentionally communicates to another so as to alter the latter's behaviour from what would have occurred otherwise'.[84] Influencing a decision means to have an effect on the final outcome. As such, expressing an opinion and taking a final decision is not the end; it is the ability of the public to have an effect on the final decision that differentiates mere consultation from participation.

Agreements emanating from a deliberative process must be reflected in governmental actions with respect to granting a project host country approval. Reflecting public opinion in decision-making engenders public trust in the entire deliberative process. For example, Article 6(8) of the Aarhus Convention provides that each party shall ensure that in the decision, due account is taken of the outcome of the public participation. The outcome must reflect the final agreed outcome such that the public will be encouraged, when next they are called upon, to give an opinion on a project proposal. The final decision should be easily traceable to the deliberative process.[85] For example, if the public vote against a project, an effective participatory process must mean that such project would not go on as designed, and will be modified to demonstrably address concerns raised during the participatory process.

In essence, not all consultative processes can be classified as meeting the threshold of an effective participatory process. The two elements of 'taking part' and 'influence' are aspects of participation that should be demonstrated and met during the design and approval of mitigation projects. When these two elements are reflected in decision-making

[84] See M. Betsill and E. Corell, 'Analytical Framework: Assessing the Influence of NGO Diplomats' in M. Betsill and E. Corell (eds.), *NGO Diplomacy: The Influence of Nongovernmental Organization in International Environmental Agreements* (Boston: Massachusetts Institute of Technology Press, 2008), 19; see also M. Betsill, 'Environmental NGOs and the Kyoto Protocol Negotiations: 1995 to 1997', in M. Betsill and E. Corell (eds.), *NGO Diplomacy*, (Boston: Massachusetts Institute of Technology Press, 2008), 177–206.

[85] See Hartz-Karp and Briand, 'Institutionalizing Deliberative Democracy', (n 70), 127.

processes, the will of the people is protected and the fundamental human rights of citizens to participate in public affairs are better preserved. Similarly, since such decisions are collectively taken, it could reduce the level of opposition and resistance facing carbon projects and actions. There is a need to place the responsibility on national governments and project proponents to demonstrate that the public have been given adequate opportunities to take part and influence the decision-making processes on whether, or how a mitigation project should be designed and executed. This should be mainstreamed as a compulsory criterion for project approval and registration.

5.2.3 Non-discrimination, equality and prioritization of vulnerable groups

One of the main concerns identified in the execution of carbon projects is the issue of siting and concentrating projects in poor and vulnerable communities.[86] These communities are particularly vulnerable because they are perceived as weak and passive citizens, who will not fight back against the poisoning of their neighbourhoods in fear that it may jeopardize jobs and economic survival.[87] In Nigeria for example, CDM projects have been sited in poor and disadvantaged communities where citizens do not have the means and resources to demand accountability or resist such projects.[88] Earlier sections of this book have discussed core international human rights instruments that show that such target selection in project siting violates human rights safeguards on equality and

[86] International Rivers, 'Submission Regarding Human Rights to the CDM' (March 25 2013), www.internationalrivers.org/resources/submission-regarding-human-rights-to-the-cdm-7899, accessed 1 April 2015; see also Inter-American Commission on Human Rights (IACHR), Hearings and other public events, 'Human Rights Situation in the Bajo Aguán Honduras', 24 October 2011; see also IACHR, 2011 Annual Report, ch. IV, 297–302; IACHR, 2010, Country Report, Honduras, 118–121; International Rivers, 'Letter to the CDM Executive Board Regarding the Barro Blanco Hydroelectric Project' (9 February 2011), www.internationalrivers.org/resources/letter-to-the-cdm-executive-board-regarding-the-barro-blanco-hydroelectric-project-3078, accessed 1 April 2015.

[87] See generally S. Owens, 'Siting, Sustainable Development and Social Priorities', *Journal of Risk Research*, 7 (2004), 101.

[88] N. Bassey, 'Foreword', in *Mired in a Fossil Trap, the Nigerian CDM Report* (Environmental Rights Action/Friends of the Earth, May 2011); Carbon Trade Watch, 'Groups slam Nigeria's submission of gas flare reductions for carbon credits' (2006), www.carbontradewatch.org/index.php?option=com_content&task=view&id=171&Itemid=36, accessed 12 October 2014; Onojiribholo, 'Kwale Chief Laments Plight of Communities', (n 77).

non-discrimination. This section will take the analyses further by exploring the scope and contents of the human rights to equality and non-discrimination and how it can be reflected in planning carbon projects.

Without a human rights perspective, carbon projects risk exacerbating the already sizeable gulf between rich and poor. As the United Nations Human Right Council recognized in its resolution 16/11, 'environmental damage is felt most acutely by those segments of the population already in vulnerable situations'.[89] Vulnerable groups include women, children, racial and ethnic minorities, migrants and non-citizens, refugees, indigenous peoples, and those living in extreme poverty, who bear the most disproportionate burden of climate actions and projects.

HRBA underscores the need for the equal treatment of people in planning. It rejects policies that marginalize people based on income, sex, ethnicity or race, age or nationality. According to the common understanding on HRBA, all people are equal as human beings and by virtue of their inherent dignity. All human beings are entitled to their human rights without discrimination of any kind, such as race, colour, sex, ethnicity, age, language, religion, political or other opinion, national or social origin, disability, property, birth or other status as explained by the human rights treaty bodies.[90] This element of the HRBA, if applied to carbon projects, would seek to remedy situations whereby projects are deliberately sited in poor and vulnerable communities. Generally, under this element, DNAs and project proponents would be required to demonstrate that extensive data segregation and impact assessment have been carried out and that a particular locality or section of society is not targeted for projects.

5.2.3.1 Elements of equality and non-discrimination

Equality of opportunity and treatment To realize, protect and fulfil non-discrimination norms of international human rights law, marginalized groups, particularly women and youths, must be accorded equal rights and opportunities to participate effectively in public life.

[89] United Nations Human Rights Council, *Resolution adopted by the Human Rights Council 16/11, Human rights and the environment*, stating in its preamble that: 'Recognizing that, while these implications affect individuals and communities around the world, environmental damage is felt most acutely by those segments of the population already in vulnerable situations.'

[90] United Nations, 'The Human Rights Based Approach to Development Cooperation', (n 1).

Furthermore, it must be ensured that no decisions directly affecting the rights and interests of such vulnerable groups are taken without their informed consent.[91] This is by ensuring the inclusion of all relevant stakeholders including the vulnerable members of the society in decision-making and by providing equal opportunity to them to oppose or support a policy or project. Inclusivity focuses on the need to provide a fair opportunity for stakeholders and representatives of diverse societal groups or interests to attend decision-making meetings. Inclusivity abhors discriminatory practices or policies that exclude vulnerable members of the society.

Equality of treatment also entails the idea of addressing practices or actions to stifle the abilities of vulnerable groups to effectively enjoy their fundamental human rights. As a Report of the Independent Expert on human rights and poverty rightly notes, due to discrimination, groups such as women, children, racial and ethnic minorities, migrants and non-citizens, refugees, indigenous peoples, persons with disabilities and older persons encounter systematic stigmatization and discrimination on the grounds of their poverty which makes it difficult for them to defend and enjoy their human rights.[92] Governments and stakeholders in carbon projects must design and adopt policies aimed at removing all barriers to inclusivity in the design and implementation of carbon projects. Inclusivity includes the notion of fairness, which emphasizes the need to protect and defend the abilities to enjoy their human rights, like other members of the society. Project implementation and planning must be geared towards removing all artificial barriers that unnecessarily stifle the abilities of vulnerable groups to access project information, participate in decision-making processes or obtain remedy when projects negatively impact them.

For example, though not legally binding, the UN 'Protect, Respect and Remedy' Framework and Guiding Principles provides detailed and persuasive explanation of how discriminatory trends can be avoided in project planning and execution.[93] It provides that business enterprises should ensure equality of opportunity and treatments as recognized in international law to ensure that all forms of discrimination are eliminated

[91] *Ibid.*, para. 4(d).

[92] Report of the independent expert on the question of human rights and extreme poverty, Magdalena Sepúlveda Carmona, on the draft guiding principles on extreme poverty and human rights, A/HRC/15/41(6 August 2010) para. 22.

[93] United Nations, Guiding Principles on Business and Human Rights: Implementing the United Nations 'Protect, Respect and Remedy', www.ohchr.org/Documents/Publications/GuidingPrinciplesBusinessHR_EN.pdf

from business related activities.[94] The Framework, in paragraph 26, specifically notes that 'States should take appropriate steps to ensure the effectiveness of domestic judicial mechanisms when addressing business-related human rights abuses, including considering ways to reduce legal, practical and other relevant barriers that could lead to a denial of access to remedy.' The Commentaries to this paragraph rightly identify that due to imbalance in power relations between the parties to business-related human rights claims; active discrimination; or unintended consequences of the way judicial mechanisms are designed and operate, individuals from groups or populations at heightened risk of vulnerability or marginalization often face additional cultural, social, physical and financial impediments to accessing, using and benefiting from justice mechanisms. The Framework notes that 'particular attention should be given to the rights and specific needs of such groups or populations at each stage of the remedial process: access, procedures and outcome'. The guiding principles provide normative guidance on the sense in redesigning project approval frameworks to ensure that the fundamental rights and freedoms of poor and vulnerable groups are protected, realized and fulfilled in the design, approval, finance and implementation of climate projects.

Governments must make conscious efforts to create a level playing field in project planning, deliberative processes and in the implementation of carbon projects so as to defend the rights of vulnerable groups. For example, barriers to participation can be removed either by choosing accessible venues for meetings or scheduling meetings in different areas of the community that are close to main centres, providing free transportation, hiring language interpreters for locals, and by reducing technicalities in discussions. Project proponents and national authorities may also utilize technology to provide an opportunity for online participation through webinars, online surveys and questionnaires.[95]

Impact assessment and burden sharing The impacts of climate actions on poor and marginalized groups also manifest in the form of lopsided projects or actions that place disproportionate burdens on poor and marginalized groups. Under a rights-based framework, governmental decisions or projects that tend to imperil the life, safety and health of

94 Guiding Principles on Business and Human Rights, *ibid.*

95 For detailed discussions on the value of e-rulemaking, see Novek, 'Electronic Revolution in Rule Making', (n 79), 434; see also Coglianese, 'Citizen Participation in Rule Making', (n 79), 943; Fung and Olin Wrights (eds.) *Deepening Democracy*, (n 79).

a section of the society must be identified and avoided. No person or group should suffer more societal harm from the project as a result of status or gender. For example, targeting poor communities as locations for carbon projects is a discriminatory practice as it subjects the poor to greater harm due to their societal status.[96] A rights-based framework will ensure that the burden of carbon projects is spread across every segment of the society and shared equally and do not excessively fall on vulnerable racial, ethnic and socio-economic groups.

Furthermore, a rights-based framework recognizes that the inequitable distribution of climate burden is a violation of international law norms on equality and non-discrimination. As such the Not-in-My-Backyard (NIMBY) syndrome must be eradicated from decision-making on mitigation projects.[97] For example, the dominant theme in the UN Guiding Principles is that policies and frameworks 'should be implemented in a non-discriminatory manner, with particular attention to the rights and needs of, as well as the challenges faced by, individuals from groups or populations that may be at heightened risk of becoming vulnerable or marginalized, and with due regard to the different risks that may be faced by women and men.'[98] It also notes that 'business enterprises should pay special attention to any particular human rights impacts on individuals

[96] A good example of discrimination in environmental planning was the alleged and highly litigated incidents of environmental racism in the United States in the late 1990s. It included the intentional siting of hazardous waste sites, landfills, incinerators, and polluting industries in communities inhabited mainly by African-American, Hispanics, Native Americans, Asians, migrant farm workers, and the working poor. For example the largest hazardous waste landfill in the United States is located in Emelle, Alabama, a poor, predominantly African-American community. It receives toxic materials from forty-five states and several foreign countries; also the South Side of Chicago, which is predominantly African-American and Hispanic, has the greatest concentration of hazardous waste sites in the nation. See C. Beasley 'Of Pollution and Poverty', *The Environmental Journal*, 2(4) (1990), 38–45; R. Bullard, *Confronting Environmental Racism: Voices from the Grassroots* (Cambridge, MA: South End Press, 1993); K. Colquette, K. Michele, A. Elizabeth, and H. Robertson 'Environmental Racism: The Causes, Consequences, and Commendations', *Tulane Environmental Law Journal*, 5(1) (1991), 153–207.

[97] NIMBY or Nimbyism is a term coined to describe the injustices of the early 1980s in the United States where poor and black neighbourhoods were selected for toxic and dangerous wastes and projects. Opposing residents are sometimes called Nimbies. See R. Bullard, *Dumping in Dixie: Race, Class, and Environmental Quality*, 3rd edn (Boulder, CO: Westview Press, 2002). According to Bullard, because low-income and minority communities in the US have had few advocates and lobbyists at the national level, social equity and distributive impacts of pollution have not been met hence the disproportionate amount of pollution and other environmental stressors in black neighbourhoods in the South.

[98] See UN Guiding Principles, page 1.

from groups or populations that may be at heightened risk of vulnerability or marginalization, and bear in mind the different risks that may be faced by women and men.[99] Though not legally binding, these norms accentuate the importance of sharing the burden of climate projects across the society to avoid exclusions, disproportionate burden and discrimination. As Robinson rightly notes, 'a climate justice approach can ensure that rights are protected and that the actions taken to, solve the climate crisis benefit, rather than burden, the vulnerable who are after all, least responsible for the causes of the global problem'.[100]

Implementing non-discriminatory policies would include laying down clear criteria for selecting project locations, conducting social and human rights impact assessment to clearly determine how a project may disproportionately affect the rights of a group. Through a Human Rights Impact Assessment (HRIA) project proponents could systematically identify, anticipate and respond to the potential human rights impact of a project on vulnerable groups.[101] As extensively analysed in other parts of this book, the key questions under a HRIA are as follows: what human rights concerns does this project raise or could it raise? Which groups are likely to be affected by these human rights concerns? Which specific rights are affected by this project or policy and what efforts would be taken to address these issues in the process of project design and implementation? A HRIA would provide clear and comprehensive answers to these questions to avoid any element of surprise or secrecy in project implementation. The HRIA goes beyond only evaluations or human rights auditing in that it identifies the areas of overlap between human rights and a project, and also identifies holistic solutions and methods for avoiding the identified human rights impacts on vulnerable groups.

HRIA would also require project proponents to demonstrate the fair distribution of projected risks. It would help to assess the implications of projects for low-income people, the representation afforded to such categories in decision-making and ways to protect the interests of the marginalized through project re-design or alternative implementation

[99] See UN Guiding Principles, para. 18.

[100] Mary Robinson Foundation, 'On Climate Justice, Human Rights and Climate Justice-A Position Paper', www.mrfcj.org/media/pdf/PositionPaperHumanRightsandClimateChange.pdf, accessed 12 December 2015.

[101] S. Walker, 'Human Rights Impact Assessments of Trade-Related Policies', in M. Gehring and M.C. Cordonier-Segger, *Sustainable Development in World Trade Law* (The Hague: Kluwer Law International, 2005) 217–256.

plans. It would make it compulsory for project proponents to demonstrate that a particular section of the society is not exceptionally disadvantaged by a project before the said project could be approved. By developing concrete risk assessment procedures to ensure better characterization of risk across populations, communities or geographic areas, measures would then be put in place to reduce high concentrations of risk among specific population groups.

Data segregation and vulnerability proofing Vulnerability proofing is a process aimed at identifying and eliminating human rights risks that a carbon project may place on sections of the community. It refers to actions taken to lessen, or perhaps eliminate, the potential negative impacts of carbon projects on poor and vulnerable communities through the life cycle of a project.[102] The aim is to reduce risks to acceptable levels through environmentally sound, economically viable and justice-based changes implemented at key stages of the project: planning, design, approval, finance and implementation.

To prevent projects and actions that discriminate based on income, status or gender, there is the need to establish and maintain an effective database for assessing risks on the basis of income, gender roles and the constraints operating against their participation in decision-making processes.[103] Project proponents and stakeholders can address the issue of non-discrimination by ensuring that official data is disaggregated, by religion, ethnicity, language, sex, migrant status, age and any other category of human rights concerns and risks. This also includes incorporating proofing as part of the wider climate governance architecture – that is proofing of all climate actions, plans and measures. Gender and income-proofing will help to assess the implications for low income people and women of any planned

[102] This definition borrows from the definition of climate proofing that has been canvassed by the ADB. See Asian Development Bank (ADB), *Climate Proofing ADB Investment in the Transport Sector: Initial Experience* (Manila: ADB, 2014), 10–15. See also Asian Development Bank (ADB), *Climate Proofing: A Risk-Based Approach to Adaptation* (Manila: ADB, 2005).

[103] F. Francioni, 'Access to Justice, Denial of Justice and International Investment Law', *EJIL*, 20 (2009) 729, ejil.oxfordjournals.org/cgi/reprint/20/3/729, accessed 2 April 2011; M. Cappelletti and B. Garth (eds.), *Access to Justice: The Worldwide Movement to Make Rights Effective* (Maurer School of Law, Indiana University: Faculty Publications Paper, 1978); J. Paulsson, *Denial of Justice in International Law* (Cambridge University Press, 2005), 134; L. Curran and M. Anne Noone 'Access to Justice: A New Approach using Human Rights Standards' *International Journal of the Legal Profession*, 15 (3) (2008), 195–229.

project, the representation afforded to such categories in decision-making and ways to protect the interests of the marginalized either by re-designing the project to avoid harmful impacts or refusing to proceed with it if such impacts cannot be effectively addressed.

An effective vulnerability proofing will rely on a HRIA to identify vulnerability and to characterize project risks across populations, communities or geographic areas. The HRIA will among other things identify the human rights concerns a project raises or could raise, Which groups are likely to be affected by these human rights concerns? Which specific rights are affected by this project or policy? What efforts would be taken to address these issues in the process of project design and implementation. Based on the identified spread of risks, measures would then be put in place to reduce high concentrations of risk among specific population groups. Vulnerability proofing provides a framework for considering and internalizing the risks and opportunities into the design, approval, finance and implementation of climate projects.

To be effective, a HRIA should be initiated as early as possible, at project planning and design stages. Conducting a HRIA should be incorporated as a pre-condition for project approvals for participants involved in climate projects ranging from project proponents, supervisory institutions, financiers, auditors, verifiers and service providers such as law firms and consulting agencies. This would ensure that any human rights impact would be put into consideration in planning and designing a project. It would also provide sufficient time for project proponents and participants to avoid the known consequences of a project. Where impact assessment indicates the possibility of potential human rights violations, it provides time for the project design to be revised to remove any incompatibility that has been found with pre-existing human rights obligations of the state concerned. This will prevent a situation whereby violations are detected only after a project has been commenced or completed. It provides an opportunity to gather relevant data, anticipate and respond to the likely impact of a project on the human rights of affected groups.

Not every possible impact of a project can be anticipated at the planning stage. As such, *ex ante* human rights impact assessments should be complemented by human rights impact assessments performed *ex post*, once the effects can be quantified. HRIA should be an iterative process, taking place on a regular basis, for instance, every three or five years. Safeguarding clauses should be included to ensure that, should such *ex post* assessments lead to the conclusion that if a project proponent is unable to comply with its human rights obligations under the project scenario, registration for such projects would be re-appraised and if necessary withdrawn.[104]

To address vulnerability, the international climate change regime must mandate project proponents to demonstrate that a particular section of the society is not disadvantaged by a project before such project could be approved. Furthermore, when considering projects for approval, project proponents should be required to demonstrate the fair and equitable distribution of projected risks.

5.2.4 Accountability

Another concern in the execution of carbon projects is the lack of accountability and transparency in the process of approving and registering projects. This is mainly due to the absence of concrete project guidelines and mechanisms that would require countries to demonstrate that all short and long term human rights impacts of a project have been considered before embarking on the project. Generally, accountability has been described as the obligation to demonstrate that a project has been conducted in accordance with agreed rules and standards and to report fairly and accurately on performance results vis-à-vis mandated roles and/or plans.[105] Accountability encompasses the obligation of the international climate change regime to review, monitor and enforce compliance with human rights standards and obligations in the design and execution of carbon projects.[106] It comprises the structural conditions, processes and indicators/outcomes through which climate change governance systems review and monitor the practical impacts of a project on human rights of the public. Consequently, an accountable regime measures, reviews and monitors the level at which human rights are respected and fulfilled in the implementation of its programmes and mandates.

104 See OHCHR, 'Guiding Principles on Human Rights Impact Assessments of Trade and Investment Agreements', www2.ohchr.org/english/issues/food/docs/20110701 Draft_Guiding_Principles_on_HRIA.pdf, accessed 23 October 2015.

105 Organization for Economic Cooperation and Development (OECD), 'Public Sector Transparency and Accountability: Making it Happen' (OECD, 2002), 7.

106 See One World Trust, *2011 Pathways to Accountability II, The Revised Global Accountability Framework*, oneworldtrust.org/publications/doc_download/470-pathways-to-accountability-ii, accessed 12 March 2015.

Accountability in human rights terms is measured by a set of indicators, which include identifying the unintended impact of laws, policies and practices on human rights; identifying which actors are having an impact on the realization of rights; revealing whether the obligations of these actors are being met; giving early warning of potential violations, prompting preventive action; enhancing social consensus on difficult trade-offs to be made in the face of resource constraints; and exposing issues that had been neglected or silenced.[107] Virtually every human rights instrument creates mechanisms for monitoring compliance and for reporting violations. The Aarhus Convention is an example of an environmental agreement that also establishes compliance mechanisms that monitor the protection and fulfilment of procedural human rights.[108]

Many human rights treaty bodies monitor implementation through the creation of additional reporting requirements for states.[109] The Commission on Human Rights also establishes special rapporteurs, or expert committees and working groups to gather human rights compliance information and to recommend actions for non-compliance. Though such recommended actions and reports are often not legally binding, they provide guidance and reference points on the mechanisms for ensuring greater accountability in the performance of human rights obligations. Similarly, the United Nations has in different non-binding resolutions identified accountability as part of the founding principles of transparency and effective public administration.[110] The UN General Assembly, for instance, has adopted Resolution 60/260 on Accountability.[111] This resolution emphasized the importance of strengthened accountability within the UN and the need for all UN

107 S. Lankford and H. Sano, *Human Rights Indicators in Development: An Introduction* (World Bank, 2010); see also UNDP, 'Using Indicators for Human Rights Accountability', hdr.undp.org/en/media/hdr_2000_ch5.pdf, accessed 12 March 2015.

108 For example, Article 15 of the Aarhus Convention provides that the Meeting of the Parties shall establish, on a consensus basis, optional arrangements of a non-confrontational, non-judicial and consultative nature for reviewing compliance with the provisions of this Convention. These arrangements shall allow for appropriate public involvement and may include the option of considering communications from members of the public on matters related to the Convention.

109 See United Nations, 'Delegation of Authority and Accountability', Report of the UN Joint Inspection Unit (2004), JIU/REP/2004/7, 13.

110 See A/RES/49/136 of 1994, A/RES/50/225 of 1996, A/RES/53/201 of 1999, A/RES/56/213 of 2002, A/RES/57/277 of 2002 and A/RES/58/231 of 2004.

111 United Nations, *Investing in the United Nations: For a Stronger Organization Worldwide*, Resolution of the United Nations General Assembly, 16 May 2006, A/RES/60/260.

agencies to ensure greater accountability within their spheres of operation for the effective and efficient implementation of legislative mandates and the best use of human and financial resources.

Despite these, however, the lack of accountability within the UN systems and within international development agencies has received scholarly attention over the last decade.[112] Accountability concerns include the prevalent development practices that fail to hold corporations liable for human rights violations in project development.[113] There is also the failure by UN organs and agencies to establish clear and compulsory human rights benchmarks that must be met before projects are approved or funded.[114] For example, the CDM EB has been variously criticized for its lack of accountability and transparency, in registering and failing to hold countries to account for projects that produce negative human rights impacts.[115] This has led to an increased call for an approach that hinges climate change mitigation on the respect for the human rights requirements on accountability.[116]

There is a need to mainstream human rights-based benchmarks or indicators into rules and procedures governing the approval and execution of carbon projects, such that compliance with human rights obligations would be measured side by side with emission reductions. This

[112] See E.B. Weiss, 'Accountability and International Law: Reflections from Water Projects' A lecture delivered at the University of Oxford on 23 February, 2015; see also O. Hathaway, 'Do Human Rights Treaties Make a Difference?', *Yale Law Journal*, 11(8) 2002); S. Kuyama and M. Fowler (eds.), *Envisioning Reform: Enhancing UN Accountability in the Twenty-first Century* (United Nations University Press, 2009), 5–8.

[113] See E.B. Weiss, 'On Being Accountable in a Kaleidoscopic World', *Asian Journal of International Law*, 1 (2011), 21–32; see also E.B. Weiss, P. Lallas and A. Herken, 'The World Bank Inspection Panel: Participation and Accountability' in Kuyama and Fowler (eds), *Envisioning Reform: Enhancing UN Accountability in the Twenty-first Century* (United Nations University Press, 2009), 271.

[114] See R. Keohane, 'Global Governance and Democratic Accountability', in D. Held and M. Koenig-Archibugi (eds.) *Global Governance and Public Accountability* (Oxford: Blackwell Publishing, 2005); see also S. Nanwani, 'Improving Accountability at the Asian Development Bank' in Kuyama and Fowler, (eds.), *Envisioning Reform*, 271.

[115] See A. Filmozer, 'CDM Projects Affect Human Rights' (February 2011), www.cdm-watch.org/wordpress/wp-content/, accessed 2 August 2014 (see also Climate Connections, 'Carbon Trade Group Backs Call to check Credits on Human Rights', climate-connections.org/2011/04/18/carbon-trade-group-backs-call-to-check-credits-on-human-rights/, accessed 2 August 2014.

[116] See Weiss, 'Accountability and International Law', (n 112); see also Hathaway, 'Do Human Rights Treaties Make a Difference?', (n 112); A. Johl, and S. Duyck, 'Promoting Human Rights in the Future Climate Regime', *Ethics, Policy and Environment*, 15(3) (2012), 300; D. Olawuyi, 'Towards a Transparent and Accountable Clean Development Mechanism Mechanism: Legal and Institutional Imperatives', *Nordic Environmental Law Journal*, 2(2) (2012), 33–52.

would also include establishing reward and sanctions for success and failure in complying with established benchmarks.[117] Accountability benchmarks would provide stakeholders with an opportunity to understand how national governments, project proponents and DNAs have discharged, or failed to discharge, their human rights obligations; and an opportunity for these public bodies to demonstrate their human rights compliance through mandatory compliance/implementation reports.[118]

5.2.4.1 Elements/indicators of accountability

Human rights indicators measure human rights realization, both qualitatively and quantitatively.[119] They help to assess the state of human rights implementation and enjoyment by public bodies. Measuring human rights can also help to allocate responsibility for human rights outcomes, that is, naming and shaming who is responsible for what.[120] Human rights measurements therefore provide clear and precise indication of areas of progress and areas where progress is lacking.

While scholars agree on the importance of measuring accountability using a set of clearly identified indicators, creating such human rights-based indicators has remained a theoretical and practical challenge.[121]

117 B. Andreassen and H.O. Sano, *Human Rights Based Approaches Indicators* (Norwegian Centre for Human Rights, 2004), 1.

118 See A. Kuper (ed.), *Global Responsibilities: Who Must Deliver on Human Rights?* (New York: Routledge, 2005).

119 Lankford and Sano, *Human Rights Indicators in Development*, (n 107); Department for International Development, 'How to Note: A Practical Guide to Assessing and Monitoring Human Rights in Country Programmes' (DFID, 2009); M. Malhotra, and N. Fasel, 'Quantitative Human Rights Indicators: A Survey of Major Initiatives' (2005) Background paper for the UN Expert Meeting on Human Rights Indicators, Turku/Abo; Norwegian Agency for Development Cooperation (NORAD), *Handbook in Human Rights Assessment: State Obligations, Awareness and Empowerment* (NORAD, 2001).

120 Transparency International's *Corruption Perceptions Index* has also had 'naming and shaming' success in Nigeria, see M. Besançon, 'Good Governance Rankings: The Art of Measurement', *World Peace Foundation Report*, 36(2) (2003).

121 Limited progress has been made towards developing commonly accepted systems of measuring human rights promotion. For example, there is an ongoing debate on whether the same set of indicators could be used to assess both civil and political, and economic and social rights. This research will not assess the nature and extent of these debates because it is not directly relevant to our scope. For more readings see J. Ackerman, 'Social Accountability for the Public Sector: A Conceptual Discussion' (Paper No 82, The World Bank, 2005); Besançon, 'Good Governance Rankings', *ibid.*; P. Ball, 'Making the Case: Investigating Large Scale Human Rights Violations Using Information Systems and Data Analysis', *Statistical Journal of the United Nations Economic Commission for Europe*, 18(3) (2001), 163–174; T. Landman, 'Measuring

The UN Special Rapporteur on the right to health, Paul Hunt, has, however, provided an instructive framework, which though not legally binding, could serve as a normative guide on how human rights accountability can be measured. He argues that human rights accountability should be measured in terms of structural conditions, processes, and outcomes.[122]

Structural conditions This measures availability of relevant legal and institutional frameworks that make it possible to monitor compliance with human rights standards. It includes availability of rules and safeguards that prevent human rights violations, and the establishment of relevant institutions to monitor and enforce such rules. At the international level, this speaks to the need to establish and empower compliance committees and review teams that would be responsible for monitoring compliance with human rights. Such institutions must be independent and should not be easily influenced by governmental interests or pressure. There is a need for independent bodies that review project implementation to ensure that established guidelines and thresholds are complied with. For example, through a monitoring committee or expert review team that would inspect project locations to ensure that projects are being executed based on approved guidelines and methodologies.

For example, to facilitate compliance with procedural rights established under the treaty, Article 15 of the Aarhus Convention establishes a compliance committee that is non-confrontational, non-judicial and consultative in approach, to review compliance with the provisions of the Convention. A strong point of the Aarhus compliance mechanism is that it allows members of the public to communicate concerns about a party's compliance directly to an international committee.[123] One or more members of the public can trigger a complaint before the compliance review mechanism concerning any party's compliance with the

Human Rights: Principle, Practice, and Policy', *Human Rights Quarterly*, 26 (2004), 906; UNDP, *The Concept and Measurement of Human Development: UNDP, Human Development Report 2002* (Oxford University Press, 2002).

[122] P. Hunt, Interim report of the Special Rapporteur of the Commission on Human Rights on the Right of Everyone to Enjoy the Highest Attainable Standard of Physical and Mental Health, United Nations General Assembly, Fifty-eighth session, Agenda item 117 (c), 10 October 2003, United Nations General Assembly, paras. 14–29.

[123] Any member of the public, that is, any natural or legal person, may submit a communication to the Committee. The person filing a communication is not required to be a citizen of the state party concerned, or, in the case of an organization, to be based in the Party concerned. See Paragraph 20 of the annex to decision I/7.

Convention.[124] A communication may concern: a general failure by a party to take the necessary legislative, regulatory or other (e.g. institutional, budgetary) measures necessary to implement the Convention; a failure of specific legislation, regulations or other measures implementing the Convention to meet specific requirements of its provisions; Specific events, acts, omissions or situations that demonstrate a failure of the public authorities to comply with or enforce the Convention. The Committee also protects the confidentiality of the complainants when requested, if the communicant is concerned that the disclosure of information submitted to the Committee could result in his or her being penalized, persecuted or harassed, he or she is entitled to request that such information, including any information relating to his or her identity, be kept confidential. The Convention's compliance committee has received several communications from the public – mostly from non-governmental organizations.

However, a major shortcoming of the Aarhus Compliance Committee is that it cannot issue binding decisions; rather, it makes recommendations to the full Meeting of the Parties (MoP). This is problematic because, in practice, MoPs occur infrequently. The independence and effectiveness of a compliance or review body would be better improved if it could implement its findings through a binding decision, such as a stop order on projects or an order for project proponents to redesign a problematic project. Despite this concern, the Aarhus Compliance Committee arguably remains a good example of how a review system could be mainstreamed into the international climate change regime to provide opportunities for receiving feedback from the public about the human rights impacts of projects.[125]

One notable accountability concern raised with the CDM over the last years is that the CDM EB has no mandate and power to monitor or consider human rights compliance; neither does the CDM EB establish human rights indicators.[126] Rather, the CDM EB examines technical

[124] Pursuant to paragraphs 18–24 of the annex to decision I/7 of the first session of the Meeting of the Parties to the Aarhus Convention, communications may be brought before the Committee by one or more members of the public concerning any Party's compliance with the Convention, unless that Party has notified the Depositary in writing by the end of the applicable period that it is unable to accept, for a period of not more than four years, the consideration of such communications by the Committee.

[125] See M. Mason, 'Information Disclosure and Environmental Rights: the Aarhus Convention', *Global Environmental Politics*, 10(3) (2010), 10–31.

[126] See The CDM Executive Board, cdm.unfccc.int/EB/index.html, accessed 12 February 2015.

requirements of projects and whether such projects have met national requirements of the host country. Similarly, the CDM EB is mainly accountable to member states in the discharge of its duties and not to the public. As such, it does have the mandate to take complaints from the public. This gap makes it impossible for individuals to approach the executive board for any projects that affect their human rights, leaving states as the only option for redress. This is a major reason why members of the public have no avenue to bring forward human rights violations linked to CDM projects.

There is a need to establish formal compliance mechanisms under the international climate change regime to measure and monitor human rights compliance in the execution of mitigation projects. This could be by expanding mandates of supervisory institutions for carbon projects, such as the CDM EB, to empower them to monitor human rights compliance. However, identifying human rights safeguards, as part of the overall international climate change regime must at first precede such efforts. Before we speak of compliance, the international climate change regime must first mainstream the obligation to protect and respect human rights as part of efforts to combat climate change. This would then provide a legal basis to expand and reform climate change governance structures. When such structures are established, they can then monitor how countries are respecting human rights in climate change actions. Such governance structures would also create synergy with national efforts by encouraging member states to monitor approved projects and report human rights compliance as part of the periodic reporting requirements which parties have under the international climate change regime.[127]

Process Process measures how established regimes and structures are functioning in practice, whether they merely exist on paper or they actually possess the tools and capacity to ensure the enforcement of human rights. International treaties might condemn lack of participation or gender discrimination and even appoint a committee to monitor the situation, but unless such committee is accessible, independent and functional, such arrangements would mean nothing. As such, it is not

[127] See, for example, the UNFCCC Reporting Guidelines on Annual Inventories require Annex I Parties, to provide by 15 April each year, annual national GHG inventories, including a National Inventory Report (NIR) on the institutional structures and quality assurance and control procedures adopted by Parties in meeting emission reduction commitments. See Articles 4 and 12 of the Convention; see also The UNFCCC Reporting Guidelines on Annual Inventories FCCC/SBSTA/2006/9 (18 August 2006).

enough to include paragraphs into a treaty to establish a compliance mechanism; 'process' measures how effective the existing compliance structures and institutions are in the enforcement of human rights. For example, it would include how functional a project monitoring committee or review team is and how transparent it is in performing its fact-finding and monitoring roles. Part of the task to be undertaken while establishing such review structures is to ensure their accessibility and independence. Such review teams should also be equipped with the resources to perform spot assessments, fact-findings and investigations, such that they could gather first-hand information on the true impacts of a project on the human rights of host communities.

Outcomes This measures progress made in human rights enforcement against set standards and goals; it asks whether, and to what extent, human rights are realized. It tests whether the structural conditions and processes are actually bringing about results in preserving the human rights of the public in project planning and execution. While the outcome may tell us whether human rights are enforced, structural conditions and processes tell us how they are enforced. Outcome indicators assess, for example, the extent to which states are complying with approved project threshold and methodologies. Through site assessments, compliance reviews and mandatory reporting obligations, the international climate change regime would be able to monitor and ensure that approved mitigation projects are executed in line with mandated guidelines. The role of a review panel in measuring outcomes cannot be overemphasized. It is easy for national governments and project proponents to lay claim to transparency and human rights protection during the design and execution of a mitigation project. However, there is a need for the review team to perform oversight functions of gathering on-the-ground facts and to investigate if projects are proceeding as approved.

The three elements of accountability should be implemented to include both compliance review and problem-solving avenues, such that they can deliver redress for human rights violations resulting from projects. Accountability moves beyond examining if a party has complied with its obligations, it also includes reviewing and monitoring the methodology and approaches adopted in doing so. An effective accountability mechanism will also be interested in resolving tensions or concerns arising from the approaches adopted by a country in meeting its emission reduction obligations, particularly in developing countries. For example, the review structure of the international legal regime on climate change should not only be interested in reviewing inventory of emission reductions, it should also consider whether such reductions are achieved in compliance with international law.[128] It is imperative to examine how the scope and mandate of emission review teams could be expanded to become full review and inspection panels with powers to examine not only inventories but also problems arising from emission reduction projects and actions. In the light of the current developments and instances where emission reduction efforts conflict with local circumstances and human rights, robust mechanisms must be put in place to inspect and gather facts from such trouble spots and apply appropriate sanctions to a defaulting party as necessary.

A good example is the problem-solving approach adopted by the World Bank Inspection Panel (WBIP). Despite the conflicting opinion on the efficiency and legacy of the WBIP with respect to transparency and accountability, the Panel is an example of how an international accountability mechanism can detect and check the excesses that arise when large-scale infrastructural projects sponsored by developed countries or international development agencies are implemented in developing countries.[129] Arguably, the WBIP example, if improved upon could provide a good example on how an international organization can monitor projects carried out in developing countries and funded by international agencies and developed countries to ensure transparency, accountability and compliance with international best practices and law.[130] The next section will examine the structural conditions, processes

[128] The Annual review of individual inventories of each Annex I Party became mandatory in 2003. Each Annex I Party is subject to at least one in-country visit during the commitment period. Expert Review Teams (ERTs) reviews compliance with emission reductions by industrialized countries through desk and centralized review and at least one visit over five years. It then raises any apparent implementation problems with the Compliance Committee. See UNFCCC Decision 19/CP.8, which empowers the UNFCCC Secretariat to select members of expert review teams from experts nominated by Parties.

[129] D. Clark, 'The World Bank and Human Rights: The Need for Greater Accountability', *Harvard Human Rights Journal*, 15 (2002), 205–226.

[130] E. Carrasco and A. Guernsey, 'The World Bank's Inspection Panel: Promoting True Accountability Through Arbitration', *Cornell International Law Journal*, 41(3) (2008); A. Ninio, 'Environmental Audit in the International Context: The Experience of the World Bank Inspection Panel' in *Ten Years after Rio 92: Sustainable Development and Law*, Proceedings of the 6th International Conference on Environmental Law, São Paulo, Brazil, (June 2002); L. Boisson de Chazournes, 'Compliance with International Standards – The Contribution of the World Bank Inspection Panel', in G. Alfredsson and R. Ring, eds., *The Inspection Panel of the World Bank: A Different Complaints Procedure* (London; Boston: Martinus Nijhoff Publishers 2001); A.R. Sureda, 'Process

and outcomes of the WBIP to draw lessons on the practical aspects of monitoring project threshold and human rights safeguards.

5.2.4.2 Monitoring project threshold: example of the World Bank Inspection Panel

The monitoring procedure of the WBIP demonstrates how an international organization can establish rules and procedures to monitor and ensure that projects carried out under any of its mandates do not conflict with approved rules and policies. Without proper monitoring, a project designed to achieve laudable goals can produce monstrous repercussions for the public.

In the 1980s, similar concerns began to emerge on how development projects sponsored by the International Bank for Reconstruction and Development (The World Bank) were producing negative environmental and social impacts in developing countries.[131] These included environmental pollution, unsafe dams and projects, mass displacement of citizens from their homes to allow for projects, concentration of projects in poor and vulnerable communities, and the lack of participation by citizens in project planning and implementation.[132] For example, the

Integrity and Institutional Independence in International Organizations: The Inspection Panel and the Sanctions Committee of the World Bank' in L Boisson de Chazournes, C.P.R. Romano and R. Mackenzie, (eds.), *International Organizations and International Dispute Settlement* (New York: Transnational Publishers, 2002); D. Hunter, 'Using the World Bank Inspection Panel to Defend the Interests of Project-Affected People', *Chicago Journal of International Law*, 4(1), 2003), 201–211.

131 The International Bank for Reconstruction and Development (The World Bank) was established in 1944 to promote the recovery and development of countries heavily affected by the impact of the Second World War. Ever since then, the World Bank has been at the forefront of providing funds for projects that assist countries to attain growth and development. It provided interest free loans and funding of over $20 billion per year to the least developed countries (LDCs) to support different development projects ranging from water dams, power stations, municipal transport systems, road infrastructures and waste disposal systems. In 2004, the World Bank provided $20.1 billion for 245 projects in developing countries worldwide. See *World Bank, About us*, web.worldbank .org/WBSITE/EXTERNAL/EXTABOUTUS/0,,contentMDK:20040558~menuPK :34559~pagePK:34542~piPK:36600,00.html, accessed 21 March 2015.

132 See for example Witness for Peace, *A People Dammed: The Impact of the World Bank Chixoy Hydroelectric Project in Guatemala*, www.witnessforpeace.org/apd.html, accessed 23 March 2015. See also D. Clark, 'Resettlement: The World Bank's Assault on the Poor' (Centre for International Environmental Law (CIEL) Brief, May 2000); F. Seymour, 'The World Bank and Environmental Sustainability' in P. Bosshard, *Lending Credibility: New Mandates and Partnerships for the World Bank* (World Bank, 1996), 43, 43–44; K. Horta, 'Rhetoric and Reality: Human Rights and the World Bank', *Harvard Human Rights Journal*, 15 (2002), 227.

Bank came under heavy criticism for financing the Sardar Sarovar Dam in India despite heavy opposition by villagers and community leaders.[133] The human rights issues at stake in this project were publicized by locally affected communities, non-governmental organizations and interest groups to highlight the need for accountability in the operations and activities of the World Bank and that international funds should not be used for projects that harm the environment or externalize development costs to the poorest members of society.[134]

The World Bank responded in September 1993 by establishing the WBIP to serve as an independent investigative forum through which individuals or communities who believe that they are or are likely to be harmed by a World Bank funded project to bring their concerns directly before the Board of Executive Directors of the World Bank.[135] The Panel serves as a fact-finding body with the power to review Bank funded projects and to determine whether the World Bank management is following its own operational policies and procedures in the design, appraisal and implementation of projects.[136] According to the Panel Operating Procedures:

> The Panel has been established for the purpose of providing people directly and adversely affected by a Bank financed project with an independent forum through which they can request the Bank to act in accordance with its own policies and procedures.[137]

The Panel has the mandate of ensuring that projects funded by the Bank are not leading to social and environmental consequences to the people in countries or communities where the projects are situated. The Inspection Panel operates as a bottom-up accountability and recourse mechanism that investigates IBRD/IDA financed projects to determine whether the Bank has complied with its operational policies

133 See K. Ramachandra, 'Sardar Sarovar: An Experience Retained', *Harvard Human Rights Journal*, 19 (2006), 275–281; see also P. Mehta, 'Internally-Displaced Persons and the Sardar Sarovar Project: A Case for Rehabilitative Reform in Rural Media', *American University International Law Review*, 20(3) (2005), 613–647; D. Clark, *The Impact of the 2002 Submergence on Housing and Land Rights in the Narmada Valley* (Habitat International Coalition, 2003).

134 See Clark, 'The World Bank and Human Rights', (n 129), 205–226; D. Bradlow, 'The World Bank, the IMF and Human Rights', *Transnational Law and Contemporary Problems*, 6 (1996), 47, 59.

135 World Bank, IBRD Resolution No. 93–10/ IDA Resolution No. 93–6 (1993), www.worldbank.org/inspectionpanel, accessed 23 February 2015.

136 *Ibid.*

137 IBRD Inspection Panel, 'Panel Operating Procedures', 19 August 1994, www.worldbank.org/inspectionpanel, accessed 23 February 2015.

and procedures (including social and environmental safeguards), and to address related issues of harm.[138] There are lessons that can be drawn in the structural set up, process and outcomes of the Panel.

Structural conditions of the World Bank Inspection Panel Structurally, the enabling resolution of this panel provides for independence in the appointment of members of the Panel.[139] The Panel comprises three members of different nationalities that are nominated by the President after consultation with the Board of Executive Directors. Similarly, no panel member may serve for more than one term; as such vacancies in the Panel are filled every five years.[140] The procedure also grants members of the Panel the security of tenure by providing that members can be removed only for cause and by a decision of the Executive Directors.[141] The manner of appointment of the Board members and tenure is very important to ensure that members maintain their neutrality and independence, and do not represent country or regional interests. Another striking provision is that the Procedure provides that a Panel member shall be disqualified from participation in the hearing and investigation of any request related to a matter in which he/she has a personal interest or had significant involvement in any capacity, equally staff members of the Bank are not eligible to serve on the Panel.[142]

These provisions are to instill some measure of structural independence, fairness and impartiality in its operations.[143] They also provide examples on how to address conflict of interest issues that often raise questions of accountability in the current international climate change regime.[144] For example, the situation in the CDM governance structure in which CDM EB members allegedly held previous positions as UNFCCC and Kyoto Protocol negotiators for their countries creates conflicts of interests and reduces

[138] See World Bank Inspection Panel, web.worldbank.org/WBSITE/EXTERNAL/EXTINSPECTIONPANEL/0,,menuPK:64132057~pagePK:64130364~piPK:64132056~theSitePK:380794,00.html, accessed 23 March 2015.

[139] *Ibid.*, see also D Clark, *A Citizen's Guide to the World Bank Inspection Panel*, 2nd edn (Washington, DC: Center for International Environmental Law, 1999), www.unece.org/fileadmin/DAM/env/pp/ppif/WB%20Inspection%20Panel%20Citizens%20Guide%201999.pdf, accessed 23 March 2015.

[140] World Bank, IBRD Resolution No. 93–10/ IDA Resolution No. 93–6 (1993), para. 4, (n 135).

[141] IBRD Inspection Panel, 'Panel Operating Procedures', (n 137). [142] *Ibid.*

[143] See World Bank, IBRD Resolution No. 93–10/ IDA Resolution No. 93–6 (1993), para. 4, *ibid.*

[144] C. Streck and J. Lin, 'Making Markets Work: A Review of CDM Performance and the Need for Reform', *European Journal of International Law*, 19(2) (2008), 409–442, where Streck and Lin question the transparency and accountability of the CDM governance structure.

the transparency of the CDM EB.[145] Similarly, the practice whereby members of expert review teams are selected by the UNFCCC Secretariat from a list of experts nominated by Parties arguably gives Parties the chance to influence the constitution and membership of ERTs.[146] To ensure transparency, only individuals with no direct link to climate change negotiations, either as country representatives, nominees of Parties or staff of the Secretariat, would be appointed to supervisory and verification bodies, such as the ERT and the CDM EB. Similarly, members of a supervisory body should be disqualified from participation in the hearing and investigation of any project in which they have personal interest or significant involvement in any capacity with the project proponents or state. Overall, the aim should be to establish an independent investigative body where members can serve without conflict of interest.

Processes of the World Bank Panel Despite the structures of the WBIP, scholars have questioned whether the Panel is truly independent and transparent in practice, particularly in its processes and procedures for admitting complaints.[147]

The World Bank Panel adopts a procedure that allows any group of individuals or community affected by a project including NGOs, to approach it for investigations. The process of investigation is triggered by the submission of a request for inspection. According to the Resolution, two or more affected people may submit a request; similarly a local organization or other duly appointed representative on behalf of the affected people; a foreign organization in exceptional circumstances if no local representative is available; or an executive director of the World Bank may submit a review request. A request may be submitted in any language and in any format, including by a mere letter, except that they must be in writing, dated and signed by the requesters.[148] The Panel also respects the confidentiality of requesters who ask that their names not be published.

According to the Panel's procedures, a request would be registered for further processing unless the Chairperson determines that it is 'without

[145] See *ibid.*, 420. See also the CDM Policy Dialogue, 'Climate Change, Carbon Markets and the CDM: A Call to Action: Recommendations of the High Level Panel on the CDM Policy Dialogue' (2012), www.cdmpolicydialogue.org/report/ues_en.pdf, accessed 10 January 2015.

[146] See UNFCCC Decision 19/CP.8.

[147] IBRD Inspection Panel, 'Panel Operating Procedures', (n 137).

[148] *Ibid.*, para. 6–15.

doubt manifestly outside the Panel's mandate'.[149] The Panel also maintains a publicly accessible register of registered requests that is available on its website.[150] This way requesters and the general public can follow the status of any project and the issues surrounding it. Other key documents such as the initial management response and the Panel eligibility reports are also posted on the website while official copies can be requested from the Panel's secretariat. Once the Panel has received and registered a Request for Inspection, the eligibility phase of the Inspection Process commences.

However, an investigation is not automatic; the Panel is required under the Operating Procedure to make a recommendation to the Board of Executive Directors of the Bank, requesting 'whether the matter should be investigated' more thoroughly. This is referred to as the Eligibility Report. Only the Board has the power to officially authorize the Panel to proceed with an investigation.[151] If the Board approves the investigation, the Panel is then able to commence the inspection process. This requirement by the Panel to seek the approval of the Board has been highly criticized as detracting from the true independence of the Panel.[152] For example, there have been instances where the Board has failed to recommend investigations in cases where it was needed.[153] This limits the Panel's ability to control what it investigates, thereby diminishing its independence.[154] To guarantee independence, a review panel should have full control over what it wishes to investigate in accordance with its own procedures after receiving a public request, without having to wait for the approval of a higher authority. The power to investigate all eligible complaints should be inherent and granted as part of the mandate of such a fact finding body, this will ensure that the

[149] *Ibid.*

[150] See World Bank Inspection Panel, www.inspectionpanel.org, accessed 23 February 2015.

[151] World Bank, Inspection Panel, Operating Procedures, V.A.31 (a)–(d).

[152] R. Bisell, 'Institutional and Procedural Aspects of the Inspection Panel', in G. Alfredsson and R. Ring (eds.), *The Inspection Panel of the World Bank: A Different Complaints Procedure* (The Hague: Martinus Nijhoff, 2001), 107.

[153] See D. Clark, 'The Rise and Fall of Accountability', *Watershed*, 6(3) (2001), 52–55; R. Bisell, 'Recent Practice of the Inspection Panel of the World Bank', *American Journal of International Law*, 91 (1997); Bradlow, 'The World Bank, the IMF and Human Rights', 47–90.

[154] In the 1999 Clarifications, the Board agreed to authorize investigations on a 'no-objection' basis except for in limited circumstances. Many other concerns on the lack of Bank independence were addressed in the Clarifications. See D. Bradlow, 'Lessons from the NGO Campaign against the Second Review of the World Bank Inspection Panel: A Participant's Perspective', *ILSA Journal of International and Comparative Law* (2001) 7, 247–257. S. Suzuki and Suresh Nanwani, 'Responsibility of International Organizations: The Accountability Mechanisms of Multilateral Development Banks', *Michigan Journal of International Law*, 27 (2005), 177, 207.

independence of the body is not eroded by official or administrative bureaucracy.

Another criticism of the WBIP is that after investigating a complaint, the Panel is required to submit its findings and resolutions to the management of the Bank. The Panel does not have the power to issue an injunction, stop a project, or award financial compensation for harm suffered. The Management of the Bank reviews the Panel's findings and if it chooses to do so, the management is able to make remedial recommendations. There have been instances whereby the management has either failed or excessively delayed implementing the recommendations of the Panel to stop a project many years after the Panel concludes its finding. This is a defect that also limits the transparency of the Panel and the level of contributions it can make.[155] The lack of mandate or power by the Panel to follow through on its own investigations and findings, limits the level of contribution the Panel can make towards ensuring timely relief to victims of bank projects. To ensure transparency and accountability, it is important for a supervisory body of this nature to have powers to provide relief and monitor the implementation of the report of its assessment and investigation. For example, review panels established to monitor carbon projects should have automatic powers to stop a project without waiting for approvals by any other governance body. To further strengthen the transparency and accountability of an investigative panel, its recommendation or findings should be automatically binding on supervisory bodies such as the CDM EB, COP or the UNFCCC Secretariat.

Outcomes of the World Bank Panel Since 1994, the Panel has received over fifty-two requests from individuals and NGOs in thirty-one countries on a wide variety of projects.[156] The Panel has recorded some progress in dealing with project requests and has recommended the suspension of projects that fail to comply with the World Bank's polices

[155] As Clark notes, most times the World Bank Board is overwhelmed with information and does not have a standing committee to track the implementation of action plans or to evaluate the effectiveness of remedial measures suggested by the Panel. Clark, 'The World Bank and Human Rights', (n 129), 220.

[156] Communities from Albania, Argentina, Bangladesh, Brazil, Burundi, Cambodia, Cameroun, Chad, Chile, China, Columbia, Congo Democratic Republic, Ecuador, Ethiopia, Ghana, Honduras, India, Kenya, Lesotho, Mexico, Nepal, Nigeria, Pakistan, Papua New Guinea, Paraguay, The Philippines, Romania, South Africa, Tanzania and Uganda have submitted project requests to the Panel. See the World Bank Inspection Panel, Cases by Country/Region, web.worldbank.org/WBSITE/EXTERNAL/EXTINSPECTIONPANEL/0,,contentMDK:21680496~pagePK:64129751~piPK:64128378~theSitePK:380794,00.html, accessed 25 July 2015.

and procedures, especially projects that involve inadequate levels of local involvement, community consultations and social and environmental impact assessments. One main strength of the WBIP is that its procedures for initiating review are simplified enough to allow two or more members of the public or representatives of a community who have concerns about a project to request a review. This opens a window of opportunities for the public to seek a review of projects that violate the rules and policies of the World Bank concerning development projects. A great number of such projects have been stopped by the Management of the World Bank based on recommendations of the Panel.[157]

The level of progress recorded by the WBIP shows that if the Panel's processes can be improved upon, it could serve as a guide for an international organization on how to effectively to monitor the implementation of projects approved under any of its programmes. The Panel has been variously described as a positive move in international law to recognize individuals and non-state actors in policy and decision-making processes and to create a 'public space' where people, states and large institutions can better interact.[158] It has also been hailed for broadening of the range of actors in international law and policymaking – a stage once restricted to national governments and the bureaucracies of international organizations.[159] In 2006, the *Global Accountability Report* recommended the Panel as the 'best practice' for accountability especially for keeping its commitments of responding to all complaints brought before it.[160] Despite its shortcomings and criticisms, the framework of the Panel can serve as a guide and template on how to mainstream accountability in the international climate change regime.

157 A.G. Gualtieri, *The Environmental Accountability of the World Bank to Non-State Actors: Insights from the Inspection Panel* (Oxford: Clarendon Press, 2002), 213.

158 See L. Boisson De Chazournes, 'The World Bank Inspection Panel; About Participation and Dispute Settlement' in T. Treves et al. (eds.), *Civil Society, International Courts and Compliance Bodies* (The Hague: T.M.C. Asser Press, 2005), 187–203; see also K. Kingsbury and Stewart, 'The Emergence of Global Administrative Law', *Law and Contemporary Problems*, 68 (2005), 15, 17; A. Bradlow, 'Private Complainants and International Organizations: A Comparative Study of the Independent Inspection Mechanisms in International Financial Institutions', *Georgia Journal of International Law*, 36 (2005), 403, at 419. See also E.B. Weiss, 'New Directions in International Environmental Law' Address Before the United Nations Congress on Public International Law (15 March 2001); L. Sohn 'The New International Law: Protection of the Rights of Individual Rather than States', *American University Law Review*, 32(1) (1982), 1–64.

159 Weiss, et al., 'The World Bank Inspection Panel: Participation and Accountability', (n 113), 271.

160 One World Trust, *2006 Global Accountability Report: Holding Power to Account* (One World Trust, 2005), 25–46.

There is a need to develop an innovative framework that builds on the structure of the World Bank Panel and corrects some of its flaws, by providing additional safeguards to ensure more accountability. For example, for a project review panel to be truly independent, it should have inherent powers to accept petitions and conduct an independent investigation without having to seek the approval of a supervisory board or 'management', such as the CDM EB, the COP or the UNFCCC secretariat. Such requirements could reduce the independence of a review panel and its abilities to carry out its oversight and investigative functions. The enabling rules and procedures establishing such a review panel should include automatic powers to carry out investigations on project review request or complaints from the public. Similarly, it is not enough to produce reports on the results of investigations, the Panel should be able to resolve the violations uncovered during its fact-finding and investigation by having the mandate to issue a stop order for a project that violates project thresholds and rules.

It is imperative for the international climate change regime to establish robust guidelines and rules that would provide opportunities to monitor and investigate the human rights impacts of carbon projects. These guidelines should include human rights indicators and a requirement for host countries and project proponents to demonstrate human rights compliance prior to project approval and during the lifecycle of a project. Guidelines alone will not be enough. The international climate change regime must also comprise of governance structures through which compliance would be substantively measured. A follow up would be to expand the mandate of expert review teams or to establish an Inspection Panel that builds on the strength of the WBIP as an independent and investigative arm within the international climate regime. The purpose of this panel will be providing people directly and adversely affected by an emission reduction project with an independent forum through which they can request a review of such projects.

5.2.5 Access to justice

Access to justice refers to the availability of mechanisms, either formal or informal, through which the public can seek remedies or redress for human rights violations.[161] Access to justice is one of the most important elements of an effective HRBA framework. In this book, access to justice

[161] UNDP reckons that access to justice encompasses not only formal institutions, such as courts and administrative tribunals, but also traditional or customary ones, such as village-level dispute resolution, and coordination mechanisms among the different components in the system. See UNDP, 'A Human Rights-based Approach to

is defined as the availability and effectiveness of complaint mechanisms through which the public may seek redress for human rights violations resulting from a carbon project. While accountability mechanisms deal with monitoring and inspecting projects to determine if and how they fail to comply with laid down rules, access to justice refers to available opportunities for the public to appeal against project approvals or procedures adopted in the planning and execution of a project. The relationship between both is that accountability mechanisms are also important to monitor the implementation of any decision handed out by a quasi-judicial body, project appeal board or review committee.

5.2.5.1 Elements of access

For the right to access to justice to be fulfilled, four essential elements are important: a normative legal framework; legal awareness; access to appropriate forums; and the effective administration of justice. There is also a need for an efficient monitoring and oversight system for the four elements.[162] A diagrammatic representation of these elements is produced below:

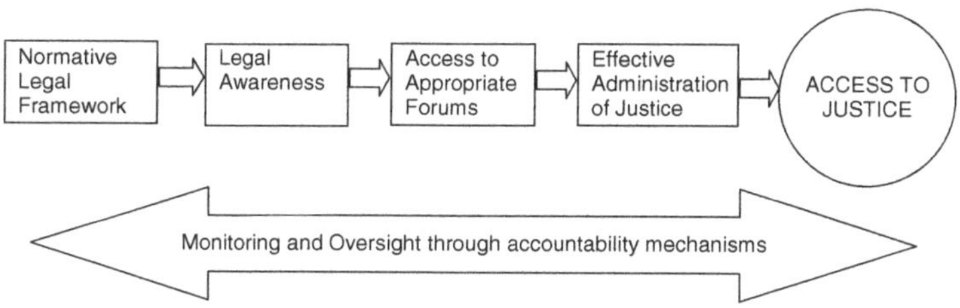

Source: Chart drawn based on information contained in UNDP: A Practitioner's Guide to a Human Rights Based Approach to Access to Justice.[163]

Figure 5.1 Elements of access to justice

These elements are discussed below.

Development Programming' in *UNDP-Adding the Missing Link* (2002), www.undp.org.fj/gold/docs/MISSINGL.PDF, accessed 10 October 2014.

162 R. Smith, 'Human Rights and Access to Justice', *International Journal of the Legal Profession*, 14(3) (2007), 261–262; see also C. Harlow, 'Access to Justice as a Human Right', in P. Alston *et al.* (eds.) *The EU and Human Rights* (Oxford University Press, 1999); S. Golub, 'Access to Human Rights: Obstacles and Issues', (International Council on Human Rights Policy, 2003), 3–5, www.ichrp.org/files/papers/99/123_-_Access_to _Human_Rights_-_Obstacles_and_Issues_Golub__Stephen__2003.pdf, accessed 10 December 2015.

163 UNDP, *Programming for All: A Practitioner's Guide to a Human Rights Based Approach to Access to Justice* (UNDP, 2005), 4–5.

Normative legal framework This refers to the set of rules, procedures, and institutions that promote access to justice. It involves the development of frameworks and governance structures to ensure that people's rights are enforced, and that remedies are provided, in cases of violations.[164] While access to justice at the national level is often tied to courts and quasi-judicial mechanisms through which citizens can seek redress for violations, access to justice at the international level is often not as straightforward. As Francioni rightly notes, there are several hindrances to individual access to international justice.[165] Primarily, access to international justice is based on specific treaty arrangements. Consequently, when a treaty does not include robust frameworks on access to justice, opportunities or mechanisms for individuals to seek redress under it become largely absent. Without specific provisions that allow individuals and non-state actors to file petitions or complaints before established complaints mechanisms, there would be no legal basis for the public to bring forward complaints on the local impacts of mitigation projects on human rights.

Secondly, even when access to justice is guaranteed as a matter of treaty obligation, other norms or doctrines of international law such as exhaustion of local remedies, requirements for standing, may effectively impede the exercise of such rights.[166] However, states may reduce the

164 See J. Cavallaro, 'The Urban Poor: Problems of Access to Human Rights', *International Council on Human Rights Policy*, (2003), 12; M. Garling, 'Enhancing Access to Human Rights: Draft Report' International Council on Human Rights Policy' (ICHRP, 2003).

165 F. Francioni, 'The Rights of Access to Justice under Customary International Law' in *Access to Justice as a Human Right* (Oxford University Press, 2007).

166 The obligation to exhaust domestic remedies has been recognized as part of customary international law. The rationale for the exhaustion rule is to afford national authorities the opportunity to prevent or put right the alleged violations. It is found in core international human rights treaties. See Article 41.1(c) of the ICCPR ' The Committee shall deal with a matter referred to it only after it has ascertained that all available domestic remedies have been invoked and exhausted in the matter in conformity with the generally recognized principles of international law. This shall not be the rule where the application of remedies is unreasonably prolonged.' The Aarhus Compliance Committee also states that any communication must specify whether steps have been taken to use the remedies available in the country in question to obtain redress in the case which is the subject of the communication (e.g. administrative or judicial review or appeals procedures operated by public authorities, courts, tribunals, ombudsperson, etc.) and if so, which steps were taken, when they were taken and what the results were. If no steps have been taken, it should be explained why not (e.g. because no remedies were available or because they were too expensive). If remedies were sought in connection with the matter, which is the subject of the communication, or in a closely related case, by a person other than the communicant, this should also be mentioned in the communication. Similarly, the communication should include information on

effects of these principles by establishing mechanisms that reduce their impacts on access to justice. For example, the requirement of exhaustion of domestic remedies is not a strict requirement under the Aarhus Convention.[167] The Committee is empowered to consider whether the application of the remedy is unreasonably prolonged or does not provide an effective or sufficient means of redress. Also, even if a domestic remedy has not been utilized, the Committee is not precluded from considering the communication. This discretionary power under Aarhus could be a fitting template under the international climate change regime to allow the public to seek remedies before an international grievance mechanism, in cases where domestic remedies are lopsided, insufficient or unnecessarily prolonged. For example, the climate change regime might water down the application of the exhaustion principle by granting its complaint resolution body the discretion to determine whether the available domestic remedies provide an effective or sufficient means of redress.

Requirements for standing might also be liberally established to widen access. For example, traditional standing requirements that insist that the complainant must establish that the challenged action will cause him/her an actual or threatened injury-in-fact act as a normative hindrance to access.[168] These requirements are often hard if not impossible to meet especially in cases brought by environmental groups or NGOs. Such stringent requirements limit access to justice. A better approach is an expansive approach that widens standing to allow environmental advocates and NGOs to bring actions against activities or projects that harm the environment, not only direct harm, but also indirect harm or the loss

whether the matter has been submitted to other international procedures (the steps taken, when they were taken and what were the results were).

[167] See Guidance Document on the Aarhus Convention Compliance Mechanism, www.unece.org/fileadmin/DAM/env/pp/compliance/CC_GuidanceDocument.pdf, accessed 12 January 2015.

[168] See *Society of Plastics Industry, Inc. v. County of Suffolk* (1991) 77 N.Y.2d 761. The court held that '[i]n land use matters. . .the plaintiff, for standing purposes, must show that it would suffer direct harm, injury that is in some way different from that of the public at large' (77 NY2d at 774). More specifically, in cases involving environmental harm, the standing of an organization could be 'established by proof that agency action will directly harm association members in their use and enjoyment of the affected natural resources. . .' According to the court, in order to have standing, a party must demonstrate an 'injury in fact' – an actual legal stake in the matter being adjudicated – which falls within the 'zone of interests, or concerns, sought to be promoted or protected by the statutory provision under which the agency has acted'. See also *Friends of the Earth, Inc. v. Laidlaw Envtl. Servs. (TOC), Inc.*, [2000]528 US 167.

of the aesthetic or recreational value of the environment. This is the approach endorsed by the Aarhus Convention.

Article 9 of Aarhus provides guidelines on what can be reviewed. Under this Article, the trigger for the review process is a complaint by any person who considers that his or her request for information and participation in environmental decision-making has been ignored, wrongfully refused, inadequately answered, or otherwise not dealt with. Article 9 places an obligation on parties to provide review procedures for any person who believes that his or her information request has not been properly dealt with. Parties must make a review procedure available when a person contends that his or her request for information has been ignored or wrongfully refused or the response is inadequate; or when he or she believes that the request was otherwise not dealt with in accordance with the provisions of Article 4. It also provides that states should ensure that such a person also has access to an expeditious procedure established by law that is free of charge or inexpensive. It further provides that final decisions shall be binding on the public authority holding the information.[169] These provisions seek to provide inexpensive and accessible means for the public concerned to seek and obtain a remedy that is final and binding in nature.

Article 9(1) also answers the contentious question of standing, that is, question on who can ask for a review. It provides that 'any person' who has requested information under Article 4 can challenge decisions. Article 4 allows 'any member of the public' to request information, while Article 4(2) defines the 'public' as natural or legal persons, and their associations, organizations or groups. This grants standing to a wide variety of the society falling within the definition of 'the public'. One of the legacies of Article 9(3) is that it provides that members of the public can file a complaint if they meet the criteria, if any, laid down in national law. This encourages a broad interpretation by parties to the Convention on the question of who has "standing" to bring a challenge.

One of the strongest criticisms against the Aarhus Convention is the discretion and flexibility it grants to Parties to apply and interpret its access to justice provisions, as well as other provisions, 'within the framework of/in accordance with national legislation'.[170] Even though this flexible approach seems understandable in view of the different legal

[169] See Aarhus Convention, (n 72).

[170] The term 'within the framework of national legislation' also appears in Articles 4(1) and 5(2) of the Aarhus Convention. See Mason, 'Information Disclosure and Environmental Rights', (n 125), 10–31.

systems and peculiarities of the various Aarhus Parties, it has not fostered a common understanding and uniform implementation of a number of the Aarhus provisions.[171] For example, the implementation of Article 9(3) has varied from one country to another. Generally, the standing requirement in the Aarhus Convention has been applied and interpreted by Parties in three different ways.[172] First is a protective norm interpretation, which requires that in order to seek and obtain remedy before a judicial or quasi-judicial body, the applicant must show that the decision or omission affects their individual or subjective right.[173] This approach is based on the requirement of standing which requires applicants to show how that their individual rights have been affected. Under this category, an NGO must demonstrate that an action results in a breach of one of its own rights or a direct injury to its financial assets or its property. For example, when an NGO's office is demolished to allow for a project. Under this protective interpretation, where an administrative decision with adverse effects on the environment does not directly affect the property or the assets of the NGOs, no judicial review will take place.

A second approach is the more expansive approach (*actio popularis*), which allows every individual and organization to challenge administrative decisions and omissions without the need to prove legal standing or individual interest in the action.[174] This provides automatic standing to

[171] See Lee and Abbot, 'The Usual Suspects?', (n 74), 80–108.

[172] See J. Darpo, 'Effective Justice? Synthesis Report of the Study on the Implementation of Articles 9.3 and 9.4 of the Aarhus Convention in Seventeen of the Member States of the European Union', (2012), ec.europa.eu/environment/aarhus/pdf/2012_access_justice_report.pdf, accessed 23 January.

[173] Popularly known as *Schutznormtheorie* in Germany where this strict form of interpretation to Article 9 of Aarhus is popular and commonly applied. According to this theory, the state is only liable when in addition to causing an injury, it also breaches a *Schutznorm* which is a legal norm protecting a subjective public right of the injured party, and which is intended not only to protect individuals generally, but also to protect a specific section of individuals to which the injured party belongs. This theory was laid down for the first time in the case of *Vloebergh v. High Authority of the ECSC* [1961] E. C. R 197. For an historical account, see H. Schermers, T. Heukels and P. Mead (ed) *Non-Contractual Liability of the European Communities* (Dordrecht; Boston; London: Nijhoff, 1988), 6–7.

[174] For example, (Article 52(3)(a) of the Portuguese Constitution provides broad access for NGOs to review decisions and to claim compensation on behalf of the aggrieved party in order to preserve the environment. This approach is also common in Latvia, Ireland, Sweden and Belgium. See Darpo, 'Effective Justice?', (n 172).

every member of the public to use in the interest of the public.[175] This approach has been criticized for its tendency to open a floodgate of litigation that could frustrate every governmental policy or project.[176] The third is an intermediate/interest-based approach which allows members of the public to demonstrate a sufficient interest and be able to challenge a decision that could result in a destruction of a resource or object to which they put to general use. A judicial review will take place where an administrative decision produces adverse effects on the environment, even though it does not directly affect the property or the assets of the individual.[177]

175 This approach has received some level of judicial recognition and application in India. Public Interest Litigation (PIL) has been used to guarantee several human rights in India, including the right to unpolluted environment, health, livelihood, free and compulsory primary education, shelter, clean drinking water, privacy, legal aid, speedy trial, and access to information. See S. Deva, 'Public Interest Litigation in India: A Critical Review', *Civil Justice Quarterly*, 28(1) (2009), 19–40.

176 A recent empirical study conducted on the impact of Articles 9(3) and (4) of the Aarhus Convention, however, concludes that the availability of litigation rights for environmental NGOs contributes to the effective enforcement of environmental law under Aarhus and has not necessarily resulted in a floodgate or rise in the number of litigations in countries that have adopted the expansive approach of *actio popularis*. See *Summary Report on the Inventory on the EU Member States' Measures on Access to Justice in Environmental Matters* (Milieu Environmental Law and Policy, Brussels 2007), 2–4, ec.europa.eu/environment/aarhus/study_access.htm., accessed 23 January 2015.

177 This approach has been adopted in several cases in the United States. See the case of *Sierra Club v. EPA* (1972) 405 US 727, at 974 (where the US Supreme Court held that a generalized 'interest' in the environment could not confer standing to challenge environmental injury, but that injury to a particular plaintiff's '[a]esthetic and environmental well-being' would be enough. See *Friends of the Earth v. Laidlaw* 528 US 167, 120 S. Ct. 693, 145 L. Ed. 2d 610, 200. In this case a Friend of the Earth (FOE) member averred in affidavits that he lived a half-mile from Laidlaw's facility which has now polluted the river; he noted that he occasionally drove over the North Tyger River, and that it looked and smelled polluted; and that he would like to fish, camp, swim, and picnic in and near the river between 3 and 15 miles downstream from the facility, as he did when he was a teenager, but would not do so because he was concerned the water was polluted. The majority held that FOE had standing to sue based on the purely aesthetic values of the river. See *Massachusetts v. EPA* (2007) 549 US 497 in which the US Supreme Court held that State of Massachusetts has a quasi-sovereign interest and standing to sue the environmental regulatory agency for failing to prevent the emission of GHGs and climate change. See also *Environmental Services United States v. Students Challenging Regulatory Agency Proceedings*, 412 US 669 (1973) (*SCRAP*), In SCRAP a student group was permitted to obtain standing based on the fact that an action involved real harm to the environment, even if there was no personal or direct harm directly traceable to the defendant's proposed action; see also *Save the Pine Bush v. Common Council of the City of Albany ('Pine Bush')* [2009] NY Slip Op 07667; *Bennett v. Spear* (1997) 520 US 154; *Lujan v. Defenders of Wildlife* (1990) 497 US 871 983.

This intermediate approach is arguably the least restrictive, yet balanced approach to standing. While it allays concerns on the likelihood of opening a floodgate of litigation under the *actio popularis* approach which grants standing to any member of the public; it also recognizes and protects instances where interests in the subject matter might go beyond direct or personal injury. For example, if a carbon project results in the destruction of a local park where I often relax every evening to gaze at nature and to meditate, even though I do not have direct interest in the park since I am not the owner, I have an indirect interest in its recreational or aesthetic use. The loss of the park could affect me emotionally and culturally; hence I should be able to challenge the continuation of such a project. This is an approach that would practically widen standing to enable individuals and NGOs to challenge a project that could violate human rights or affect the environment, even where there is no actual personal injury.

Access to appropriate forums Access to justice does not stop at establishing a framework; there is also the need to provide mechanisms aimed at aiding or assisting members of the society that cannot afford the cost of administrative review. Article 9(4) of Aarhus provides detailed guidelines on the minimum requirements and standards to ensure adequate access to justice. These include: equity, fairness, timeliness, cost effectiveness, requirements to record decisions in writing and to make decisions publicly accessible. According to Article 9(4), the lack of cost-friendly access to justice is a major hindrance to adequate access. An effective justice system is that which provides access to justice even for the poor such that everyone can hold their government accountable and redress power imbalances. For example, in many developing countries, even in the face of pollution, most citizens are either unable to afford a court case or are denied standing to bring such cases. This produces a multiple effect of denying justice to the oppressed members of the society. It also leads to a continuation of societal oppression and a loss of interest in the judiciary as the home for justice. An accessible grievance mechanism can enable the public to hold governments accountable where domestic remedies are insufficient or unavailable.

Though not legally binding, the Ruggie Guiding Principles on Business and Human Rights Framework also provides comprehensive definitions of an 'effective grievance mechanism'.[178] The Framework advocates

[178] Ruggie, 'United Nations Guiding Principles on Business and Human Rights', (n 64).

greater access to effective remedy, both judicial and non-judicial for victims of human rights violations arising from businesses. According to the Framework, in order to ensure effectiveness, non-judicial grievance mechanisms, both state-based and non-state-based, should be: legitimate, accessible, predictable, equitable, transparent, rights compatible, a source of continuous learning and should be based on engagement and dialogue.

The Commentaries to the Framework provide useful definitions for these terms, which are adopted for our discussions here. For a grievance mechanism to be rights-compatible, it must ensure that outcomes and remedies accord with internationally recognized human rights. Human rights principles must be well reflected in the procedures and modalities of a grievance mechanism such that it could effectively admit and address human rights complaints from the public. Accessibility is defined in the Framework in terms of being known to all stakeholder groups for whose use they are intended, and providing adequate assistance for those who may face particular barriers to access. Factors that hinder access, such as standing, prohibitive cost, language barriers and other procedural hindrances, must be addressed in order to provide the public adequate opportunities to come before the body. Legitimacy means the mechanism should engender trust from stakeholder groups for whose use they are intended, and should be accountable for the fair conduct of grievance processes. Predictable relates to providing a clear and known procedure with an indicative time frame for each stage, and clarity on the types of process and outcome available and means of monitoring implementation. Equitable refers to seeking to ensure that aggrieved parties have reasonable access to sources of information, advice and expertise necessary to engage in a grievance process on fair, informed and respectful terms; while transparency refers to keeping parties to a grievance informed about its progress, and providing sufficient information about the mechanism's performance to build confidence in its effectiveness and meet any public interest at stake. It should also be a source of continuous learning by drawing on relevant measures to identify lessons for improving the mechanism and preventing future grievances and harms; it should also foster engagement and dialogue by consulting the stakeholder groups for whose use they are intended on their design and performance, as well as focusing on dialogue as the means to address and resolve grievances.[179]

[179] *Ibid.*

Most recently, scholars have suggested international investment arbitration as the 'most appropriate mode of settlement' that can provide a fair, timely, cost effective and adequate access to the public to challenge foreign investments or development projects that violate human rights.[180] Arbitration could provide non-confrontational opportunities for communities, civil societies and governments to sit together to address the intersections between an investment or project and human rights of the community. Simma argues that international investment protection and human rights are not 'separate worlds' and that investment arbitrators can be empowered to consider human rights issues arising from international investments by foreign corporations. However, there are problems associated with applying human rights within foreign investment arbitration, which are not sufficiently addressed by the present architecture of international investment arbitration. Problems could arise when a government refuses to submit itself to arbitration, which could be the dead end for victims. There are also problems related to the fact that most times, arbitral awards are compensatory in nature, they do not anticipate problems or provide opportunities to prevent disputes from arising or to solve future problems. An anticipatory approach would provide in-built mechanisms that would enable the public to block a project that could violate human rights even before it produces any harm.

There are also epistemic concerns which Simma describes as 'genetic problems' associated with the fact that, the large majority of investment arbitrators have private or commercial law, rather than a public law or public international law background and might thus tend to see international human rights as a potential, or probable, cause of political disturbances, intruding in their 'purely legal', autonomous field.[181] As Simma notes, one way to avoid such difficulties is through a robust framework that systematically harmonizes human rights obligations into treaties to address areas of overlap. Such treaty process must also be supported by concrete institutional reforms that provide adequate human resources and capacity development for investment arbitrators or non human rights experts to be able to interpret and enforce human

[180] B. Simma, 'Foreign Investment Arbitration: A Place for Human Rights?', *International and Comparative Law Quarterly*, 60 (2011), 584; A. Tanzi, 'On Balancing Foreign Investment Interests with Public Interests in Recent Arbitration Case Law in the Public Utilities Sector', *The Law and Practice of International Courts and Tribunals*, 11(47) (2012).

[181] See Simma, 'Foreign Investment Arbitration', *ibid.*

rights obligations in cases brought before them. Through such harmonization, the problems of treating investment and human rights as two separate worlds can be addressed; it would also help provide opportunities for access to justice and the availability of remedies for civil society in the host state of foreign investments that affect human rights.[182] This is the nature of reform proposed by this book. There is a need to integrate human rights and the international climate change regime through a robust and holistic framework that harmonizes human rights obligations into the international climate change regime to address areas of overlap; allow both communities to share recourse and expertise, establish legal partnerships between both communities and also equip administrators with the requisite skills to interpret and enforce human rights standards.

There is also a need to establish appropriate governance structures to implement such harmonization. Arguably, establishing public complaint mechanisms as part of the international climate change regime could allow the public to seek remedies or redress for human rights violations. Such a body could be created to allow stakeholders to request review of a carbon project prior to registration for emission credits. Providing the opportunity for individuals to request a review of projects before a panel prior to approval would provide an opportunity for individuals to raise and highlight the human rights issues arising from a specific project. It would provide a preventive medium through which issues can be addressed before they produce negative effects.

There is equally the need for a project appeal board within the international climate regime through which individuals could appeal against the decisions to continue on-going projects. Some projects might meet all the requirements and the threshold for registration during the pre-approval stages, while the processes of their execution might turn out to produce negative environmental and human rights effects. This raises the need for post approval appeal panels, which could consider such unforeseen effects of projects. Such a panel would provide an opportunity for victims of these projects to seek their status review. The appeal panel should be created to consider such petitions arising from the continued existence of approved projects.

[182] See also F. Francioni, 'Access to Justice, Denial of Justice, and International Investment Law' in P.-M. Dupuy, F. Francioni and E.-U. Petersmann (eds.), *Human Rights in International Investment Arbitration* (Oxford University Press, 2009).

Legal awareness This refers to knowing the law, rights, obligations and how to access channels to resolve grievances. It has been rightly described as the degree of people's knowledge of the opportunity to seek redress through the justice system, whom to demand it from, and how to start a formal or traditional justice process.[183] Without basic knowledge of the laws and the legal system, the public cannot demand justice even in the face of brazen and brutal injustice. As Butegwa rightly notes, 'education and training are needed in the HRBA business as people have to know their rights so as to be able to hold duty-bearers to account'.[184] The preamble to the UDHR, also declares that 'every individual and every organ of society, keeping this declaration constantly in mind, shall strive by teaching and education to promote respect for these rights and freedoms'.

When the public know their rights, and the level of protection that they can get under the human rights framework, it will provide the critical mass to fight violations. If people do not know about their rights and cannot exercise them, the fact that the human rights treaties have enshrined them will be of little practical importance. Legal awareness provides people with the knowledge and understanding of rights and legal issues, together with the skills they need to gain access to the justice systems.

There is thus a need for the design of a Human Rights Education (HRE) programme, primarily aimed at members of the public especially rural dwellers, to increase legal awareness, improve understanding of the judicial system and inform them of the process of seeking justice. The HRE should provide members of the public with free and accessible information on the laws, legal systems and the methods for seeking help when their human rights are violated. HRE as an element of an HRBA system will assist in spreading basic information that can equip a critical mass of the population with knowledge of their human rights and how to demand government action when such rights are violated.

Under the HRBA to carbon finance, the focus would not only be on informing people about their rights, but would also be on sensitizing people on the nature and causes of environmental problems, how such problems violate their human rights and the options available to them for redress. As of today, a number of people, especially in developing countries, do not know the nature and causes of a number of environmental

183 UNDP, *Programming for All*, (n 161), 4–5.

184 C. Butegwa, *Enterprise Development and Poverty Reduction: The Case for a Rights-based Approach in Non-Governmental Organizations and Business* (Dakar, Senegal: ENDA, 2005), 2.

problems. For example, in Nigeria, a number of citizens are not pressing for governmental policies that can enhance climate change mitigation and adaptation because they believe climate change is an 'Act of God'.[185] In many parts of Nigeria, sections of the community do not know which Ministry to hold responsible for what. Such ignorance and lack of awareness can only deny the public opportunities to explore and access institutions designed to empower them.

Human rights education could also be beneficial to policymakers and administrators. In order to effectively adopt the HRBA, HRE is essential so that staff can be familiar with human rights standards, including human rights laws, their monitoring and implementation mechanisms.[186] For example, climate change administrators that have little or no experience or training on human rights would need a high level degree of training and education on human rights in order to understand how to apply and implement human rights in their areas of work.[187]

Effective administration of justice For the public to feel protected, there is a need to establish a process that effectively guarantees access to justice.[188] Effective administration of justice relates to removing every identifiable barrier to justice when projects clearly violate human rights, and providing remedies or compensations for victims of such violations. The UNDP notes that the HRBA would aim to remove barriers that have been observed to hinder the enjoyment of human rights.[189] These barriers include long delays; prohibitive costs of using the system; lack of available and affordable legal representation that is reliable and has integrity; formalistic and expensive legal procedures; abuse of authority

[185] A great number of Nigerians also understand climate change in terms of change in weather pattern, thus they complain of 'climate change' when rain does not fall or when they get too much rainfall. See This Day, 'Nigeria: Citizens See Climate Change as an Act of God' allafrica.com/stories/200903120056.html, accessed 11 March 2015.

[186] Hamm illustrates this with the example of an expert from the German Ministry of Economic Cooperation and Development (BMZ), who notes that development experts in his ministry lack knowledge of human rights and thus they are not in a position to implement the HRBA. See Hamm, 'A Human Rights Approach to Development', (n 1), 1005–1031.

[187] See D. Olawuyi, 'Recognizing the Intersections between Human Rights and the Environment in Legal Education and Training', *Asian Journal of Legal Education*, 1(2) (2014), 103–113.

[188] See S. Golub, 'Beyond Rule of Law Orthodoxy: The Legal Empowerment Alternative', *Carnegie Endowment for International Peace*, (2003), 2–3.

[189] UNDP, *Programming for All*, (n 161), 4–5.

and powers, resulting in unlawful searches, seizures, detention and imprisonment; and weak enforcement of laws and implementation of orders and decrees. It also includes the severe limitations in existing remedies provided either by law or in practice as most legal systems fail to provide remedies that are preventive, timely, non-discriminatory, adequate, just and deterrent.

To ensure the effectiveness of established complaint mechanisms, they must be accessible to everyone including the poor and marginalized in the society, such as women, children, poor and uneducated people who are often marginalized in decision-making processes for carbon projects in many developing countries. The international climate regime could place an obligation on parties to demonstrate that adequate legal assistance is readily available to members of the public that cannot afford the cost of enforcing their rights before the public complaint boards. The emerging regime could also allow non-governmental organizations to bring joint actions on behalf of the affected communities. Finally, there is need for the creation of an international monitoring group to ensure that after registration and approval, the execution of projects are in line with approved standards and do not create unanticipated problems.

5.3 Conclusion

A detailed understanding of the normative elements of the HRBA is a *sine qua non* for designing rights-based risk management frameworks that will ensure that climate actions and projects do not exacerbate human rights violations, poverty, societal exclusions and discrimination. To address human rights risks in carbon projects, the international climate change regime must integrate procedural safeguards that mandate project proponents to demonstrate full adherence to human rights norms and principles, particularly those on ATI; participation; accountability; equality and non-discrimination; and access to justice. These elements form the crux of an approach that could help anticipate, prevent and avoid the human rights consequences produced by current and future carbon projects.

Adopting the HRBA to carbon finance would mean providing project approval guidelines and thresholds that would stipulate that any project proposal that does not guarantee or satisfy human rights requirements would not be approved no matter how many tonnes of emission reductions it could generate. This could ensure that carbon projects are implemented without violating the rights of the public. To mainstream these human rights requirements in the international climate change regime, there is a need for legal and institutional reforms that provide legal basis for supervisory bodies for carbon projects to adopt and apply these human rights requirements in the first place.

The next chapter examines the question of 'how', that is, how can these human rights norms be mainstreamed into the international climate change regime, what are the legal and institutional frameworks required? It discusses a legal framework that could assist in moving the idea of integrating human rights requirements and principles into the international climate change regime from theory to practice.

6

Legal framework for implementing the human rights-based approach to carbon finance

6.1 Introduction

Developing conceptual formulations and standards on human mainstreaming is only a part of the task. Other essential questions relate to the mechanisms to be put in place to practically implement a human rights-based approach (HRBA) to carbon finance. How theoretical concepts would be operationalized and translated into concrete action plans; how progress would be monitored and evaluated; and what further action would be taken in the case of unsatisfactory progress; are all key questions that must be answered. These questions are examined in this chapter. This chapter discusses a legal framework for implementing the HRBA to carbon finance.

6.2 Minimum legal threshold for project approval

To mainstream human rights norms into the international climate change regime, there is need for a project approval framework that would ensure that actions, measures or projects that do not guarantee or satisfy the requirements of accountability, access to information (ATI), participation, equality and access to justice would not be approved no matter the volume or value of carbon credits from such action or project. Therefore, there is need for a legal framework that sets out the modalities for enforcing the linkages between carbon finance and human rights. This framework would provide in concrete terms rights-based modalities for project approval and the human rights conditions that must be met before projects can be approved. This would be achieved by establishing a minimum human rights threshold for project approval and registration. This legal threshold would lay down the levels of protection for individual human rights, which would be regarded as the minimum acceptable outcome under a given project or policy scenario.[1]

[1] On the importance of a legal threshold, see S. Caney, 'Cosmopolitan Justice, Responsibility and Global Climate Change', *Leiden Journal of International Law*, 18 (2005), 747–775;

The framework would also set out modalities that contain these threshold levels that should not be breached either directly by the climate change itself or as a result of policy measures or emission reduction projects. The framework would also be the basis to grant supervisory bodies, such as the CDM EB and the UNFCCC Secretariat, powers and a mandate to consider human rights issues and to enforce this threshold.

For a rights-based reform of the international climate change regime, there is a need to establish a rights-based legal threshold or performance standards that project proponents must meet before project approval and registration. Such performance standards (threshold) would emphasize the importance of an integrated human rights impact assessment to identify human rights risks of projects, the importance of effective community engagement through disclosure of project-related information; consultation with local communities on matters that directly affect them; and the need for project proponents to manage and disclose the human rights concerns.[2] Establishing a rights-based legal threshold under the international climate change regime would provide clear guidelines that projects are required to meet before they could be approved, financed, registered and recognized for the purpose of gaining emission credits under the international climate governance architecture. A clear legal threshold at the international level could also spur reform at the national level as it could serve as guidance on what policymakers should look out for in order to design a balanced and rights-based policy response and when designing carbon projects.

This section discusses a six-part legal threshold that could serve as the minimum core content of a rights-based international climate change regime on carbon finance: the project eligibility screen, sustainable development (SD) screen, human rights impact assessment (HRIA), stakeholder identification and consultation, information disclosure system and project review mechanism. The threshold addresses the need for a pre-project analysis of any project to

S. Caney, 'Global Justice, Rights and Climate Change', *Canadian Journal of Law and Jurisprudence*, 18(2) (2006), 255–278; see also ICHRP, *Climate Change and Human Rights: A Rough Guide* (Geneva: International Council on Human Rights Policy, 2008), 18.

[2] See, for example, social and environmental performance standards of the International Finance Corporation, 'Performance Standards on Social and Environmental Sustainability' 30 April 2006, Performance Standard 1, Social and Environmental Assessment and Management Systems, para. 2.24.

ascertain its nature and the manner in which it would be planned and executed.

6.2.1 Pre-project eligibility screen

This test illustrates the need for a detailed description and analysis of a project proposal, to ascertain the nature, objectives and likely impact of the project. Preliminary questions here would include whether the nature and scope of the proposed project raise any conflict with local priorities, the nature of technology involved in the proposal the proposed location of the project, including the criteria for site selection; and also whether the project could potentially affect the quality of the environment of the proposed location. To effectively answer these questions, a detailed environmental impact assessment is important early on in the project planning stages. For example, one key concern with the implementation of the CDM is that the CDM Rules do not provide for a compulsory EIA if the host country does not have specified requirements for an EIA.[3] Specifically, under the CDM Rules, EIA is only imperative 'if the project has the potential for a negative environmental impact'. The gap in this requirement is that it leaves the definition of a 'project that has a potential negative impact' in the hands of national authorities. This has resulted in a situation whereby if the DNA determines that a project does not have any negative environmental impact it would go ahead, leaving room for short- and long-term human rights risks. To ensure transparency and accountability in project design and implementation, there is a need for mandatory EIA for all projects at the planning stages. This would help shed light on the nature of the project and its likely social and environmental effects.

Furthermore, results of the pre-project analysis should be included in a comprehensive project design document (PDD), which would, among other things, detail the purpose and nature of the project, the likely short- and long-term environmental impact and alternative measures to reduce the impact. The nature of information to be disclosed will include a detailed description of physical and technical characteristics of the proposed activity, including an estimate of the expected residues and emissions; the size, pace, reversibility and scope of any proposed

[3] See CDM Modalities & Procedures, 3/CMP.1 (2005), cdm.unfccc.int/Reference/COPMOP/08a01.pdf, accessed 12 June 2015.

project or activity; the duration of the project and/or activity; the locality of areas that will be affected; a preliminary assessment of the likely economic, social, cultural and environmental impacts, including potential risks; a description of the measures envisaged to prevent and/or reduce the effects, including emissions in line with the precautionary principle; a description of personnel likely to be involved in the execution of the proposed project (for example indigenous peoples, private sector staff, research institutions, and government, and employees); procedures that the project may entail; information as to whether the activity is subject to a national or transboundary environmental impact; a non-technical summary of the foregoing information and an outline of the main alternatives studied by the applicant; as well as the main report and advice issued to the public authority and the time when the public concerned shall be informed of the proposed activity.[4] Information on projects and policies would also need to be completely disclosed.

The pre-project analysis of project location would also help identify, with precision, the various stakeholders that might be affected by a project. This would help to identify options for redesigning such project, so as to minimize negative impacts and risks. For example, if this analysis is done early on at the planning stage, it becomes easier to find out how many people would lose their homes due to this project. This data would help with the decision as to whether there is a need for an alternative site location for the project and how to resettle or compensate those who would be inevitably affected.

6.2.2 Sustainable development screen

With renewed global focus on ending poverty, fighting inequality and tackling climate change by 2030, as parts of the United Nations Sustainable Development Goals, the need to balance development projects or goals with the attainment of sustainable development has gained

[4] See Article 6(6) of the Aarhus Convention on Access to Information, Public Participation in Decision-Making and Access to Justice in Environmental Matters (Adopted 25 June 1998, entered into force 30 October 2001) No. 37770, treaties.un.org/doc/Publication/MTDSG/Volume%20II/Chapter%20XXVII/XXVII-13.en.pdf, accessed 19 March 2015; see also Report of the International Workshop on Methodologies Regarding Free Prior and Informed Consent E/C.19/2005/3, endorsed by the *United Nations Permanent Forum on Indigenous Issues* (UNPFII) at its Fourth Session in 2005.

renewed vigour.[5] To achieve sustainable development goals in climate mitigation, there is a need to develop robust legal and institutional frameworks to ensure that climate actions and projects do not hamper sustainable development, that is, combating climate change in a way that aids and not hinders sustainable development goals.

Under this requirement, the goal would be to balance the goals of emission reduction and sustainable development so as to avoid projects that lead to trade-offs and the sacrifice of sustainable development for emission reduction.[6] Carbon projects have come under intense scrutiny for not achieving the dual goals of emission reductions and sustainable development.[7] Examples include the cases highlighted in this book. Implementation of the CDM has shown that sustainable solutions are not necessarily the most cost-effective as they often require new and more expensive technologies. Thus, projects that may lead to cleaner air, fewer emissions and better health require new and costly technologies, while the affordable and less costly projects may not have meaningful and sustainable impacts on the host country of the project.[8] Consequently, developed countries have so far had to choose between investing in projects involving expensive and new technologies which could lead to greater sustainability or settling for cost-effective projects that lead to emission reductions but that do not lead to real and lasting sustainable development for the host country. Many developed countries have settled for the latter.[9] For example, wind power, although more expensive, has

[5] United Nations Development Program, Sustainable Development Goals, www.undp.org/content/undp/en/home/mdgoverview/post-2015-development-agenda.html, accessed 13 December 2015.

[6] The Kyoto Protocol set out to achieve two main objectives with the project-based mechanisms: emission reduction and sustainable development (SD). For example Article 12(2) of the Kyoto Protocol identifies two specific goals for the CDM: (a) to assist developing countries in achieving sustainable development, (b) to assist industrialized countries in complying with their emissions reduction commitments, and to contribute to the attainment of the environmental goals of the Framework Convention. See Conference of the Parties to the Framework Convention on Climate Change, *Kyoto Protocol*, 10 December 1997, UN Doc FCCC/CP/1997/L.7/add.1, 37 ILM 22 (1998) [Kyoto Protocol].

[7] See Olsen, 'The Clean Development Mechanism's Contribution to Sustainable Development: A Review of the Literature', cd4cdm.org/Publication/, accessed 21 January 2014.

[8] The term 'Host Country' refers to the country where a project is located or carried out while the industrialized country carrying out such a project is described as the project proponent. See Myung-Kyoon (ed.), CDM Information and Guide Book.

[9] See A. Olhoff, *CDM Sustainable Development Impacts* (UNEP Risoe Center on Energy, Climate and Sustainable Development), 10.

no significant environmental side effects and is as such more attractive in terms of sustainable development. However, it is not often pursued because of its cost.[10]

Implementation of the CDM has therefore seen a rise in the spate of projects that deliver carbon credits and emission reductions but do not contribute to the host country's sustainable development. As Seroa da Motta rightly noted, when a cleaner technology, with more cost and more sustainable development prospects, will result in the same level of emission reduction as a dirtier and cheaper technology that has adverse environmental effects and lesser sustainability, there is a tendency for most developed countries to settle for the most cost-effective project, irrespective of its sustainable development prospects and its long-term effect to the host country.[11] Commenting on this trade-off, Sutter argued that no registered CDM project is likely to fulfil the Kyoto Protocol's two-fold objectives of delivering GHG emission reduction and contributing to sustainable development at the same time.[12]

Arguably, the current trade-offs are attributable to the fact that a number of developed countries attempt to generate high volumes of carbon credits at the least cost possible, without paying proper attention to associated sustainable development and human rights implications. Taking the Panama project as an example, while the project is appealing from a cost-benefit perspective, it is not as appealing from an environmental and human rights perspective. The negative environmental impacts of the project, coupled with the attendant human rights impacts, make this project the least attractive in terms of sustainable development. These trade-offs are also manifest in the other examples described in this book.

Similarly, failure to enforce sustainability in CDM projects is attributable to lax understanding and application of sustainable development criteria at both national and international levels. This can be attributed to the lack of capacity and technical knowledge in some countries on how to effectively set standards for SD and how to measure progress.[13]

10 *Ibid.*

11 R. Seroa de Motta, 'Brazil: CDM Opportunities and Benefits', in D. Austin and P. Faeth (eds.), *Financing Sustainable Development with the Clean Development Mechanism* (Washington DC: World Resource Institute, 2000), 2–4.

12 See C. Sutter, 'Does the Current Clean Development Mechanisms Deliver Its Sustainable Development Claims', (2005) *HWWA Report*, 1.

13 H. Kimura, A. Srinivasan, and K. Iyadomi, 'Clean Development Mechanism' in A. Srinivasan *et al.* (eds.), *Asian Aspirations for Climate Regime beyond 2012* (Institute for Global Environmental Strategies, 2006), 36.

A number of projects that have short- and long-term human rights impacts are endorsed as sustainable at national levels due to limited understanding of the practical application of the concept of sustainable development. In many developing countries, national governments have failed to establish clear and detailed national SD agenda that avoids trade-offs. The result is a culture of approvals in which projects that do not promote sustainable development gain host country approval.

Furthermore, relying on SD criteria set by national governments has been a major weakness of the accountability of the CDM process. Relying on decisions and assessment by the host country that a project meets its sustainable development standards has led to a situation whereby states in a bid to attract projects lower sustainable development standards and approve all manner of projects as meeting their sustainable development standards, including projects that produce adverse environmental and human rights effects. The need for avoiding this sort of trade-off in the international climate change regime cannot be over-emphasized.

State efforts and projects aimed at achieving emission reductions should not conflict with sustainable development and respect for human rights.[14] Generating credits through projects that harm the environment and violate fundamental human rights is tantamount to transferring problems from one sector to another. There is therefore a need to carry out a thorough analysis of current experiences gained with CDM projects to inform and establish detailed sustainable development criteria that would serve as common guidelines on how to measure the SD impact of a project. This unified standard would impose a form of quality control on all market-based projects to avoid the current situation whereby such decisions are taken solely by the host governments. A common sustainability standard must be placed at the heart of a new international climate change regime as a condition that must be met before projects can be registered for credits.

Indicators for assessing sustainability, as discussed in this book, encompass balancing two complementary pursuits that are often portrayed as opposing: emission reduction and respect for human rights. It is

[14] See C. Figueres, 'Sectoral CDM: Opening the CDM to the Yet Unrealized Goal of Sustainable Development', *McGill International Journal of Sustainable Development Law and Policy*, 2(1) (2006), 5–25.; see also C. Streck and J. Lin, 'Making Markets Work: A Review of CDM Performance and the Need for Reform', *European Journal of International Law*, 19(2) (2008), 409–442; S. Cohen, 'Climate Change and Sustainable Development: Towards a Dialogue', *Global Environmental Change*, 8 (4), (1999), 341.

important to expound these indicators as necessary criteria that a project must meet before it can be approved under the sustainable development threshold. By exploring the main linkages between carbon projects and their impact on the economic, environmental and social pillars of SD (which includes respect for human rights), countries can design and select projects that are associated with the largest development benefits.[15] Examples of factors to be demonstrated under each indicator include the following:

i) Economic criteria

For a carbon project to be considered sustainable, it must lead to financial returns and profits to local entities, increased investments in priority sectors, generation of local employment opportunities, a positive balance of payments and technology transfer to the host country.[16] A project must result in improved economic activities in the host country and should add to the economic development of the host country. Projects that do not encourage positive economic returns in the host country are unsustainable.[17] A project should also result in the transfer of clean and cost-effective technologies that replace dirty technologies in the host country.[18]

ii) Social criteria

Carbon projects must provide increased social benefits, equity and improved quality of life in the host country. Sustainable carbon projects should alleviate social concerns and problems in the host country, by providing relevant infrastructure that is either non-existent or inefficient in the host country. Examples include improved access to clean, affordable and sustainable water supply; improved transportation through projects that emphasize efficient transportation and less reliance on fossil fuels; improved health facilities services through transfer of up-to-date and energy efficient technologies; and improved power supply through the transfer of alternative energy sources such as solar panels,

[15] See UNFCCC, 'Benefits of the Clean Development Mechanism 2012', cdm.unfccc.int/about/dev_ben/ABC_2012.pdf, accessed 21 March 2015; Olhoff, *CDM Sustainable Development Impacts*, (n 9), 17.

[16] See Olsen, 'The Clean Development Mechanism's Contribution to Sustainable Development', (n 7).

[17] Olhoff, *CDM Sustainable Development Impacts*, (n 9), 17–18.

[18] R. Bruneau, 'Maximizing Benefits from the Clean Development Mechanism: Developing Country Supply of Carbon Sinks and Technological Transfer Projects', *International Environmental Affairs* (2006), 6.

wind projects, clean cook stoves and hydro-electric power projects. It must also foster greater community participation, reduction of wealth disparities, increased capacity building and trainings.

Similarly, the UNFCCC has recently identified respect for human rights as part of the social criteria of development. The UNFCCC takes an expansive view of sustainable development, by classifying respect and consideration of the rights of local/indigenous people and community or local/regional involvement in decision-making, as part of the social indicators of sustainable development.[19] This interpretation, which embeds respect for human rights as part of social development, must be reflected in the international climate change regime.[20] Under this test, any carbon project that produces human rights impacts or does not positively improve the quality of life of the public in a host country cannot be adjudged as leading to sustainable development.[21]

iii) Environmental criteria

This is arguably the most important criterion of all. Carbon projects should lead to progress in environmental issues, including reduction of GHGs and the use of fossil fuels, reduced air and water pollution, conservation of local resources, reduced soil erosion from deforestation, improved health and reduced pressure on local environments.[22] Carbon projects must result in genuine emission reductions and an improvement in the overall quality of the environment in the host country.[23]

A number of organizations and project verification agencies have also developed helpful sustainability standards that provide guidelines on what

[19] UNFCCC, 'Benefits of the Clean Development Mechanism 2012', (n 19).

[20] See the 2009 United General Assembly Resolution (A/RES/64/236), which also describes economic development, social development and environmental protection as interdependent and mutually reinforcing pillars of sustainable development, daccess-dds-ny.un .org/doc/UNDOC/GEN/N09/475/99/IMG/N0947599.pdf?OpenElement, accessed 25 May 2015.

[21] *Ibid.* On how CDM cook stoves can help households in Nigeria, see UNFCCC, 'New Proposed Methodology: Switching to Biomass Residues as the Fuel Source for Individual Stoves' (2009), cdm.unfccc.int/methodologies/PAmethodologies/publicview.html?meth_ref=NM0307, accessed 12 August 2015. See also UNFCCC, 'Improved Cooking Stove for Nigeria', cdm.unfccc.int/ProgrammeOfActivities/poa_db/7R1B09HSJV3FKIZYCA4D6XQOETP5GN/view, accessed 12 August 2015.

[22] Olsen, 'The Clean Development Mechanism's Contribution to Sustainable Development', (n 7).

[23] *Ibid.* See also L. Schneider, 'Is the CDM Fulfiling Its Environmental and Sustainable Development Objectives? An Evaluation of the CDM and Options for Improvement', a report prepared for the World Wildlife Fund (Berlin: Öko-Institut, 2007).

project proponents should demonstrate under each indicator. For example, the Gold Standard Foundation – an internationally recognized project verification agency – developed a comprehensive set of indicators for a sustainable development screen.[24] To obtain a Gold Standard Certificate, all project proponents are 'required to demonstrate that their project activities will have clear sustainable development benefits through a Detailed Impact Assessment'. All project proponents are required to demonstrate that their project activities will have no severe negative environmental, social and/or economic impact, through a 'Do No Harm' Assessment.[25] Projects that fail to satisfy the Gold Standard requirements on sustainability are rejected by the foundation. The Gold Standard achieves a high-quality standard by using the three sustainable development indicators for assessing projects in what it calls a 'sustainable development matrix'.[26]

A sustainable development screen examines how a project would contribute to social, economic and environmental progress. Developing and integrating sustainability standards into the design, approval, finance and implementation of carbon projects would ensure that projects that do not demonstrate sustainability under each indicator are not approved. It is pertinent for the international climate change regime to incorporate a sustainable development screen as a mandatory threshold to be met by project proponents. Criteria for sustainable development are linked with human rights. The linkages between human rights and sustainable development have been increasingly recognized.[27] As such, many of the

[24] The Gold Standard Foundation was established by the World Wildlife Fund (WWF) in 2003. It provides certification standards for carbon mitigation projects and is recognized internationally as the benchmark for quality and rigour in both the compliance and voluntary carbon markets. The Foundation certifies renewable energy and energy efficiency carbon offset projects to ensure that they all demonstrate real and permanent greenhouse gas (GHG) reductions and sustainable development benefits in local communities that are measured, reported and verified. The Gold Standard certification has been endorsed by more than 80 NGOs worldwide, including the United Nations. To be eligible for Gold Standard certification, projects must: employ renewable energy and/or energy efficiency technologies; adhere to the strictest standards on additionality and positively impact the economy, health, welfare and environment of the local community hosting the project. See the Gold Standard Foundation, www.cdmgoldstandard.org/about-us/who-we-are, accessed 12 March 2015.

[25] The Gold Standard Foundation, *Gold Standard Eligibility Requirements*, (2009), www.cdmgoldstandard.org/wp-content/uploads/2011/10/GSv2.1_Requirements-11.pdf, accessed 12 March 2015.

[26] *Ibid.*

[27] For a most recent discussion of the linkages between human rights and sustainable development, see the Statement of the UN High Commissioner where she notes that 'Strategies based on the narrow pursuit of economic growth without due regard for equity

Table 6.1 *Sustainable development criteria*

Social dimensions	Economic dimensions	Environmental dimensions
Improvement in quality of life	Provision of financial returns to local entities, employment	Reduction of GHG emissions and the use of fossil fuels; reduced air and water pollution
Poverty reduction	Positive impact on balance of payments	Conservation of local resources and reduced pressure on local environments
Improvement in community participation and equity issues, capacity building, health access	Transfer of environmentally sound technologies	Meeting local renewable energy portfolio standards and other environmental standards

current inequities generated by carbon projects often point to the fact that they are not sustainable projects and do not meet the sustainable development threshold level set out in the foregoing. If a project satisfies the entire elements of the sustainable development threshold, there would arguably be a diminished probability for such a project to generate human rights concerns.

Table 6.1 summarizes the key indicators of sustainable development under the social, economic and environmental pillars. These key indicators could guide governments, project proponents and relevant stakeholders in screening a carbon project for its sustainable development prospects.

The sustainable development screen would ensure that projects that could potentially have an adverse social, economic and/or environmental

and related environmental, social and human rights considerations, will both fail in their economic objectives, and risk damaging the planet and the fundamental rights of people' See OHCHR, 'Pillay urges states to inject human rights into Rio+20' (18 April 2012), www.ohchr.org/en/NewsEvents/Pages/DisplayNews.aspx?NewsID=12070&LangID=E, accessed 21 April 2015. See also N. Pillay, 'Open Letter by the United Nations Human Rights Commissioner' (30 March 2012), www.ohchr.org/Documents/Issues/Development/OpenLetterHC.pdf, accessed 14 April 2012, 2; UNFCCC, 'Benefits of the Clean Development Mechanism 2012', (n 19).

impact are not approved or registered for emission reduction credits under the international climate change regime.

6.2.3 Human rights impact assessment

In order to anticipate and address human rights risks in carbon projects, project proponents must, at the very early stages of a project, assess, quantify and evaluate the likely impact of the project on fundamental human rights. They must also demonstrate or describe efforts put in place to mitigate or avoid such human rights risks. The Ruggie Framework expatiates on the need for business enterprises to carry out 'human rights due diligence' or a 'human rights risk assessment' before executing projects.[28] HRIA is the process for systematically identifying, assessing, predicting and responding to potential human rights that might be affected or put at risk by an activity, operation or project.[29] It includes assessing actual and potential human rights impacts, evaluating human rights risks data, developing risk management or mitigation plans, integrating and implementing such plans, tracking responses and communicating how impacts are addressed.

Different carbon projects are likely to affect certain rights more than others, and in different ways. For example, hydropower projects may result in flooding of farmlands and neighbouring houses. This could violate several human rights, including subsistence rights where crops are destroyed, a violation of the right to property where the reservoir floods homes or the right to health where the flooding causes transfer of

[28] According to the Framework, in order to identify, prevent, mitigate and account for how they address their adverse human rights impacts, business enterprises should carry out human rights due diligence which: (a) Should cover adverse human rights impacts that the business enterprise may cause or contribute to through its own activities, or which may be directly linked to its operations, products or services by its business relationships; (b) Will vary in complexity with the size of the business enterprise, the risk of severe human rights impacts, and the nature and context of its operations; (c) Should be ongoing, recognizing that the human rights risks may change over time as the business enterprise's operations and operating context evolve. J. Ruggie, 'United Nations Guiding Principles on Business and Human Rights: Implementing the United Nations "Protect, Respect and Remedy" Framework' (2011) UN Document A/HRC/17/31 (21 March 2011), para. 17.

[29] See OHCHR, *Human Rights Translated: A Business Reference Guide* (Geneva: OHCHR, 2008), xvii, www2.ohchr.org/english/issues/globalization/business/docs/Human_Rights_Translated_web.pdf, accessed 12 December 2015. See also, generally, OHCHR Briefing Paper, *The Global Compact and Human Rights: Understanding Sphere of Influence and Complicity* (Geneva: OHCHR, 2004), 18–19.

diseases. A REDD+ project may not necessarily produce the same human rights consequences. It is more likely to violate the rights of indigenous communities to use lands for cultural and religious purposes and the right to food. It is therefore important to identify with precision through a self-reporting system the human rights that would likely be affected by a project and the measures that would be taken to avoid these impacts.

Through a HRIA, project proponents and other stakeholders could systematically identify, anticipate and respond to the potential human rights impacts or risks of a carbon project.[30] Adopting explicit human rights policies, along with mechanisms for their implementation and reporting, provides adequate opportunities for an investor or project proponent to prevent human rights risks and to deal effectively with any allegations of human rights wrongdoing that may arise.[31]

The HRIA is different and more comprehensive than 'human rights auditing' that has been suggested in the literature.[32] The key questions under a HRIA are which human rights concern does this project raise or could it raise? Which groups are likely to be affected by these human rights concerns? Which specific rights are affected by this project or policy? And what efforts would be taken to address these issues in the process of project design and implementation? A HRIA would provide

[30] S. Walker, 'Human Rights Impact Assessments of Trade-Related Policies', in M. Gehring and M.C. Cordonier-Segger (eds.), *Sustainable Development in World Trade Law* (The Hague: Kluwer Law International, 2005), 217–256.

[31] OHCHR, *Human Rights Translated: A Business Reference Guide*, (n 29), viii.

[32] Simma suggests a system of Human Rights Audit, that would be a joint undertaking of foreign investors and host States, to survey the host State's human rights treaty commitments especially in the field of economic and social rights and methods for implementing such commitments (such as doctrines of incorporation or transformation). According to Simma, the ultimate result of conducting a 'human rights audit' would be a better definition of the landscape of the foreign investor's 'legitimate expectations' in a way that would not leave excessive *ex post* discretion to arbitrators, should investor–host State disputes arise in the future. On the one hand, foreign investors would be able to better estimate and prepare for alternative scenarios of regulatory measures which the host State might take to vindicate its economic and social human rights obligations, and on the other, host States would not be unduly constrained from defining their public policy agenda as a result of investment protection guarantees within foreign investment contracts and their corresponding treaties. As Simma notes, 'human rights audit would not be the same as "human rights impact assessments", which is of a much broader scope, very detailed and fact-intensive and spanning inter-disciplinary approaches'. HRIA goes beyond evaluating human rights instruments or measuring compliance, it includes on-the-ground assessments of human rights issues in a project and providing holistic solutions on how to avoid unintended impacts. B. Simma, 'Foreign Investment Arbitration: A Place for Human Rights?' *International and Comparative Law Quarterly*, 60 (2011), 595–596.

clear and comprehensive answers to these questions to avoid any element of surprise or secrecy in project implementation. The HRIA goes beyond only evaluations or human rights auditing, in that it identifies areas of overlap between human rights and a project and also holistic solutions and methods for avoiding the identified human rights impacts.

The earliest call for HRIA emerged in 2001 with respect to trade agreements. In its resolution 2000/7, the Sub-Commission for the Promotion and Protection of Human Rights requested the High Commissioner for Human Rights to undertake an analysis of the human rights impact of the Agreement on Trade-Related Aspects of Intellectual Property Rights (the TRIPS Agreement) of the WTO.[33] The High Commissioner was called upon to examine how the TRIPS Agreement could affect the enjoyment of several rights – in particular, the right to health, food, development and the human rights of indigenous peoples. The report of the High Commissioner demonstrated how the TRIPS agreement could affect human rights, especially the right to health, and discussed the need for HRIA mechanisms to be mainstreamed into the TRIPS and into national trade policies. In 2011, the United Nations adopted the Ruggie Framework, which among other things gives a mandate to business enterprises and multinational corporations to ensure non-infringement of human rights principles by conducting human rights 'due diligence'.[34] This due diligence approach has been developed to include the conduct of a compulsory HRIA for project proponents to 'know and show' that they respect human rights. In March 2012, the United Nations Special Rapporteur on the right to food, Olivier De Schutter, unveiled a set of seven guiding principles on HRIAs, in relation to trade and investment agreements.[35]

The evolution and increased recognition of HRIAs underscores the importance of ensuring that the international climate change regime is consistent with obligations under international human rights

[33] United Nations, 'The Impact of the Agreement on Trade-Related Aspects of Intellectual Property Rights on human rights', Report of the High Commissioner E/CN.4/Sub.2/2001/13.

[34] United Nations, 'Guiding Principles on Business and Human Rights: Implementing the United Nations "Protect, Respect and Remedy" Framework' (2011) UN Res. A/HRC/17/31, para. 17.

[35] OHCHR, 'Guiding Principles on Human Rights Impact Assessments of Trade and Investment Agreements', www2.ohchr.org/english/issues/food/docs/20110701Draft_Guiding_Principles_on_HRIA.pdf, accessed 23 February 2015.

instruments.[36] HRIAs would help measure the impact of carbon projects, which are generally not designed with human rights in mind, to reduce potential negative human rights effects and increase the positive gains of a project.[37] They would complement other impact assessments such as the EIA and the SD screen. The main difference is that it would be framed by appropriate international human rights principles and conventions to demonstrate how a project could affect the recognized human rights of stakeholders and how best to directly provide a leeway to those people whose rights may be at risk.[38] A HRIA would help to produce comprehensive analyses that show a direct link between a project activity and how it might affect human rights. It takes a project-by-project approach to aid a quick understanding of the current situation, issues and the human rights impacts both now and in the future. It also helps to shed light on measures that could be taken to avoid any anticipated impact.

HRIA would also require project proponents to demonstrate the fair distribution of projected risks. It would help to assess the implications of projects for low-income people, the representation afforded to such categories in decision-making and ways to protect the interests of the marginalized through project re-design or alternative implementation plans. It would make it compulsory for project proponents to demonstrate that a particular section of the society is not exceptionally disadvantaged by a project before the said project could be approved. By developing concrete risk assessment procedures to ensure better characterization of risk across populations, communities or geographic areas, measures would then be put in place to reduce high concentrations of risk among specific population groups.

Table 6.2 summarizes the core contents of a HRIA. As seen in the table, the HRIA is a fact-based assessment of any human rights impact or risk that may result from the implementation of a project. By identifying and aggregating all of the risks, project proponents, supervisory bodies and other stakeholders are able to mitigate and eliminate such risks.

To be effective, a HRIA should be initiated as early as possible, at project planning and design stages. Conducting a HRIA should be incorporated as a pre-condition for project approvals for participants involved in carbon projects, ranging from project proponents, supervisory institutions, financiers, auditors, verifiers to service providers such as law firms and

[36] See Human Rights Impact of Trade and Investment Pacts (27 February 2012), www.twnside.org.sg/title2/wto.info/2012/twninfo120211.htm, accessed 23 March 2015.

[37] *Ibid.* [38] OHCHR, *Human Rights Translated: A Business Reference Guide*, (n 29).

Table 6.2 *HRIA due diligence matrix*

Which human rights concern does this project raise or could it raise?	Example: Are there possibilities of flooding, air and water pollution, increased noise levels in the project construction and operational phase, mass displacements of people from their homes, loss of access to drinking water, loss of arable farmlands, risks of spillage or explosions?
Which groups are likely to be affected by these human rights concerns?	Which municipality, town or local government would feel these impacts the most? Which members of this locality stand to suffer the most consequences from this project, for example, farmers, fishermen, traders or school teachers?
	Evaluate and examine specifically that if the above occur, one or more of the following rights could be affected in the process: the right to life, the right to adequate food, the right to the highest attainable standard of health, the right to adequate housing, the right of indigenous peoples to the conservation and protection of the environment of their lands and resources, the right to self-determination, the right to safe drinking water and sanitation, and the right to subsistence
Which specific rights are affected by this project or policy?	Have the public concerned been consulted through the FPIC process?
What efforts would be taken to address these issues in the process of project design and implementation?	Have alternative routes or designs, that avoid key human rights risks, been considered for the project?
	What procedures have been put in place to provide redress or compensations for victims of the inevitable effects of the project

consulting agencies. This would ensure that any human rights impact would be put into consideration in planning and designing a project. It would also provide sufficient time for project proponents and participants to avoid the known consequences of a project. Where impact assessment indicates the possibility of potential human rights violations, it provides time for the project design to be revised to remove any incompatibility that has been found with pre-existing human rights obligations of the state concerned. This will prevent a situation whereby violations are detected only after a project has been commenced or completed. It provides an opportunity to anticipate and respond to the likely impact of a project on the human rights of citizens and stakeholders.

Not every possible impact of a project can be anticipated at the planning stage. As such, *ex ante* HRIAs should be complemented by HRIAs performed *ex post* once the effects can be quantified. HRIA should be regarded as an iterative process, taking place on a regular basis, for instance, every three or five years. Safeguarding clauses should be included to ensure that, should such *ex post* assessments lead to the conclusion that the state is unable to comply with its human rights obligations under the project scenario, registration for such projects would be re-appraised and if necessary withdrawn.[39]

It is pertinent for the international climate change regime to establish HRIA as a mandatory participation requirement for the design, approval, finance and registration of carbon projects. HRIA would enable project proponents to identify who may be affected; to catalogue the relevant human rights issues; to project how the proposed mitigation project and activity could have adverse human rights impacts on those identified; and to develop alternative approaches to avoid such problems. An effective HRIA would also enable project proponents to pay special attention to any particular human rights impacts on individuals from groups or populations that may be at heightened risk of vulnerability or marginalization and to bear in mind the different risks that may be faced by the poor and marginalized communities, women and other vulnerable groups.

6.2.4 *Information disclosure system*

This threshold would make it mandatory for project developers, national authorities and other key stakeholders, to establish public disclosure

[39] See OHCHR, 'Guiding Principles on Human Rights Impact Assessments of Trade and Investment Agreements', (n 35).

programmes that enhance easy and cost-effective dissemination of project information. This includes the need to place mandatory public disclosure requirements on project proponents and on national governments to publish on a regular basis all information related to a project, including new risks and challenges, even when citizens do not specifically make requests for such information. Information disclosure does not end at the planning stages, because as mitigation projects move from one stage of development to the other, information is most likely to change with new facts emerging and some previous project information becoming outdated or updated. There is thus a need to establish a comprehensive information management system that allows the public to receive the latest and most up-to-date information about approved projects. This information would enable the public to assess if projects are environmentally sound or if they have become harmful at some point along the line.

Disclosure of relevant project information on an ongoing basis helps affected communities understand the risks, impact and opportunities of the project. Consequently, this threshold would require public bodies to do more than accede to requests for information. They must also actively publish and disseminate key categories of information that are of significant public interest. Public bodies are to publish and disseminate widely documents that are of significant public value and interest, for example, information about how the public body functions and the content of any decision or policy affecting the public.[40] This places a duty on governments and project developers to run disclosure programmes that allow individuals to gain regular access to publicly held information. This could be by keeping a public database or by releasing periodic documents on governmental affairs. The aim is to reduce the culture of secrecy in governance and to ensure that the right to information is not only reactive but also proactive. Project developers must demonstrate that they have established a system of publishing information *suo motu* (proactively) on their own volition. The requirement here is to publish and inform citizens about known hazards relating to a project as soon as such information is available, even when members of the public have made no request. It takes access to information one step further by placing an obligation on custodians of publicly held information which is vital and important to the people to publish it *suo motu*. Examples of project information that should

[40] See Report of the Special Rapporteur, 'Promotion and Protection of the Right to Freedom of Opinion and Expression', UN Doc E/CN.4/2000/63, 18 January 2000.

be constantly released through public disclosure programmes include operational information, budgeting and costs, information on complaints, procedures for public input and the details of decisions taken regarding issues affecting the public.[41]

Furthermore, the process of obtaining information should not be tedious and complex, neither should it be designed to frustrate and discourage individuals from seeking publicly held information. Public bodies responsible for handling information requests must process all requests for information quickly and fairly. Unnecessary and duplicative procedures and administrative bottlenecks that complicate the process should be removed to make access easier and less demanding. Similarly, cost of access to information should not be so high as to deter requests. By establishing cost-effective disclosure programmes, the public are encouraged to seek and obtain publicly held information.

At national levels, it is important for countries to put in place institutions and procedures that facilitate and simplify the processes of obtaining public information. This can be done by publishing project updates on dedicated websites, sending periodic emails and newsletters to subscribing members of the public concerned, through radio, television documentaries and photos. It also includes delivering project information in the appropriate language and format understood by locals and in a way that such information reaches the most remote and rural communities.

Part of the information management system is for project developers to promote a culture of openness by providing the public information about their ATI rights, scope of information available and the manner in which the right can be exercised. This is different from the requirement to publish as it makes clear the need to promote public awareness on the existence of this right and how it can be enforced. Establishing an information system would be meaningless if people are not aware of their rights, the existence of ongoing projects and policies, as well as how to access more information about them. This element imposes an obligation on national governments to provide relevant information and awareness on institutions that are in charge of public information and the processes for accessing such information. Tailored educational and public awareness programmes are to be developed especially for vulnerable groups such as women, children and the least educated so as to

[41] Article 19, 'The Public's Right to Know: Principles on Freedom of Information Legislation' (1999): www.article19.org/pdfs/standards/righttoknow.pdf, accessed 12 May 2015.

provide them equal opportunities to be aware of important issues concerning projects and actions that might affect them.

As discussed earlier, one practical challenge to administering public disclosure programmes based on these standards is the cost and resources involved. To reduce the costs and resources required to establish and maintain public disclosure programmes, it is pertinent for national authorities to leverage on already existing structures and resources of different government programmes and institutions. For example, many countries already have freedom of information laws, which establish information access offices and programmes. There is a need to reflect on the flaws of these national programmes and how to improve them, such that ATI rights can be protected during the execution of mitigation projects. This will include strengthening or reforming inefficient institutions in line with the discussed ATI standards to make them better prepared to administer public disclosure programmes. DNAs would also serve as the focal point to coordinate information disclosure for carbon projects from different overlapping national institutions that are custodians of pertinent project information.

6.2.5 *Stakeholder identification and consultation*

The question that project proponents must answer under this threshold is 'have all key stakeholders been identified and consulted?' The first part of this test highlights the need for project proponents to demonstrate that all key stakeholders have been identified during the project analysis stage. This goes back to the question of expanding the definition of 'the public' in the international climate change regime to include two categories of the public who should be considered as key stakeholders and who should participate in decision-making, project planning and implementation.[42] They are the *specific public concerned* (individuals, landowners, community leaders, historically marginalized groups of women and youths, corporations and other business organizations) and *interested stakeholders* (representatives from non-governmental organizations (NGOs) and interests groups).[43] For example, the Aarhus Convention defines the public as 'one or more natural or legal persons and, in accordance with national legislation or practice, their associations, organizations or

[42] See C. Nwapi, 'A Legislative Proposal for Public Participation in Oil and Gas Decision-Making in Nigeria', *Journal of African Law*, 54 (2010), 184–211.

[43] See Arts. 2 and 4 of the Aarhus Convention, (n 4). See also Nwapi, 'A Legislative Proposal for Public', *ibid.*, 184–211.

groups'.[44] Article 2(5) of the Aarhus Convention also states that the public concerned means 'the public affected or likely to be affected by, or having an interest in, the environmental decision-making'.[45] An expansive definition of the public should include both the 'specific public concerned' by a project and 'interested stakeholders'. This will comprise individuals, landowners, community leaders, corporations and other business organizations, and representatives from NGOs and interests groups. Specific public concerned and interested stakeholders must be accorded opportunities to take part in and influence decision-making processes on the design and approval of carbon projects.

Project proponents must demonstrate that they have identified the public, that is, specific public concerned that will be affected by the impact of a project, and 'interested stakeholders' who have interests in or are concerned about the project or activity in question. The need to properly identify the key stakeholders cannot be over-emphasized. Some of the complaints in the implementation of CDM projects are that project proponents only liaise and discuss with some stakeholders, leaving out others. There is also the issue that project proponents and governments at times liaise with people who are not considered representatives of the community or who do not have the mandate of the community in decision-making. When dealing with representatives of the community, it is pertinent to verify their status and mandate as elected representatives of the public. This test also requires the participation of all members of the public affected by a decision irrespective of social status or societal strata. It would mandate the representation of all significant sectors of the society at the decision table.[46] It requires that citizens collectively and cooperatively analyse and discuss a problem with the view of proposing solutions.[47]

The second test here is whether the identified members of the public have been duly consulted and whether their Free Prior Informed Consents (FPIC) have been obtained. This should constitute a minimum legal threshold that must be demonstrated before a project is approved or registered. The FPIC standard on participation must be fully adopted as the requirement for good participation as opposed to mere ticking of a box or consultation. The FPIC would require that

[44] See Article 2(4) of Aarhus Convention, *ibid.* [45] Aarhus Convention, *ibid.*

[46] See J. Poiser, 'A Civic Republican Perspective on the National Environmental Policy Act's Process for Citizen Participation', *Environmental Law*, 26 (1996), 53 at 83.

[47] See J. Hartz-Karp and M. Briand, 'Institutionalizing Deliberative Democracy', *Journal of Public Affairs*, 9 (2009), 127.

consent be freely given, obtained prior to implementation of activities and be founded upon an understanding of the full range of issues implicated by the activity or decision in question.[48]

For FPIC to be effective, the affected public must be given timely information about the project and must be provided adequate opportunities to take part in and to influence the decision-making process. This would include providing a common platform for everyone to participate, either by choosing a suitable venue or strategic locations for meetings, by providing free mass transportation for the poor, ensuring participation by members of vulnerable groups such as women and minorities, hiring language interpreters for locals who cannot speak the common language and by reducing technicalities in discussions. Technological advancements could also be used to provide opportunities for online participation through webinars, online surveys and questionnaires.[49]

To avoid practical challenges to implementing a genuine participatory process, national authorities must streamline the processes in line with national peculiarities. For example, to avoid a situation whereby powerful NGOs and interest groups hijack or frustrate a consultative process, a way forward is to carry out an initial survey of relevant questions and information that are to be discussed and to arrive at an agenda for consultative sessions as early as possible in the process. This could be by identifying key members within the community to assist with collecting information. The criteria decided by the community would form an agenda for a participatory process. The agenda will provide a template for focused discussions and would lay down fair and transparent rules on the processes and voting methods to be adopted for decision-making. This would guide against deviations or unnecessary bickering. Selecting the appropriate facilitative role for the DNA, in line with local circumstances, would also be important. It must be decided whether the DNA

[48] See UNEP, *UN-REDD Programme Guidelines on Free Prior Informed Consent* (UNEP, 2012), 7. See also M. Colchester and F. MacKay, *In Search of Middle Ground: Indigenous Peoples, Collective Representation and the Right to Free, Prior and Informed Consent* (Moreton-in-Marsh: Forest Peoples Programme, 2004), 8–14; A. Perrault, K. Herbertson and O. Lynch, 'Partnerships for Success in Protected Areas: The Public Interest and Local Community Rights to Prior Informed Consent (PIC)', *Georgetown International Environmental Law Review*, 19(3) (2007), 477.

[49] For detailed discussions on the value of e-rulemaking, see S. Novek, 'Electronic Revolution in Rule Making', *Emory Law Journal*, 53 (2004), 434; see also C. Coglianese, 'Citizen Participation in Rule Making: Past, Present and Future', *Duke Law Journal*, 55 (2006), 943; A. Fung and E. Olin Wrights (eds.), *Deepening Democracy: Institutional Innovations in Empowered Participatory Governance* (London; New York: Verso, 2003).

should play an interactive, inclusive or interpretive role.[50] Selecting the appropriate role for the DNA would go a long way in reducing tensions with NGOs who due to previous experiences and mistrust might also have mutual suspicions that the DNA might hijack and frustrate the process. Generally, the most important point is that the decision to go ahead with a carbon project or not must reflect the popular opinion as expressed in the deliberative process.

6.2.6 *Project review mechanisms*

There is a need to establish project review and grievance redress mechanisms at national and international levels to provide the public opportunities to obtain redress or review of non-compliant projects. Practically, this will include establishing a project review committee or panel prior to project approval and registration and establishing a project appeal committee or board to address emerging complaints about approved or registered projects.

A project review mechanism is necessary to provide a chance for stakeholders, who have legitimate concerns or whose rights might be affected by a project, to raise their concerns and have them addressed prior to registration. It would enable project proponents to address any claims that affected persons may have early in the project planning stages. This would provide an opportunity for a remedy before disputes become inflamed. A review mechanism would make it possible for grievances to be identified, addressed and remediated early before they result in human rights violations.

[50] A participatory session is *interactive* if the facilitator plays only the role of gathering the stakeholders to a neutral venue and observing the decisions taken. In this approach, the facilitator is not involved in the process, it only listens, gathers facts and takes note of the final outcome and the decisions taken. It is *inclusive* if the facilitator becomes a member of the group and provides options and inputs on outcomes that could be adopted by the group. The interpretive mode is the middle ground where the facilitator presents its idea to the group and then becomes an outsider in order to allow the group to explore it, review other options and then make up its own mind on how to proceed. Each of these approaches has its peculiar strengths and suitability. It is important for the DNA to select its roles in deliberative sessions in line with local circumstances. Over the course of a deliberative process, different approaches could also be adopted at different stages. For example, at the beginning, the DNA might play an inclusive role and then step aside at the decision making stage to reduce pressures on local communities with respect to the final outcome and decisions to be taken. See generally, N. Sen, *Beyond Participatory Rituals: Strengthening Citizen Monitoring in Dharali* (New Delhi: UNESCO, 2012), 11–12.

Since it is practically impossible to precisely anticipate all the likely human rights impacts of a project, there is a need to establish redress or grievance mechanisms that could provide remedies for impacts that appear after approval or registration. Grievance mechanisms would also provide redress for members of the public who have been denied access to project information, participatory opportunities or whose other human rights have been violated by project proponents. Once grievances are identified, grievance mechanisms make it possible for such complaints to be addressed at the earliest opportunity before it escalates.

Generally, as new project information emerges, new human rights issues could crop up too. As such, it is not enough to only provide updated information on projects; it is pertinent for project proponents to provide a project complaint and review platform for stakeholders to establish complaints that have arisen especially after the initial consultations. Lack of procedural opportunities for the public to request review of projects before or after registration has been a reoccurring concern in the implementation of several carbon projects. This has created a culture of 'approvals' whereby different projects that evidently infringe human rights of host communities still inexplicably get approved and registered. This gap underscores the need for a rights-based reform of the international climate change regime.

A practical approach would be to establish project review panels that would obtain feedback from stakeholders on projects and to consider if and how these projects could affect them. For example, an investigative arm could be established under the international climate change regime to evaluate and examine how projects designed to attain emission reductions are affecting host communities. Such assessments can be through household perceptions, opinion surveys or a simple questionnaire; for example, 'do you think the Kwale project violates or could violate human rights?' Such feedback would allow the UNFCCC to monitor climate change actions and projects designed by parties to ensure they do not result in human rights consequences. It would also provide a chance to demand human rights enforcements from parties and to prevent human rights violations.

Apart from reforms to governance structures within the international climate change regime, project proponents and host countries should also establish their independent grievance mechanisms that would allow them to assess and understand the concerns of host communities and to provide adequate compensation as soon as such complaints occur. For example, the World Bank introduced a requirement for project proponents and clients to establish a grievance mechanism as an element of community engagement by the project sponsor.[51] This mechanism is part of the requirements of the Performance Standards on Social and Environmental Sustainability. According to this Standard, if the project proponents anticipate ongoing risks to or adverse impacts on affected communities, the client will establish a grievance mechanism to receive and facilitate resolution of the affected communities' concerns and grievances about the client's environmental and social performance. According to the World Bank, the grievance mechanism should be scaled to the risks and adverse impacts of the project. It should address concerns promptly using an understandable and transparent process that is culturally appropriate, readily accessible to all segments of the affected communities, and at no cost and without retribution. The mechanism should not impede access to judicial or administrative remedies and the project proponents are to inform the affected communities about the mechanism in the course of its community engagement process.[52] This Standard established by the World Bank is a good example of how project proponents and host countries can establish internal control mechanisms to address human rights impacts of projects. The international climate change regime could make such internal control mechanisms part of the threshold to be met or demonstrated by project proponents before project registration and approval.

6.3 Conclusion

Two complementary types of reforms are required to integrate human rights norms into the design, approval, implementation and finance of carbon projects: legal and institutional reforms. The necessary legal reforms include the need to entrench and adopt a minimum legal threshold that must be met and demonstrated before projects are approved and registered. The six-part legal threshold discussed in this chapter provides practical guidelines on the minimum core content of a rights-based international climate change regime, namely project eligibility screen, sustainable development screen,

[51] International Finance Corporation, 'Performance Standards on Social and Environmental Sustainability' 30 April 2006, Performance Standard 1, Social and Environmental Assessment and Management Systems, para. 23, www.ifc.org/ifcext/enviro.nsf/AttachmentsByTitle/pol_PerformanceStandards2006_full/$FILE/IFC+Performance+Standards.pdf, accessed 12 March 2015.

[52] *Ibid.*

human rights impacts assessment, stakeholder identification and consultation, information disclosure system and project review mechanism. This six-part legal threshold sets the levels of protection for individual human rights that should be regarded as the minimum acceptable outcome to be demonstrated by project proponents under any given project scenario. Taking human rights seriously in the international climate change regime would mean that any project that does not meet this threshold should not be approved or registered under the international climate change regime. In addition to meeting the requirements of this threshold, project proponents would be strongly encouraged to comply with applicable national laws, including those laws that implement host country obligations under international law. It is important for the international climate regime to adopt these requirements in order to effectively harmonize human rights obligations as part of climate change responses.

This threshold would represent a move from the current needs-based approach to carbon finance, under which human rights only come to the table when there is a problem with a project, to a transformative, empowering and process-oriented approach under which human rights are considered before and during project implementation. By requiring project proponents to demonstrate that this threshold has been met, a far-reaching obligation is placed on project proponents to consider the direct and indirect impacts of a particular project on the enjoyment of existing human rights. Project proponents are placed under the spotlight to demonstrate how they have anticipated and avoided the likely human rights risks of a given project. This threshold also provides a platform for stakeholders to demand their rights and to be involved in project planning and decision-making. This is the heart of a human rights-based approach - an opportunity to redress inequitable power relations by reconciling human rights and development activities.

To ensure a successful application of the above legal threshold in the design, approval, finance and implementation of carbon projects, there is a need to establish or empower institutions within the international climate regime to spearhead and practically apply the legal framework. The next chapter will examine institutional requirements and governance structures needed to implement the rights-based approach to carbon finance.

7

Mobilizing structures: institutional framework for implementing the human rights-based approach to carbon finance

7.1 Introduction

Identifying a legal framework through which human rights principles may be mainstreamed in the design, approval, finance and implementation of carbon projects is only part of the mainstreaming process. There is also a need to examine the institutional requirements and governance structures needed to implement these human rights mandates and deal with human rights concerns and risks arising from carbon projects. This chapter discusses the nature of institutional reforms required to implement the rights-based approach to carbon finance.

Institutions would go a long way to determine how the legal threshold discussed in the last chapter would be operationalized and enforced, how progress will be monitored and evaluated, and what further action could be taken in case of unsatisfactory progress. Four complementary institutional reforms are required to move the idea of human rights mainstreaming within the context of the international climate change regime from theory to practice. They are the creation of a specific human rights and climate change unit (HRCU) within the climate change governance structure to be responsible for the overall mainstreaming strategy; expansion of the terms of reference and powers of supervisory boards; establishing a compliance team to ensure that after registration the execution of projects are in line with approved standards and that it does not create unanticipated problems; and the setting up of a public complaint branch that could hear appeals by stakeholders who may be adversely affected.

This institutional framework mirrors the shift away from top-down planning towards more participatory forms of development and recognition for stakeholders in climate actions and projects. It provides a window of opportunity and a platform for stakeholders to bring their complaints about a project before an international supervisory body at every stage of project planning and execution, that is, prior to registration and post-registration. Mobilizing and empowering institutions would, arguably, enable individuals, communities and non-governmental organizations (NGOs) to have a voice in the processes of approving carbon projects. Furthermore, enhancing the mandates of climate governance institutions will provide a legal basis for them to reject projects that raise human rights risks. This would go a long way in addressing the current pervasive culture of approving emission reduction projects without adequate attention to their potential impact on human rights.

7.2 Institutional framework

7.2.1 Human rights and climate change lead agency

One of the very first institutional prerequisites for effective mainstreaming is the need to appoint and mandate a lead agency that will be the focal point for identifying and safeguarding human rights issues in climate actions or projects.[1] The lead agency would be mandated to spearhead essential reforms necessary for integrating human rights norms into the approval, finance and registration of carbon projects.

To forge a stronger alliance between human rights and climate change epistemic communities, and to enhance cooperation between human rights and climate change institutions, there is a need for a body at the international level that would bring together approaches, concerns and issues from human rights and climate change spheres.[2] Institutional coherence and harmony could be achieved with the international climate change regime by leveraging institutional support, services and

[1] See United Nations, 'The Human Rights Based Approach to Development Cooperation: Towards a Common Understanding among UN Agencies' (2003) www.undg.org/archive_docs/6959The_Human_Rights_Based_Approach_to_Development_Cooperation_Towards_a_Common_Understanding_among_UN.pdf, accessed 12 October 2015; UNESCO, 'Guidelines for Human Rights-Based Programming Approach' Executive Directive 1998-004; UNFPA, 'Core Concepts of a Human Rights Based Approach' (2010), www.unfpa.org/webdav/site/global/shared/documents/publications/2010/hrba/introduction.pdf, accessed 15 July 2014.

[2] See, for example, Article (7)(2)(l) of the Convention provides that the COP shall seek and utilize, where appropriate, the services and cooperation of, and information provided by, competent international organizations and intergovernmental and non-governmental bodies. This provision is also contained in Article 13(4)(i) of the Kyoto Protocol, which empowers the CMP to seek and utilize, where appropriate, the services and cooperation of, and information provided by, competent international organizations and intergovernmental and non-governmental bodies.

guidelines from UN bodies such as the Office of the High Commissioner for Human Rights (OHCHR) and the United Nations Human Rights Council (UNHRC) to strengthen human rights interpretation and application within the international climate change regime.[3]

A HRCU would serve as a linking point to reinforce the linkages between climate change and human rights regimes and to create a platform that allows for easy and coordinated sharing of good practices with these bodies. It will also provide an opportunity at the international level for local communities that are negatively affected by carbon projects to seek accountability and redress.[4] In light of the emerging evidence that a number of developing countries either have failed to address the impacts of carbon projects on the human rights of their citizens or, worse still, have sometimes cooperated with project proponents to entrench such violations, the responsibility to address these concerns and to protect the rights violated by climate actions and projects must fall to the international climate governance architecture and institutions.[5] When a national government fails to carry out its duty to screen, identify and address human rights risks in a carbon project, international supervisory bodies can step in to provide remedies for individuals by failing to approve or register such project.

[3] As the UN Secretary General rightly noted:

> There still remains a need for a body that brings together the key developed and developing countries to address the critical inter linkages between human rights and the environment, trade, finance, the handling of pandemic diseases and economic and social development. To be effective, such a body must operate at the level of national leader.

See Report of the UN Secretary General's High-Level Panel on Threats, Challenges and Climate Change, (2004), www.un.org/secureworld/report2.pdf, accessed 12 July 2014.

[4] H. Shue, 'Human Rights, Climate Change, and the Trillionth Ton', in D. Arnold (ed.), *The Ethics of Global Climate Change* (Cambridge: Cambridge University Press, 2011), 292.

[5] Shue's argument is premised on the emerging Responsibility to Protect (R2P) framework that has been popularised by the United Nations since 2005. It consists of an emerging norm, or set of principles, based on the idea that sovereignty is not a right, but a responsibility and that if a state fails to protect its citizens from human rights violations and mass atrocities, and peaceful measures have failed, the international community has the responsibility to intervene through coercive measures. Even though the R2P focuses on preventing and halting four crimes: genocide, war crimes, crimes against humanity, and ethnic cleansing, its underpinning notion of international intervention when national governments fail to protect their citizens from human rights violations may provide support for the need to incorporate protective measures and mandatory threshold into the international climate change regime to prevent human rights violations and repressions by national governments.

It is therefore pertinent for a special HRCU to be established as the focal point or lead agency that will spearhead and implement rights-based reforms under the international climate change regime. The HRCU would be an international focal point that would spearhead the necessary reforms aimed at bringing human rights and climate change issues to the fore. It would lead efforts to ensure that policy measures and projects aimed at mitigating climate change do not violate fundamental human rights. The HRCU would serve as the bridge that would bring together key stakeholders from the different international organizations, NGOs and human rights groups relevant to climate change mitigation and adaptation.

Mainstreaming human rights in the international climate change regime would require coordination between the OHCHR and the Human Rights Council, on one hand, and UNDP, UNEP, UNFCCC and the COP/MOP to the climate change regime on the other hand. This coordination is important to maximize the available resources, expertise and experience of these specific agencies and to avoid duplication of roles. The HRCU would bring together these key agencies, identify areas of priorities and distribute resources accordingly. It would also identify the key structures or resources that can be tapped from existing institutions and ways to avoid an institutional overlap in the areas covered by these agencies.

The HRCU would also design and establish an evaluation procedure that adopts the minimum legal threshold to assess the contribution of the prospective projects to the overall goals of the international climate change regime and to international human rights. It would ensure that all the relevant impact assessments are carried out for all projects designed and implemented for emissions reduction credits. It would spearhead further and continuous research on how climate change policies intersect with human rights and how these intersections can be recognized and managed within the international climate change regime.

The HRCU would be positioned to provide capacity development and training required for the human rights mainstreaming process. It would provide means for the dissemination of information and building the essential constituency of support for the mainstreaming effort. It would also provide information and advice for national governments and project proponents on how to mainstream human rights norms to project and policy design at the national level. It would be responsible for communicating the process required for a rights-based approach to national authorities and project proponents and other stakeholders responsible for designing climate change policies and projects.

To achieve these, the HRCU would create special reporting mechanisms such that parties can demonstrate in their climate change reports that actions, measures and projects approved at national levels do not violate human rights. The HRCU would also serve as the monitoring agency that would oversee the work of supervisory bodies under the international climate change regime, such as the CDM EB and the compliance committee, to ensure that human rights are respected and enforced within their scope of activities. It would ensure that only carbon projects that respect, protect and fulfil human rights are approved and registered under the international climate change regime.

For the HRCU to function effectively there is a need for proper staffing and capacity development. Since the mandate of the HRCU would include evaluating human rights impacts of climate actions and projects and designing policies on human rights and climate change, there is a need to appoint international human rights experts and individuals who have experience and training in human rights compliance and enforcement to lead the work of such agency. The international climate change governance structure, as it currently stands, largely comprises environmental scientists and international development experts who cannot (for the purpose of accuracy) be described as human rights experts or members of the human rights epistemic community. It is therefore doubtful if existing governance and supervisory institutions have the capacity to accommodate such complex transformation that requires deliberating on human rights issues. Consequently, to properly integrate human rights issues into the international climate change regime, it is pertinent to appoint staff with training and experience in human rights. This would ensure that human rights and human rights instruments are expertly understood, internalized, implemented and enforced.[6]

There is also a need for continuous human rights education (HRE) for climate change administrators. As the UN General Assembly recognizes, if human rights norms are to be reflected in the works of all UN entities, then HRE is needed to serve as a bridge in knowledge for all UN staff and development practitioners in general.[7] Though not legally binding, the United Nations Declaration on Human Rights Education and Training

[6] See J. Knox, 'Linking Human Rights and Climate Change at the United Nations', *Harvard Environmental Law Review*, 33 (2009), 477–498.

[7] A/HRC/RES/16/1.

emphasizes that human rights education and training are essential for the 'promotion of universal respect for and observance of all human rights and fundamental freedoms for all'; it is pertinent for HRE to be emphasized as part of the overall human rights mainstreaming strategy at national and international levels.[8] Article 7(4) of the Declaration provides that states, and where applicable relevant governmental authorities, should ensure adequate training in human rights and, where appropriate, international humanitarian law and international criminal law, of state officials, civil servants, judges and law enforcement officials, as well as promote adequate training in human rights for teachers, trainers and other educators and private personnel acting on behalf of the state. Article 11 also provides:

> The United Nations and international and regional organizations should provide human rights education and training for their civilian personnel, and military and police personnel serving under their mandates.[9]

In order to effectively adopt the HRBA, HRE is essential so that climate change administrators can be familiar with human rights standards, including human rights laws and their monitoring and implementation mechanisms.[10] Through a high-level 'Training-the-Trainers Program', staffs that have not acquired direct and specific training on human rights could be enlisted for short courses and advanced training on human rights instruments. This would provide some basic knowledge necessary to understand and handle mundane human rights issues arising within the secretariat and also equip them with the knowledge to train national authorities on human rights mainstreaming. For example, the advanced international certificate course on 'Mainstreaming Human Rights', which has gained increased recognition in recent years, could be a good background course for such practitioners.[11] It could provide some level of training and general education on human rights and on how to apply human rights norms. It would help to prepare 'outsiders' within climate change institutions for the tasks and challenges of mainstreaming human

[8] *Ibid.* [9] *Ibid.*

[10] Hamm illustrates this with the example of an expert from the German Ministry of Economic Cooperation and Development (BMZ), who notes that development experts in his ministry lack knowledge of human rights and thus they are not in a position to implement the HRBA. See B.I. Hamm, 'A Human Rights Approach to Development', *Human Rights Quarterly*, 23 (2001), 1005–1031.

[11] Course 1T09: Mainstreaming Human Rights, www.hrea.org/index.php?doc_id=676, accessed 12 May 2015.

rights. Other short courses, in-house training, workshops and webinars can equally help foster a deeper understanding of the contours of human rights mainstreaming, thereby reducing the gap in knowledge between the human rights epistemic communities and the perceived outsiders.

Generally, a lead agency, in the mould of the HRCU, would be important for human rights mainstreaming to be effective. Without establishing a lead agency with the specific mandate of coordinating climate change actions and projects in a way that identifies the striking linkages between human rights and climate change, there is the risk that efforts aimed at recognizing these linkages could remain uncoordinated, overlapping and conflicting. Establishing this lead agency at the international level would also provide guidance to national authorities to equally address possibilities of achieving strong inter-governmental linkages and ministerial coordination between agencies, departments, ministries and parastatals that are responsible for designing and approving climate change actions and projects, so as to ensure a legal partnership between human rights and climate change communities.

7.2.2 *Reformed supervisory boards*

As earlier noted, one critical human rights concern in the approval of carbon projects is that supervisory institutions at the international level, such as the CDM EB and the JI Supervisory Committee, do not have sufficient legal authority and mandate to consider human rights questions in project approval processes. This concern must be addressed if carbon markets are to respect, protect and fulfil human rights. To achieve this, the starting point would be to reform supervisory bodies responsible for overseeing and registering carbon projects under the international climate change regime ('Supervisory Boards') to grant them legal authority, mandates and power to consider human rights risks in carbon projects.[12]

For example, the CDM EB has attracted enormous criticisms for not doing enough to protect human rights in project approval processes. This is mainly because current CDM Rules place the responsibility to address human rights and sustainable development issues on the host

[12] For example, the Joint Implementation Supervisory Committee (JISC), which is responsible for supervising that emission reductions projects and activities under Joint Implementation (JI), meets the relevant requirements of Article 6 of the Kyoto Protocol and the Joint Implementation Guidelines.

government of the project.[13] Placing such a crucial decision in the hands of a country that is interested in such projects has led to the approval of all sorts of projects, even the ones that lead to loss of life and the violation of human rights. Countries have been more concerned with hosting CDM projects at all costs, irrespective of the human rights consequences of such projects.[14] As such, there have been increased tendencies to lower sustainability standards and to encourage foreign CDM projects despite their potential short- and long-term effects on human rights enjoyment.[15]

The crux of the rights-based approach is to prevent human rights violations in project implementation. As such, the most effective entry point would be through the body that approves and registers these projects for carbon credits in the first place. A preventive approach is to empower Supervisory Boards to reject projects that do not demonstrate compliance with international human rights law irrespective of the demonstrated emission reduction potentials of the project. This would require Supervisory Boards to play more important roles in providing avenues for challenging projects that do not adequately fulfil the requirements of the identified minimum legal threshold.

A starting point therefore would be to expand the mandate of Supervisory Boards, within the international climate change regime, by vesting them with powers to refuse registration for projects that violate human rights. Supervisory Boards for carbon projects should be empowered to exclude projects that result in human rights violations from registration. Adopting the six-part legal threshold identified in this book into project approval guidelines and rules could provide a legal and normative basis or checklist that could guide Supervisory Boards in performing this role. One of the elements of the discussed legal threshold is the need for mandatory human rights impact assessment (HRIA)

[13] See The Response of the CDM Executive Board to the Petition, www.cdm-watch.org/wordpress/wp-content/uploads/2011/02/EB59–14_CDM-Watch_Concerns-on-CDM-project-in-Honduras-Ref.-3197_Response.pdf, accessed 02 August 2014.

[14] See CDM Watch Group News Letter, (February 2011), www.cdm-watch.org/wordpress/wp-content/uploads/2011/02/cdm_watch_newsletter_15_february_2011.pdf, accessed 23 July 2014.

[15] See M. Jung, 'Host Country Attractiveness for CDM Non-Sink Projects', *Energy Policy* (2006), 2174; A. Silayan, 'Equitable Distribution of CDM Projects Among Developing Countries', *HWWA-Institut für Wirtschaftsforschung-Hamburg (HWWA) Report*, 255 (2006), 1; K. Olsen, 'The Clean Development Mechanism's Contribution to Sustainable Development: A Review of the Literature', cd4cdm.org/Publication/, accessed 21 January 2014.

without which a project would not be registered. A rights-based Supervisory Board should have the mandate and powers to reject projects that fail to meet any of the identified criteria under the HRIA.

Similarly, project eligibility rules must be modified and expanded to include human rights impacts as part of the validation requirements. For example, the CDM Rules provide that the review by the Executive Board shall be made when the review request is related to issues associated with the validation requirements.[16] However, HRIA is not listed as one of the validation requirements. For rights-based reform of supervisory institutions, the rights-based threshold discussed in this book must be included as core elements of project validation requirements.

Validation rules and requirements should also expand the terms of reference of project auditors and designated operational entities, such that only projects that meet the human rights criteria are validated and brought forward to Supervisory Boards for registration. It is important to integrate human rights due diligence and compliance into project validation processes to ensure that human rights issues are raised at all key phases and cycles of a carbon project. HRIAs should be incorporated as one of the indicators to assess while validating projects. In accrediting project auditors, Supervisory Boards must ensure that only institutions and firms that have capacity and human resources to validate HRIA are accredited to perform project validation functions.

There is also the need to provide elaborate rules on the constitution and membership of Supervisory Boards to safeguard and ensure impartiality, transparency and accountability. For example, the CDM EB has been frequently criticized for its lopsided membership and constitution in which certain CDM EB members also hold positions as UNFCCC and Kyoto Protocol negotiators for their countries.[17] This creates conflicts of interests and reduces the transparency of the CDM EB. To ensure transparency, individuals with no link to member states should be appointed to project Supervisory Boards. It is especially important for board members to be appointed and remunerated as international civil servants during the term of their membership of the board to prevent

[16] CDM Modalities & Procedures, 3/CMP.1 (2005), cdm.unfccc.int/Reference/COPMOP/08a01.pdf, para. 41, accessed 12 June 2015.

[17] See C. Streck and J. Lin, 'Making Markets Work: A Review of CDM Performance and the Need for Reform', *European Journal of International Law*, 19(2) (2008), 409–442.

conflict of interest.[18] Similarly, a Supervisory Board member should be disqualified from participation in the hearing and investigation of any project in which they have a personal interest or a significant involvement in any capacity with the project proponents or state.

Furthermore, there is great value and sense in creating a balanced mix in environmental and human rights expertise in membership of Supervisory Boards for carbon projects. For example, CDM Rules only provide that members of the EB shall possess 'appropriate technical and/or policy expertise' without specific mention of human rights expertise or skills.[19] For a Supervisory Board to effectively mainstream human rights principles into project registration and eligibility requirements, it must be properly constituted to include human rights experts, not only technical or policy experts. It would be helpful to clarify what the exact constitution of the panel would be in terms of expertise, how many human rights experts, as well as how many policy and technical experts. This would ensure that at every sitting of the Board, all areas of expertise would be represented.

Similarly, Supervisory Boards for carbon projects must be empowered to receive direct project complaints and grievances from the public. Supervisory Boards must establish grievance mechanisms or procedures for receiving and dealing with requests for project review prior to registration and approval. This would enable the Board to receive and facilitate the resolution of the affected stakeholders or communities' concerns and grievances about the human rights, environmental and social impacts of a project prior to the registration of the project. This could be achieved by appointing members of the Board into an ad-hoc project review committee to deal with, and make decisions on, project review requests. The nature and role of the panel are discussed in what follows.

7.2.3 Project review panel

Equally important is the need to establish complaint mechanisms and procedures for stakeholders or private individuals whose consultative rights or human rights in general have been infringed during project approval processes. This would enable them to block the approval of such projects or seek review of already approved projects. For example, one of

[18] See, for example, the World Bank Inspectional Panel, para. 10 of the Operating Resolution. See Resolution No. IBRD 93-10, Resolution No. IDA 93-6 'The World Bank Inspection Panel'.

[19] See CDM Modalities & Procedures, para. 8 (c), (n 16).

the key concerns under the CDM is that only governments or three CDM EB members may request a review of projects under CDM Rules.[20] Such a requirement limits the authority of the CDM EB to hear appeals from members of the public whose rights might be affected by a project, even in cases such as the Aguan gas project, where several petitions were received, which indicated significant infringements of human rights. This creates a one-way mechanism under which the decision of the state is final and most often rubber stamped by the international supervisory body. It freezes out the common citizen whose fundamental human rights might be repressed by the home state and the repression endorsed by the international climate change regime. The lack of participation by citizens and stakeholders in project planning and implementation is a violation of one of the most important international human rights standards and principles. Project rules and guidelines for carbon projects must be reformed to allow stakeholders to request a review of projects before and after registration.

Apart from establishing project review panels to hear human rights complaints respecting carbon projects, it is equally important to simplify the rules of procedure for such appeal body to ensure easy access. Rules of procedure to be established for the review panel must be simple and expansive so as to ensure that any group of individuals or community affected by a project (including NGOs) can approach it for investigation simply by submitting a request or complaint. To simplify the procedure of the project review panel, its rules should avoid any technicality or excessive requirements on the form of a petition. To make the panel accessible to the less educated people in many developing countries, a handwritten petition or report should be sufficient to trigger a review process.

Furthermore, rules on admissibility of petition should be widened and simplified. A simplified procedure would be able to admit a petition whenever the following three requirements are met. An individual or NGO alleges that:

(a) A carbon project or action would cause actual or threatened injury or human rights violations to the public. This would make it possible

[20] Para. 41 of the CDM Modalities and Procedures provides that 'The registration by the Executive Board shall be deemed final eight weeks after the date of receipt by the Executive Board of the request for registration, unless a Party involved in the project activity or at least three members of the Executive Board request a review of the proposed CDM project activity.' See CDM Modalities & Procedures, *ibid.*

for key stakeholders, which includes NGOs, to be able to file a petition even if they suffer no direct or actual injury.

(b) The injury or violation is traceable to the project or action. Linking the project to a public harm or human rights violations would satisfy this condition; for example, emission reduction projects that displace citizens or lead to loss of life, arable land or loss of income. Once complainants can establish that violations can be traced to an emission reduction project, it should create sufficient grounds for the public complaint branch to investigate and penalize the parties involved.

(c) The injury is avoidable or redressable by the intervention of the project review panel. This third condition would establish the importance of a complaint. It must be shown that the intervention of the public complaint branch would assist in preventing or redressing the violations. This would require the complainants to demonstrate that the project is one that is carried out under the international climate change regime and how the complaint branch can assist either by redress or prevention.

Redress would be the best remedy if the project had already been completed. In such cases, the public complaint branch could apply consequences that can redress or mitigate the harm suffered by the public, for example by requesting the party to pay compensation or to ensure resettlement where necessary or face the risk of losing emission reduction credits obtained as a result of such a project. For an ongoing project, complainants can request for the project to be stopped by the complaint branch to prevent any likely impact on human rights or environmental issues. The panel can then compel the parties to stop the project pending a comprehensive investigation on the scale and magnitude of human rights violations involved and how the violations can be prevented.

Most of these tasks would be to deal with petitions and complaints regarding any human right impact. To establish these claims, there is a need for proper investigation; this would include visiting countries where projects are initiated to investigate the claims. Mandates of constituted expert review teams that typically verify emission reduction claims or reports can either be expanded to serve as the investigative or fact finding arm of the international climate change regime, or the project review panel could periodically constitute an inspection committee made up of its own members to investigate claims brought before it and to produce a comprehensive report on which the panel could base its decisions.[21] The committee would serve as an ad-hoc investigative organ of the review panel and would be charged with producing a comprehensive report on its findings. The purpose of the investigation is to establish the facts and determine whether the complaint has merit.

For the inspection committee to function effectively, part of its terms of reference would be to hold consultations, public hearings and meetings with directly affected people and general members of the affected community. There would also be the need to visit project sites and to speak with as many interested NGOs on the field who have vital information. The report of the inspection committee should, among other things, indicate the relevant facts and steps taken to conduct the investigation; committee's findings on whether a project proponent has complied with relevant policies and procedures; and the recommendations of the committee to the project review panel.

The team's report would form the basis for a decision by the project review panel on whether or not to proceed with the registration of a project. This comprehensive process would make it possible to exclude projects that violate human rights, environmental and social standards prior to their registration. It represents a preventive and anticipatory approach that makes use of human rights safeguards to avoid the sources of some of the problems that have impacted the integrity of CDM projects. This dispute resolution mechanism could also help to avoid litigation arising from carbon projects as it provides an opportunity for concerns to be addressed at an early stage before they escalate. This could prevent violations and litigation.

However, since this process focuses on review prior to registration, a question that arises is, what of projects that satisfy all the requirements and were registered but subsequently turned out to violate some of the minimum legal threshold, that is, what remedy would be available for victims in the case of projects that violate the legal threshold post-registration or in the long term? Arguably, the compliance committee of the Kyoto Protocol (which is currently vested with powers to facilitate

[21] Under the UNFCCC and the Kyoto Protocol, Expert Review Teams are constituted by the UNFCCC Secretariat to check annual inventories and to make sure they are complete, accurate and conform to the Guidelines under Article 8 of the Kyoto Protocol. Members of Expert Review Teams are selected by the Secretariat from experts nominated by parties. However, they currently perform only technical functions, which is the technical review of greenhouse gas inventories from parties included in Annex I to the Convention.

and enforce compliance with the Protocol and its rules) could be reformed to take up such post-registration compliance issues.

7.2.4 Monitoring institution or enforcement committee

A project might be human rights compliant in its design, and during the approval process, but then raise human rights issues during implementation. It is therefore important to put in place institutions or committees to monitor and address post-registration issues relating to human rights violations and non-compliance arising from carbon projects. This involves monitoring carbon projects throughout their lifecycles to detect non-compliance and human rights violation as early as possible.

Many multilateral environmental agreements contain compliance mechanisms that monitor transparency and compliance by parties with commitments under the regime.[22] The important question, however, is whether the compliance mechanism established under the international climate change regime has robust mandate and power to consider human rights issues. Without integrating human rights oversight functions into the scope and mandate of the compliance mechanism, issues relating to how policy measures aimed at achieving emission reductions may affect human rights would be left without adequate protection and safeguard.

Post-registration compliance issues may be addressed either by reforming compliance mechanisms of the international climate change regime to include human rights oversight functions or by establishing a new public complaints branch that will be empowered to deal with human rights issues arising post-registration.

7.2.4.1 Reformed compliance architecture

Under the first approach, the scope and mandate of the compliance mechanism under the international climate change regime may be expanded to empower compliance institutions to prescribe consequences or penalties for countries implementing carbon projects that violate human rights. The advantage of this approach is that it would minimize the costs of mainstreaming human rights in overall compliance architecture of the international climate regime. The use of existing institutions is

[22] For example, the Kyoto Protocol Compliance Committee was established to provide advice and assistance to parties, promote their compliance and implementation of the Protocol and to determine consequences for parties not meeting their commitment. See UNFCCC, 'An Introduction to the Kyoto Protocol Compliance Committee', unfccc.int/kyoto_protocol/compliance/items/3024.php, accessed 21 April 2015.

one way to reduce the exorbitant costs, in terms of structure and resources, as it builds on existing capacities and resources. Since the climate regime already has a compliance mechanism for reviewing compliance, the existing compliance architecture may leverage its existing expertise, experience, best practices and resources to assess and review if, and to what extent, countries are implementing human rights risk management frameworks in the design, approval, finance and implementation of carbon projects.

However, expanding the mandate of extant compliance architecture would also raise the need to re-structure and re-constitute constituent committees to include human rights experts. To ensure that human rights and human rights instruments are understood, internalized, implemented and enforced under the compliance mechanism and framework, there is a need to recruit individuals from the human rights epistemic communities, that is people with training and expertise in human rights law.

It is also pertinent for the compliance regime to define comprehensive methodologies or criteria for measuring if and to what extent a project is in compliance with the approved procedures and standards. This could be achieved through the use of qualitative data such as surveys, questionnaires and findings of the inspection panels, as supplements to quantitative data generated from the implementation of the discussed threshold, to determine whether a project complies with international human rights and environmental standards and to estimate to what extent it does or fails to do so.

Furthermore, the monitoring process adopted by the compliance regime must be participatory, that is, committees should be set up to provide opportunities for stakeholders to comment and give feedback on the noticeable impact of a project and to report any inconsistency or deviation from the terms of the consent issued to proponents of a project during the pre-registration consultation or project design stages. Rules or requirements that prevent the public from filing direct complaints must be removed. Stakeholders who are affected by a policy measure or project are the most credible sources of information when it comes to the environmental or human rights threats they suffer.[23] By providing them with an

[23] As noted in Principle 10 of the Rio Declaration:

> Environmental issues are best handled with the participation of all concerned citizens, at the relevant level. At the national level, each individual shall have appropriate access to information concerning the environment that is held by public authorities, including information on hazardous

opportunity to bring attention to the existence of such threats, the compliance regime can better serve as a true 'watch dog' for non-compliance. It would be better placed to perform its overseeing functions in ensuring that parties comply with approved methodological requirements and procedures for carbon projects. This emphasizes the need for a public complaint mechanism to be introduced into the compliance architecture, through which stakeholders can have the space and chance to report violations and non-compliance with approved methodologies and procedures.

7.2.4.2 Public complaints branch

The second approach is to establish a 'public complaints branch' as part of the compliance mechanism of the climate regime. The drive for accountability under international law has attained a momentum that has led several organizations to establish feedback units or a branch through which public petitions and protests could be received or dealt with. For example, virtually all international human rights treaties establish individual complaint procedures that allow private citizens to bring an action at an international forum for human rights abuses.[24] Other international organizations have also established the Office of the Compliance Advisor or an Ombudsperson to receive complaints and comments from the public.[25] Such is the reform that is needed under the international climate regime to enhance accountability.

A public complaints branch (PCB) could serve as a sort of 'appellate body' where post-registration issues arising from projects could be raised by stakeholders. The PCB could be established as a specialized branch responsible for receiving and reviewing claims by individuals, private

> materials and activities in their communities, and the opportunity to participate in decision-making processes. States shall facilitate and encourage public awareness and participation by making information widely available. Effective access to judicial and administrative proceedings, including redress and remedy, shall be provided.

24 At the UN level, individual complaint mechanisms are found under eight conventions. See Chapter 2 above.

25 See, for example, the International Finance Corporation's Compliance Advisor Ombudsman (CAO) an independent recourse mechanism for projects supported by the International Finance Corporation (IFC) and Multilateral Investment Guarantee Agency (MIGA) – the private sector lending arms of the World Bank Group. The CAO was established in 1999 to address the concerns of individuals or communities affected by IFC/MIGA funded projects, to enhance the social and environmental outcomes of IFC/MIGA projects; and to foster greater public accountability of IFC and MIGA. See The Compliance Advisor Ombudsman (CAO), www.cao-ombudsman.org, accessed 23 April 2015.

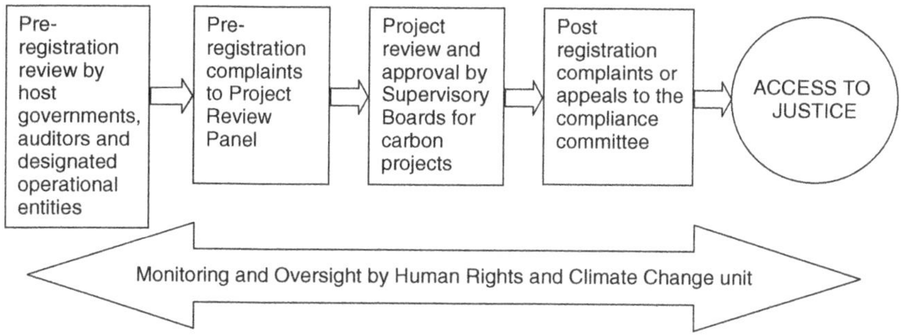

Figure 7.1 Project review process

entities and NGOs against policy measures adopted by states for meeting their obligations if such measures produce a negative impact on human rights or social and environmental concerns. Human rights issues arising from projects that have already been registered could be brought under the jurisdiction of the PCB in line with approved guidelines and rules. A public complaint branch would create a public space or forum through which individuals can access the international climate change regime and provide opinion, comments or complaints about policy measures and ongoing projects that violate international human rights norms and standards. This would offer individuals adequate opportunity to hold their governments directly accountable for adverse policy measures before an international supervisory body. The PCB would also determine appropriate remedies and consequences for non-compliance with approved project procedures or methodologies.

To function effectively, the PCB must be appropriately staffed and provided the space, resources and structures to discharge its mandate and responsibility, as a stand-alone branch of the compliance regime. For the PCB to effectively consider human rights issues and petitions, it is pertinent for its members to include international human rights staff who have experience and training in human rights compliance and enforcement (Figure 7.1).

7.3 Conclusion

Actions, measures and projects designed to combat climate change must address complex and fundamental causes of inequality and human rights violations. These inequalities and violations arise from lack of consultation in decision-making, discrimination and inequality; non-disclosure of vital project information; and the absence of grievance and feedback mechanisms through which stakeholders could seek and obtain redress. The institutional framework discussed in this chapter provides practical guidelines on how human rights risks in carbon projects may be addressed under a human rights-based approach to carbon finance. By reforming governance structures under the international climate change regime to become more rights-based, transparent and accountable, many of the negative impacts associated with carbon projects can be effectively anticipated and prevented.

Apart from establishing a lead agency to oversee the overall integration of human rights into the design, approval, finance and implementation of carbon projects, establishing a project review panel would ensure that pre-registration disputes and complaints are addressed before the project can be registered as a carbon project, while establishing a compliance regime with mandate and authority to measure and monitor human rights compliance post-registration would ensure that appropriate consequences and penalties are prescribed for parties that fail to execute carbon projects in line with approved plans.

To practically realize the key reforms and changes required under a human rights-based approach to carbon finance, theoretical viewpoints and public interest agitations must be matched with commitments at national, regional and international levels to integrate human rights standards and requirements into the international climate governance architecture. The next chapter examines procedural requirements that must be met and logistical challenges that must be overcome to move the human rights-based approach to carbon finance from theory to practice.

PART IV

From theory to practice: practical challenges, paradoxes and potentials of a human rights-based approach to carbon finance

8

Making mainstreaming work

A three-step approach to implementation

8.1 Introduction

One of the key barriers to integrating human rights principles into global climate change governance is that proponents of the human rights mainstream paradigm have, arguably, not devoted sufficient attention to examining relationships between its theoretical importance and the practical implications. Laying down a theoretical framework is one thing, understanding its practical implications, possible real-life limitations and the requisite steps for moving the concepts from the realm of theory to practice is another.

This chapter examines legal and practical potentials, the required processes and challenges that may arise in implementing the human rights-based approach to carbon finance. It discusses a step-by-step approach for mainstreaming human rights approach in the international legal regime on carbon finance. This approach includes three complementary steps: reflecting human rights language in the international climate change regime, adopting concrete legal guidelines or documents on carbon finance and human rights and enforcing a carbon finance certification scheme.

8.2 Three-step approach for mainstreaming human rights norms into the international climate change regime

Adopting a human rights-based approach to carbon finance requires a comprehensive review and reform of the international climate change regime to make it more accountable, participatory and rights-based. Implementing a rights-based reform of the international climate change regime is not a task that could be achieved merely by editing texts of a convention or treaty. Rather, it requires a systematic process that will ensure that projects approved and registered for carbon credits do not violate or infringe human rights. This section discusses a three-step approach through which a rights-based approach could be practically implemented in the international climate change regime. This approach includes two complementary steps: reflecting human rights language in the international climate change regime and adopting concrete legal guidelines or documents on carbon finance and human rights. This section analyses legal processes involved under each of the steps. This three-step approach could provide guidance on how to reform climate change agreements to reflect human rights provisions and the procedures for adopting new or additional protocols that stipulate human rights standards in the design, approval, finance and implementation of carbon projects.

8.2.1 Reflecting human rights language in the international climate change regime

A first step in implementing a rights-based approach to carbon finance is to mainstream human rights standards and language into global climate change instruments. For example, as noted earlier in this book, one of the key reasons for the increased neglect of human rights in the approval and execution of CDM projects is that the UNFCCC and the Kyoto Protocol do not establish clear recognitions of the linkages between human rights and climate change; neither do they empower the CDM EB to consider human rights issues in the process of reviewing and approving projects. Concerns of human rights violations in CDM projects, and the perceived inertia and lack of legal authority for the CDM EB to address such concerns as they arose show that efforts to address human rights concerns in carbon projects must proceed first with a strong recognition of human rights in the international climate change regime. Examples of the recognition required is the decision adopted by the Conference of the Parties to the UNFCCC in Cancun (COP 16) that 'Parties should in all climate change actions respect human rights'; and in the Paris climate agreement that 'parties should, when taking action to address climate change, respect, promote and consider their respective obligations on human rights, the right to health, the rights of indigenous peoples, local communities, migrants, children, persons with disabilities and people in vulnerable situations and the right to development, as well as gender equality, empowerment of women and intergenerational equity'.[1]

[1] See Preamble to the Paris Agreement under the United Nations Framework Convention on Climate Change, Conference of the Parties Twenty-first session Paris, 30 November to

These agreements reflect growing consensus on the need to reinforce the international climate change regime with human rights provisions and standards to prevent overlap and lack of coherence between climate mitigation goals and international human rights. However, the international climate change regime must move beyond mere preambular and textual recognitions of human rights concerns and risks in climate actions and projects, to establishing concrete legal authority and mandate for supervisory boards of carbon projects to investigate, examine and address these concerns.

The authority to periodically review the implementation of multilateral environmental instruments, amend any related legal instruments and to make decisions necessary to promote the effective implementation of the convention typically falls on the COP.[2] The COP to the international climate change regime has powers to examine commitments of parties in light of the Convention's objective, new scientific findings and experience gained in implementing climate change policies and to facilitate the coordination of measures adopted by parties to address climate change and its effects.[3]

The general authority of the COP can be invoked by parties to the international climate change regime to recommend additional actions and to adopt rules for the implementation of a rights-based approach to carbon finance. It could be the basis for the COP to review the implementation of carbon projects and to make the decisions necessary to promote the effective implementation of the international climate change regime. Parties can therefore agree to incorporate more robust human rights provisions and language into texts of climate change instruments to specifically identify the critical linkages between human rights and climate change. For example, pursuant to Article 15 of the UNFCCC, any

11 December 2015, FCCC/CP/2015/L.9, unfccc.int/resource/docs/2015/cop21/eng/l09.pdf, accessed 18 December 2015.

[2] See Multilateral Environmental Agreements for example Article 7.2 of the United Nations Framework Convention on Climate Change; see also Article 9 of the Kyoto Protocol, Conference of the Parties to the Framework Convention on Climate Change, Kyoto Protocol, 10 December 1997, UN Doc FCCC/CP/1997/L.7/add.1, 37 ILM 22 (1998) [Kyoto Protocol] which empowers the Conference of the Parties serving as the meeting of the parties to this Protocol (CMP) to periodically review the Protocol in the light of the best available scientific information and assessments on climate change and its impacts, as well as relevant technical, social and economic information. Such reviews are to be coordinated with amendments to the Protocol and the Convention.

[3] On the legal status, powers and functions of the Conference of the Parties under multilateral environmental agreements, see B. Desai, *Multilateral Environmental Agreements: Legal Status of the Secretariats* (Cambridge University Press, 2010), 133–169.

party may propose amendments to the Convention.[4] The Convention does not specify or impose any substantive limitations on what might be adopted as an amendment. Thus, an amendment could specify new commitments, change or vary existing commitments, or establish new institutions or procedures if supported by at least a three-quarters majority vote of the parties present and voting. Parties to the international climate regime therefore have strong roles to play in ensuring that human rights principles and language are adopted and integrated as part of the international climate change regime, specifically guidelines and rules governing the approval and registration of carbon projects: first, by developing detailed rights-based proposals for inclusion in the international climate change regime and, second, by mobilizing broad-based support of other parties for such proposals to be adopted and included in the international climate change regime. Questions on the likelihood of securing a majority vote for a rights-based reform are considered in later parts of this chapter.

An arguably more straightforward option could be to incorporate human rights requirements in annexes of climate change instruments to reflect human rights screening processes for carbon projects. Annexes are an integral part of multilateral environmental agreements that allow parties to attach, append or add implementation rules and requirements to treaties.[5] They contain lists, forms and other scientific, technical, procedural or administrative rules or requirements.[6] For example, in contrast to amending texts of the UNFCCC, new annexes and amendments to annexes do not need to be ratified; they come into force automatically, except in the case where parties lodge written objections.[7] Therefore, to integrate human rights

[4] See also Article 20(1) of the Kyoto Protocol, which provides that 'any Party may propose amendments to this Protocol'. Article 20 establishes the requirement of a three-quarters affirmative vote for any amendment to the Protocol.

[5] See, for example, Article 21 of the Kyoto Protocol, *ibid.*, which provides that annexes to the Protocol shall form an integral part thereof. It also provides that any additional annexes shall be restricted to lists, forms and any other material of a descriptive nature that is of a scientific, technical, procedural or administrative character. Article 21(2) of the KP provides that 'any Party may make proposals for an annex to this Protocol and may propose amendments to annexes to this Protocol'.

[6] See, for example, Article 16 of the United Nations Framework Convention on Climate Change adopted by the United Nations General Assembly, 20 January 1994, A/RES/48/189.

[7] This process was adopted with relative success under the UNFCCC. For example, Article 4.2(f) called on the COP to review the lists in Annexes I and II not later than 31 December 1998 and amend them as appropriate, with the approval of the party concerned. In line with this, Annex I was amended at COP 3 in 1997, by consensus, to add the names of Croatia, Liechtenstein, Monaco and Slovenia, and to replace Czechoslovakia with the

into the international climate change regime, an available in-route is to insert procedural or administrative rules that describe a list of human rights items to look out for in a carbon project as annexes to climate change instruments. An annex that establishes a human rights impact assessment procedure and project review rules could be included as part of the international climate change regime to provide guidelines for countries on how to anticipate and avoid human rights problems in the design, approval, finance and implementation of carbon projects. This amendment procedure could also be evoked to propose an additional Annex that contains a comprehensive list of prohibited mitigation activities or projects, or a list that itemizes mitigation activities and projects that must only proceed with a public disclosure programme or public consultation and prior informed consents. A list of prohibited mitigation activities could include activities that violate human rights or that do not meet the human rights threshold described earlier. This approach has been adopted in a number of multilateral environmental agreements. For example, the Annexes to the Rotterdam Convention on the Prior Informed Consent Procedure for Certain Hazardous Chemicals and Pesticides in International Trade provide comprehensive lists of chemicals subject to prior informed consent procedure and a list of information disclosure requirements in its Annexes; same for the Montreal Protocol on Substances that Deplete the Ozone Layer, which provides a list of controlled substances.[8]

8.2.2 *Rights-based modalities and rules for project approval*

A required second step, in integrating human rights requirements in the design, approval, finance and implementation of carbon projects, is the adoption of concrete guidelines, rules and modalities that set out human

Czech Republic and Slovakia (decision 4/CP.3). The amendment came into force on 13 August 1998. Annex II was amended at COP 7 to delete the name of Turkey from Annex II (decision 26/CP.7). COP 7 also invited the parties to recognize the special circumstances of Turkey. This amendment entered into force on 28 June 2002. Turkey became a party on 24 May 2004. In accordance with Article 4.2(g), any non-Annex I party may notify the Depositary that it intends to be bound by the commitments contained in Article 4.2(a) and (b) relating to the mitigation of climate change (see chapter 11.B). Kazakhstan took such action in 2000.

[8] See Montreal Protocol on Substances that Deplete the Ozone Layer (Adopted 16 September 1987, effective 1 January, 1989); Rotterdam Convention on the Prior Informed Consent Procedure for Certain Hazardous Chemicals and Pesticides in International Trade (Adopted 10 September 1998, effective 24 February 2004).

rights due diligence steps that must be taken or demonstrated by a project proponent before the project can be registered.

One of the key functions of the COP is to establish decisions, modalities and rules to enhance the effective implementation of climate change agreements.[9] For example, under the UNFCCC, decisions are adopted by a two-third majority vote at COP meetings, which according to Rule 4.1 of the COP Rules of Procedure are to be held once every year. Most COP/CMP meetings culminate in decisions that clarify and specify rules for the practical and effective implementation of aspects of the Convention and the Protocol. For example, the Marrakech Accord was adopted at COP 7 as a set of decisions and rules for the implementation of the CDM among other things. Similarly, at the COP 16 in Cancun, the COP adopted a decision which included for the very first time a recognition of the linkages between human rights and climate change by stating that 'Parties should in all climate change actions respect human rights.' This shows that a COP decision could be a practicable in-route for recognizing how carbon projects affect human rights.

The COP could in another decision build on existing recognitions of the linkages between carbon mitigation activities and human rights by establishing detailed guidelines, rules and modalities on how parties can respect human rights in carbon projects. The six-part legal framework proposed in this book could form the crux of such additional decision, modalities, rules or guidelines. It would provide in concrete terms, rights-based modalities for project approval and the human rights conditions that must be met before carbon projects can be approved, registered and recognized for the purpose of gaining emission credits. A clear legal threshold at the international level could also spur reform at the national level as it could serve as guidance on what policymakers, project proponents, verifiers, technical partners and service providers such as law firms should look out for to design balanced and rights-based policy responses and climate change mitigation projects.

8.2.2.1 Participation guidelines or directives for project participants

Integrating human rights within decision-making processes on carbon projects is not a task for governments and international institutions alone. As discussed in Chapter 2, corporations and business enterprises are key actors in designing and scoping carbon projects. Engineering and

[9] Desai, *Multilateral Environmental Agreements: Legal Status of the Secretariats*, (n3), 133–135.

technical firms assist countries to design project design documents and to implement them at the project implementation stages. Third-party verifiers such as auditors, asset management firms and validators play pivotal roles in validating project design and confirming carbon credits; law firms, accounting and consulting firms provide legal and policy services to document the allocation of risks and transfer of carbon credits; while international finance institutions, development banks and commercial lenders fund carbon projects.

Due to the high incidence of human rights petitions associated with carbon project development, private sector participants have come under scrutiny from civil society leaders and indigenous rights groups for failing to provide sufficient human rights safeguards to uphold land and forest rights of local and indigenous peoples. For example, a German development bank had to suspend its approved loan funding for the Aguan CDM Project due to several human rights violations.[10] While the response of the German bank is laudable, more could be done to ensure that financing or funding for carbon projects is not approved, in the first place, without detailed human rights impact assessment. The situation whereby governments and development banks, such as the World Bank, finance or facilitate trade in carbon credits that have been generated from carbon projects that engender land grabs and forced displacements undermines international law and threatens local rights. This underscores the importance of integrating human rights safeguards as requirements for accessing global carbon finance and funds for carbon projects.

A practical example of this approach is the Equator Principles (EP) already adopted by over eighty commercial banks worldwide for determining, assessing and managing environmental and social risk in projects and to ensure that projects that violate environmental and social goals are not funded.[11] One drawback is that Equator Principles currently apply only to banks that have voluntarily become signatories, which excludes several development banks and commercial banks in developing countries that fund carbon projects but have not become signatories

[10] CDM Watch, 'German Bank Won't Lend to Honduran CO2 Project' (April 2011), www.bloomberg.com/news/2011-04-18/german-bank-won-t-lend-to-honduran-co2-project-cdm-watch-s. The Kimberley Process Certification Scheme (KPCS) imposes extensive requirements on its members to enable them to certify shipments of rough diamonds as 'conflict-free' and prevent conflict diamonds from entering the legitimate market. See, tradeays.html, accessed 1 August 2014.

[11] The Equator Principles: A financial industry benchmark for determining, assessing and managing environmental and social risk in projects (June 2013), www.equator-principles.com/resources/equator_principles_III.pdf, accessed 1 August 2014.

to the Equator Principles. Furthermore, the Equator Principles apply only to projects of over $10 million. This also excludes several small-scale carbon projects that raise human rights concerns. Similarly, the Equator Principles also do not apply human rights threshold as its focus and guiding principle; it seeks to address compliance with relevant host country laws, regulations and permits that pertain to environmental and social risks.

Despite the limitations of the Equator Principles, it is a practical example of how the international climate change regime can mainstream human rights impact assessments and due diligence processes as pre-conditions for accessing financing for carbon projects.

At all levels of involvement, the international climate change regime must provide guiding principles that mandate corporations and business enterprises involved in carbon projects to measure and address human rights risks in carbon investments and projects, in accordance with the overall aim of the human rights-based approach to carbon finance: that is, respect, protection and fulfilment of human rights in the design, approval, finance and implementation of carbon projects. The guiding principles would set out the required human rights procedures that companies must follow when financing, participating in or executing project activities.

All projects are different as each firm is unique; however, the international climate change regime could establish minimum procedures based on the legal framework discussed in this book, which verifiers, service firms and financial institutions must complete either before issuing a verification report or issuing funding or finance for carbon projects. Financial institutions must invite project proponents to demonstrate compliance with the legal threshold and to demonstrate that all human rights vulnerabilities and risks of a carbon project have been examined and addressed prior to financing such projects. Project proponents and participating entities must also put in place internal corporate risk management frameworks and risk mitigation strategies that improve business practices and reinforce human rights compliance systems in the execution of a carbon project. Only projects that have been certified as meeting compliance thresholds should be eligible for carbon finance. Firms that fail to adhere to such participation guidelines should be delisted or sanctioned from further participation in carbon projects. Establishing fair play rules will not only send a strong statement of intent to participants on the need to safeguard human rights in project design and implementation but will also go a long way in addressing allegations of collusion, oppression, bribery and neglect that have trailed verifiers and corporate participants in carbon projects.

As for the role of lawyers representing companies and participants in carbon projects and investments, it is important to emphasize the significance of human rights risk assessment when providing counsel. Law firms and consulting firms, with expertise in legal and regulatory risk management, can work with project proponents and carbon investors to provide holistic support with assessing, managing and mitigating human rights risks in carbon projects. As key participants in the carbon markets, law firms can provide advice on how project participants can redesign and enhance corporate compliance and governance systems and processes to conform with international human rights requirements. This will include compliance advice on how business entities and organizations can review and update their internal control guidelines, project design and implementation processes, including carbon procurement practices to screen and exclude carbon projects and credits that violate human rights. Regulatory risk analysis and management could also provide project participants with tools to manage human rights risks to minimize disruptions to ongoing carbon projects.[12]

Such corporate risk management processes would save a business entity from sanctions, penalties, negative publicity and public scrutiny. It will also provide ready toolkits through which investors, stakeholders and employees can understand and ascertain the company's strategy for implementing human rights in carbon projects or initiatives. The next section discusses key elements of corporate risk management processes that business may adopt when investing in carbon projects.

8.2.2.2 Internal control and due diligence guidance for project participants

Businesses face increasingly complex challenges associated with the management of human rights risks in carbon projects and investments. The ascendance of human rights norms and language in the international climate regime impose stricter obligations on corporations to evolve rights-based risk management and corporate responsibility policies and frameworks for their carbon investments and projects.[13] The speed with

[12] See Norton Rose Fulbright, *Regulations and Investigations: Dealing with legal and regulatory risk*, www.nortonrosefulbright.com/files/regulation-and-investigations-112016 .pdf, accessed 12 December 2015.

[13] See Z. El Hassan and P. Saintilan, 'The Paris Agreement: Moving from Climate Conscious to Climate Compliance' (Norton Rose Fulbright, 2015), www.regulationtomorrow.com/

which human rights risks in carbon projects can materialize and spread requires business enterprises to have a firm understanding of the practical requirements, institutional demands and business implication of implementing, monitoring and enforcing human rights obligations in carbon investments and project portfolios, over the short, medium and long term.

To stay compliant under a rights-based international climate change regime, there is a need for business enterprises to adopt and implement a four-step approach that could enable them to effectively: understand their human rights risk exposure, evaluate internal compliance systems, and ensure that robust governance, risk management and human rights compliance strategies are put in place to manage and minimize risk. The first step is to conduct internal gap analysis to determine what is missing within company policies on climate change, carbon credits procurements, project design and project implementation. The human rights impact assessment matrix discussed in Chapter 5 could provide a helpful assessment tool in this regard to determine whether carbon investments and projects affect human rights and to identify the scope of recognition given to human rights issues in corporate policies on climate change.

If there are no internal policies in place on emission reduction and carbon finance, the required follow up step is to develop robust and coherent internal policy or statements that adopt a rights-based approach to carbon finance. This might take the form of one policy document or a series of statements that reference and enshrine the company's decision to integrate international human rights standards and conventions into business practices to prevent the direct and indirect impacts of climate change. The policy will state specifically the company's decision to ensure that all its actions, policies and projects respect, protect and fulfil human rights. This will include restating the company's commitment to design, finance; implement carbon projects in line with local, regional and international human rights laws; and to exclude projects that do not safeguard local rights from their portfolio of carbon investments.

A third step is to integrate the human rights-based approach to carbon finance in management decision-making processes. This is by appointing top management level officers of the business to be responsible for

au/the-paris-agreement-moving-from-climate-conscious-to-climate-compliance/, accessed 18 December, 2015, discussing how new obligations under the Paris Agreement will influence the decisions of financial institutions and fiduciary investors (particularly pension funds) as they become more climate change conscious and sensitive to the risk of stranded assets on their investment portfolios over the short, medium and long term.

implementing the policies designed to screen out project proposals that do not protect, promote or fulfil human rights of local communities. Institutional questions on capacity development and human rights training can be addressed in line with proposals described earlier in Chapter 7.

The final step is to establish monitoring mechanisms through which management can systematically review progress made in line with goals set in the corporate risk management policy. The goal is to review and check if the policies have been implemented in alignment with the business's overall human rights strategy.

8.2.3 *Carbon finance certification scheme*

A follow up to the adoption of participation modalities and guidelines is for the international climate regime and finance agencies to establish and implement a carbon finance certification scheme. The aim of this scheme will be to ensure that only projects that have met prescribed human rights threshold are approved for the exchange of carbon credits. The international carbon finance architecture must establish guidelines that would ensure that only carbon credits generated in compliance with prescribed human rights threshold would be sold and exchanged in international carbon markets. Under a rights-based emission reduction scheme, exchange of carbon credits should be accompanied by a compliance certificate issued by the responsible international supervisory body, guaranteeing that carbon credits have been generated in accordance with human rights and have met the minimum requirements of the prescribed legal threshold.

A number of relatively successful examples of certifications schemes abound in international law. Over the past decade, certification schemes have grown in popularity as mechanisms for encouraging and enforcing the sustainable production and sourcing of commodities, such as natural and mineral resources, including agricultural products.[14] For example,

[14] See, for example, the International Sustainability and Carbon Certification (ISCC), a system for certifying that biomass and bioenergy industries apply social and environmental standards in the production of bio energy. ISCC certification applies legal requirements in the bioenergy markets to ensure sustainable use of land, protection of natural biospheres, reduction of greenhouse gas emissions, and social sustainability in bioenergy projects. Additionally, ISCC PLUS has been developed for food, feed, technical/chemical (e.g. bioplastics) and other bioenergy applications (e.g. solid biomass). ISCC received the world's first official state recognition through the German government's biomass sustainability ordinance (BioNachV) in 2010 and has since been recognised by the European Commission as one of the first certification schemes to demonstrate compliance with the EU Renewable Energy Directive's (RED) requirements. See

due to perennial concerns that resources were extracted from African countries and sold in violation of several human rights, the United Nations in 2002 established the Kimberley Process Certification Scheme (KPCS), to ensure that sale and transfer of rough diamonds are consistent with international law governing international trade.[15] The KPCS establishes guidelines that ensure that sale and exchange of diamonds is 'conflict-free'. Under the terms of the KPCS, participating states must meet 'minimum requirements' and must put in place national legislation, institutions and export, import and internal controls that ensure transparency, accountability, and adequate exchange of information. Participants can only legally trade with other participants who have also met the minimum requirements of the scheme, and international shipments of rough diamonds must be accompanied by a KPCS certificate guaranteeing that they are conflict-free. For example, the Republic of the Congo, the Democratic Republic of the Congo and Côte d'Ivoire were expelled from the diamond market at some point due to non-compliance. Implementation is monitored through 'review visits' and annual reports, as well as by regular exchange and analysis of statistical data.

Despite criticisms on the effectiveness of the KPCS scheme, most especially due to its voluntary and non-binding nature, statistics show that incidence of sale and exchange of conflict diamonds has reduced when compared to pre-2002 levels.[16] This shows that with proper

International Sustainability and Carbon Certification (ISCC), www.iscc-system.org/en/, accessed 23 June 2015, also, www.sustainablepalmoil.org/standards-certfication/certification-schemes/international-sustainability-carbon-certification-iscc/, accessed 23 June 2015. Although the ISCC is designed for biomass activities with focus on social and environmental impacts and not necessarily human rights, it provides practical examples on how project participants can be required within the international carbon finance architecture to demonstrate compliance with the prescribed human rights legal framework under the climate change regime.

[15] See United Nations General Assembly Session 55 Resolution 56. *The role of diamonds in fuelling conflict: breaking the link between the illicit transaction of rough diamonds and armed conflict as a contribution to prevention and settlement of conflicts* A/RES/55/56 1 December 2000. See also Preamble to the Kimberley Process Certification Scheme Core Document, www.kimberleyprocess.com/en/kpcs-core-document, accessed 23 May 2015.

[16] The effectiveness of the KPCS has been questioned by organizations such as Global Witness, which pulled out of the scheme on 5 December 2011, claiming it has failed in its purpose and does not provide markets with assurance that the diamonds are not conflict diamond. See Elisha Sword, 'A Flawed Initiative: The Kimberley Process Certification Scheme and the Growing Conflict Diamond Trade' where the author criticized the KPCS as lacking the backbone necessary to prevent KP members from contributing to the conflict diamond industry. According to the author, KPCS's status as a

monitoring, a binding carbon certification scheme could stem the tide of generating and exchanging carbon credits at all costs, even in the face of wanton human rights violations and repressions. Even though the magnitude and scale of human rights problems associated with carbon projects are, arguably, smaller in scale when compared to circumstances that led to the KPCS, human rights violations, forced land grabs and repressions, no matter how small, require stringent international legal response and redress. Through a carbon certification scheme, the international legal regime on climate change can assign certification requirements to different stakeholders in the carbon finance cycle, based on their role in the supply chain, to monitor and ensure that carbon credits generated through any form of human rights violations are not accepted for trade in the international carbon market. Practically, this will mean that prior to any sale or transfer of carbon credits, the selling party must obtain and provide a carbon certificate from supervisory boards, guaranteeing that carbon credits are human rights compliant and violations-free.

Furthermore, in addition to approving projects for sustainable development goals, the country of origin of carbon credits (the country where carbon mitigation project or activities took place) must be required to grant host country approval and certification to projects to demonstrate that they comply with the legal and human rights thresholds discussed in this book. Countries must demonstrate that they have put in place appropriate legislation and a local enforcement framework to ensure that only projects that comply with the threshold are granted host country approval and certification. Countries that do not put in place internal mechanisms, legislation and policies to ensure that carbon projects are designed, executed and implemented in accordance with human rights framework should, as a punitive measure, be excluded from carbon markets.

voluntary programme means any country dissatisfied by the KPCS's provisions for whatever reason may drop out and trade unethical diamonds. See also Global Witness, 'Loopholes in the Kimberley Process: Illegal Trade Undermines Efforts to Combat Conflict Diamonds', (October 2007), www.globalwitness.org/sites/default/files/import/loopholes_in_the_kimberley_process.pdf, accessed 21 May 2015; also I. Smillie, 'The Kimberley Process: The Case for Proper Monitoring', in I. Smillie (ed.), *The Diamonds and Human Security Project* (Ottawa, Canada: Partnership Africa Canada, 2002), 1. For a positive assessment, see V. Haufler, 'The Kimberley Process Certification Scheme: An Innovation in Global Governance and Conflict Prevention', *Journal of Business Ethics*, 89 (2010), 403–416.

8.3 Conclusion

Effective human rights risks management requires a holistic approach. The HRBA places obligations on governments, project participants, finance institutions, business enterprises, project auditors, carbon verifiers and importantly international governance institutions, to ensure that processes to anticipate, manage and address human rights risks are embedded at all levels and stages of project design, approval, finance and implementation. This chapter has discussed three complementary steps that could support and reinforce the practical adoption and integration of a human rights-based approach to carbon finance: the reform or expansion of the scope of the international climate change regime to reflect human rights standards, the adoption of rights-based modalities and rules for project participation and approval; and the implementation of a carbon finance certification scheme at international and organizational levels. These human rights risks management steps will not only ensure that only projects that have met prescribed human rights threshold will be financed but they also provide guidance and legal frameworks on how participants in the carbon market can anticipate human rights risks and design robust risk management strategies aimed at screening out carbon projects that may violate human rights.

Implementing human rights risk management frameworks arguably provides a win-win scenario for all parties involved in the life cycle of a carbon project. While it will provide indigenous communities an increased level of confidence in the international carbon market, knowing that repressive projects will not be allowed to trade in or benefit from carbon credits generated, it also provides participating countries and business enterprises additional assurance and confidence that carbon credits sold or obtained from the international carbon markets are human rights compliant and violations-free. For international supervisory bodies, practical options such as the carbon certification scheme provide opportunities to address post-registration issues for projects that satisfy all the requirements and were registered but subsequently turn out to violate minimum legal threshold. In such cases, carbon credit certification can be withheld for such projects.

Designing risk management frameworks is only one essential aspect of the task of implementing a human rights-based approach to carbon finance. A further step is to review and examine practical barriers and limitations that such a proposal may face along the way. The next chapter provides real-life checks on practical challenges that may limit the success of any effort aimed at promoting a rights-based reform of the international climate change regime.

9

Minding the gap

Practical paradoxes and barriers to the adoption of a human rights-based approach to carbon finance

9.1 Introduction

The last chapter examined legal scenarios, requirements and procedures for integrating the human rights-based approach to carbon finance by participants across the international carbon markets. The three-step approach discussed provides potential options and scenarios for integrating human rights safeguards into the design, approval, finance and implementation of carbon projects. A practical question, however, is whether the international community will achieve the consensus required to move such rights-based reform of climate change agreements from theory to practice. Three practical challenges that might limit the success of any effort aimed at invoking a rights-based reform or amendments of the international climate regime are examined in this chapter.

This chapter examines how lack of political will, polarization and political alignments, as well as negotiation blocs may act as practical barriers to a rights-based reform of the international climate change regime to integrate human rights consideration in the design, approval, finance and implementation of carbon projects.

9.2 Practical barriers to a rights-based reform of the international climate change regime

9.2.1 Political will

To integrate robust human rights risks management procedures and guidelines under any of the amendment scenarios described in the last chapter, the first step would be for one of the parties to the international climate change regime to individually, or as part of a group, propose an amendment or updates to the climate regime. This raises the question whether countries will generally agree to bind themselves or vote in support of a rights-based reform that could holistically empower a large section of the public to block carbon projects, demand accountability, request project information and to even challenge project decisions before international supervisory bodies. Thus arises the question whether such a rights-based proposal will not be viewed as an attempt to grant the public a cudgel with which to beat the state into submission or to empower NGOs to habitually oppose governmental projects. These concerns are in fact reasonable, particularly the fear that NGOs and interest groups could capture and frustrate mitigation projects and plans through the proposed rights-based processes. However, an evolved understanding of the basis and value of the proposals in this book could arguably address these concerns.

Primarily, it is important to understand that the HRBA does not necessarily introduce new obligations that many states do not already have under several international human rights treaties. The rights-based framework builds on existing human rights obligations under international law, which virtually all parties to the international climate change regime already agreed to protect, respect and fulfil. As such, the legal framework proposed would not grant new or revolutionary rights that NGOs and the public do not already possess.[1]

What a rights-based legal framework would do is to provide an opportunity for countries to incorporate these already existing obligations and considerations, while planning and designing carbon projects in order to prevent violations and tensions. Through the framework proposed in this book, parties could meet their obligations under both human rights and the international climate change regimes in a holistic, coherent and mutually supportive manner. It would allow states to respect, protect and fulfil human rights while meeting their climate change obligations. The reform proposed would provide an avenue to package a collection of international obligations under both human rights treaties and the international climate change regime into one toolkit to enhance implementation. Similarly, it would provide an avenue to draw upon human rights obligations to underpin and inform the design and execution of emission reduction goals provided in climate change agreements. This evolved understanding of the

[1] There are 166 parties to the ICCPR, 160 to the ICESCR. The majority of these parties have signed the UN Framework Convention on Climate Change and the Kyoto Protocol. See L. Rajamani, 'The Increasing Currency and Relevance of Rights-Based Perspectives in the International Negotiations on Climate Change', *Journal of Environmental Law*, 22(3) (2010), 391.

reforms proposed would foster a balanced recognition of why it is important to incorporate human rights obligations into climate change mitigation actions and projects. It would also provide an opportunity for governments to work with stakeholders to anticipate, address and prevent human rights problems that carbon projects could generate, such that projects do not result in petitions, protests and violence.

A second aspect of the political will question is whether any of the developing countries that count on financial and economic benefits from project-based mechanisms would have the political will or motivation to support a rights-based reform of carbon markets. A reverse side of the question is whether any of the industrialized countries that are arguably beneficiaries of the *status quo* would vote in support of a reform that could substantially alter current market trends, to make project approvals more streamlined and process-based.[2] Taking a very close look at trends in the implementation of the CDM, there are very few reasons to be optimistic about this happening. With respect to the CDM, several developing countries competitively struggle to attract as many CDM projects as possible, including by lowering sustainability standards so as to attract CDM opportunities.[3] This accounts for why national authorities at times, arguably, look away from human rights complaints or even collude to seize communal lands without compensation, suppress protests using military personnel and even kill protesting citizens. Project proponents from industrialized countries have equally looked away from these incidents, as demonstrated by the Panama, Honduras and Nigerian examples. The question then is whether there would be sufficient political support for a reform of business-as-usual models and practices of extant market systems.

The answer to this could, to a large extent, depend on the level of awareness that is created at national levels on the need for action. Considering the nature of problems generated by carbon projects, particularly the dislocation of people from ancestral lands and homes, it is important to raise awareness on why robust rights-based reforms to

[2] For an excellent analysis of the impacts of economic standpoints on climate negotiations, see S. Bailer and F. Weiler, 'A Political Economy of Positions in Climate Change Negotiations: Economic, Structural, Domestic, and Strategic Explanations', *The Review of International Organizations*, 10 (1), (2015), 1–12.

[3] For a discussion of this, see D. Olawuyi, 'Beautifying Africa for the Clean Development Mechanism: Legal and Institutional Issues Considered', in B. Richardson, Y. Bouthillier, H. Mcleod-Kilmurray and S. Wood (eds.), *Climate Law and Developing Countries: Legal and Policy Challenges for the World Economy* (Cheltenham, UK: Edward Elgar Publishing, 2009), chp. 11.

extant project approval processes are not only important but also required. International diplomacy has been largely influenced by the abilities of NGOs and environmental interest groups to raise awareness and to put pressure on national governments to effect change or reform at the international level.[4] As Sterk rightly notes, international politics does not happen in a vacuum; the positions countries take internationally are determined mainly by their domestic political situations.[5] As such,

[4] Although NGOs are formally only observers at a number of United Nations conferences including climate change conferences, with no right to vote and with restricted access to the corridors, plenary sessions and some contact groups; many scholars agree that NGOs still make a great difference in global environmental politics. NGOs and interests groups have played active roles in putting pressures on international gatekeepers to support environmental negotiation processes. One frequently cited example is how Environmental NGOs (ENGOs) contributed to drafting UNFCCC at the Rio Earth Summit in 1992 by participating in government delegations, lobbying, building public pressure and contributing to content and structure of the negotiation text. The influence and roles of NGOs in international law making can therefore not be sidelined. NGOs such as Greenpeace International, Amnesty International, Down to Earth Group, Earth Right and Earth Justice can play extensive roles in creating international awareness on the need to address the human rights impacts of climate change mitigation schemes. For example, Greenpeace International and Down to Earth Group have already launched a series of projects that highlight these concerns. If stakeholders or government in a developing country, say Nigeria, can develop a detailed amendment proposal based on the legal framework discussed in this book and seek the support of these NGOs on the importance of protecting the respect for human rights in climate change actions, a coordinated awareness drive would be generated that could result in widespread support for countries to integrate human rights principles into the international climate change regime. See K. Rietig, 'Public Pressure versus Lobbying – How do Environmental NGOs Matter Most in Climate Negotiations?', Centre for Climate Change Economics and Policy Working Paper, 79 (2011); M. Betsill, 'Environmental NGOs and the Kyoto Protocol Negotiations: 1995 to 1997', in Betsill and Corell (eds.), *NGO Diplomacy* (Massachusetts Institute of Technology Press, 2008); M. Betsill and E. Corell, 'NGO Influence in International Environmental Negotiations: A Framework for Analysis', *Global Environmental Politics*, 1(4) (2001), 65–85; C. Carpenter, 'Business, Green Groups and the Media: The Role of Non-governmental Organizations in the Climate Change Debate', *International Affairs*, 77(2) (2001), 313–328; E. Corell and M. Betsill, 'A Comparative Look at NGO Influence in International Environmental Negotiations', 86–107. B. Arts, *The Political Influence of Global NGOs, Case Studies on the Climate and Biodiversity Conventions* (Utrecht, Netherlands: International Books, 1998); K. Raustiala, 'Nonstate Actors in the Global Climate Regime' in U. Luterbacher and D. Sprinz (eds.), *International Relations and Global Climate Change* (Massachusetts Institute of Technology Press, 2001); S. Oberthür, et al. *Participation of Non-Governmental Organisations in International Environmental Governance: Legal Basis and Practical Experience* (Berlin: Erich Schmidt Verlag, 2002); S. Andresen and L. Gulbrandsen 'The Roles of NGOs in Promoting Climate Compliance' in J. Hovi, O. Stokke, and G. Ulfstein, *Implementing the Climate Regime: International Compliance* (London: Earthscan, 2005), 169.

[5] W. Sterk, 'House Cleaning in Doha: UN Climate Summit Delivers Second Life for Kyoto but no Deal to Revive Carbon Market', *Carbon Mechanisms Review*, 1(1) (2013), 10, where

progress in reforming current international climate negotiations to make them more rights-based would be enhanced if pro-climate advocacy coalitions can be brought together in the key countries and across borders to demand for change. The fear of possible backlash from the press or environmental NGOs could elevate the need for negotiators to push for reforms. With an intensive awareness drive by NGOs and interest groups, particularly NGOs in developed countries, there could be increased pressure on negotiators, policymakers and international gatekeepers to support amendments aimed at addressing the human rights impacts of climate change measures and projects on communities in developing countries. Such awareness drives have enormous potential to culminate in the eventual rights-based reform and amendment of key provisions of the international climate change regime to make it more rights-based, transparent and efficient. They might also culminate in integrating some of the human rights reforms and governance structures proposed in this book into a new guideline or modalities for carbon projects.

9.2.2 Negotiation blocs and alignments

Getting a party to spearhead the amendment process by submitting a proposal is one task, obtaining the requisite majority vote to push through the proposal is yet another concern.[6] The question is whether enough political support could be garnered within the international climate change negotiation framework to allow such a rights-based reform proposal. For example, over the years, the COP to the UNFCCC has been criticized for its 'managerialism' approach in dealing with serious legal issues.[7] Due to political divisions, entrenched alignments and bargaining coalitions, the COP has also been described as a threat to accountability, procedural fairness, climate justice, and the

the author argues that international negotiations can rarely take decisions that have not been previously prepared nationally, and that the current situation is that most key countries have no appetite to undergo the fundamental economic and ecologic transformation that is necessary to revive carbon markets under current climate change regimes.

[6] Guzman describes this as the problem of getting to 'yes' in international law. See A. Guzman, 'Against Consent', *Virginia Journal of International Law*, 52(4) (2012), 1–2.

[7] See A. Nollkaemper, J. Wouters and N. Hachez, 'Accountability and the Rule of Law at International Level', *Policy Brief of Leuven Centre for Global Governance Studies*, 8 (2009), 8–10; J. Wouters, A. Marx and N. Hachez, 'Managerial Accountability in International Organizations – Comparative Perspectives', *Policy Brief of Leuven Centre for Global Governance Studies*, 3 (January 2008), 2–7.

rule of law.[8] These characterizations rightly underscore how entrenched alignments and posturing within the international climate change regime could act as barriers to holistic rights-based reform of carbon markets.[9] Practically, robust rights-based reforms to the international climate regime, as proposed in this book, could be snuffed out in a second by the solidarity and support from countries and negotiation blocs with similar economic and political interests and ties that might consider such a reform as inimical to the growth of carbon markets.[10]

[8] Nollkaemper, Wouters and Hachez, 'Accountability and the Rule of Law at International Level', *ibid.*; Guzman, 'Against Consent', (n 6), 1–2. See also R. Audet, 'Climate Justice and Bargaining Coalitions: A Discourse Analysis', *International Environmental Agreements: Politics, Law and Economics*, 13 (3), (2013), 369. L Tomlinson, *Procedural Justice in the United Nations Framework Convention on Climate Change: Negotiating Fairness* (Springer 2015), 118–120.

[9] S. Bailer and F. Weiler, 'A Political Economy of Positions in Climate Change Negotiations: Economic, Structural, Domestic, and Strategic Explanations', (n 2), *The Review of International Organizations* 1 (2015), 1–43. See also S. Alam, S. Atapattu, J Razzaque, & C. Gonzalez (eds.) *International Environmental Law and the Global South* (Cambridge: Cambridge University Press, 2015) discussing how bifurcations and conflicts between rich and poor nations (the North–South divide) have compromised international environmental law.

[10] For example, there are two main negotiation groups within the COP to the UNFCCC: developing countries and developed countries. Within these groups there are further negotiating blocs. These blocs and groups take important decisions on whether to support a proposal or not. The decisions by the influential blocks often determine how far a proposal can make it within the COP. *Developing countries:* (G77 and China), which comprises all developing countries that are members of the climate change convention (Africa, Alliance of Small Island States (AOSIS), Latin American countries, The Arab Group, Asia-Pacific Group, Least Developed countries and Brazil, South Africa, India, and China (BASIC) Group). Generally they argue for legally binding emission reduction commitments for industrialized countries, market mechanisms, substantial international financial, capacity-building and technological support. Specifically, the *Alliance of Small Island States* argue for action as strong and swift as possible, due to their extreme vulnerability to sea level rise, while the BASIC group are GHG emitters that advocate that developed countries should allow developing countries 'equitable space for development' unencumbered by emission reduction obligations. *Developed countries: EU block* (twenty-seven member countries), *Environmental Integrity Group* (Mexico, the Republic of Korea and Switzerland), *Umbrella Group/JUSSCANZ* (Japan, Switzerland, Canada, Australia, New Zealand, Iceland, Norway, the Russian Federation, Ukraine and the United States). While the EU advocates a strong global agreement with stringent economy-wide emissions reductions, substantial financial support, and robust compliance mechanisms; the Umbrella Group clamours for a legally binding treaty with market mechanisms and with emission reduction commitments for large emitters, including BASIC countries such as India and China. See UNFCCC, Regional Groups and Negotiating Blocks, unfccc.int/cop6/parties/regiogroup.html, accessed 12 May 2015. See also, X. Zhu, 'International Ethics and Governance for Climate Change Amid the Rise of BASIC countries', *International Social Science Journal*, 64 (1), (2013), 211–212.

A good example is how the comprehensive proposal submitted by Tuvalu at the COP 15 in Copenhagen, which called for the adoption of a legally binding Copenhagen Protocol, was snuffed out. More surprising was the process that led to the adoption of the Copenhagen Accord by a small group of powerful developed and developing countries - The Friends of the Chair Group - despite the fact that there was no consensus as required by the COP Rules of Procedure.[11] The question therefore is whether a transformative proposal such as implementing a rights-based approach to carbon finance would pass through the eye of the COP needle in the face of these alignments and intrigues.

The answer would depend on whether the proposal to mainstream human rights norms in climate change agreements is understood as important to: address some of the human rights problems that currently bedevil the carbon markets (as exemplified, for example, in the report of the CDM

[11] While it was widely reported that the AWG-LCA Chair in Plenary, Michael Zammit-Cutajar dismissively remarked about the Tuvalu proposal that 'Tonight when I go to bed, I will take the proposal with me for bedtime reading', the Friends of the Chair Group was able to provoke more than a bedtime reading for its proposal. The Group was created on the last day of the conference and consisted of twenty-five countries, including the United States, the United Kingdom, Sweden, Spain, Saudi Arabia, Russia, Norway, Mexico, the Maldives, Lesotho, South Africa, Bangladesh, Algeria, Denmark, Germany, France, India, Ethiopia, Colombia, Korea, China and Brazil. Led by US President Barrack Obama, the group involved mostly Heads of State. The end product from the Friends of the Chair Group was a two-page document labelled as the Copenhagen Accord. In the document, reference to 'legally binding' outcome of future climate negotiations was dropped and replaced with allowing countries to determine their own national emission targets and policy actions. It, however, pledged substantial financial commitments by industrialized countries for 30 billion dollars for 2010–2012 and long-term finance of 100 billion dollars per year by 2020. What followed was a chain of rejections by Tuvalu, Nicaragua, Bolivia, Cuba, Venezuela, Sudan, Pakistan, and Saudi Arabia. *Bolivia:* 'We say this emphatically and as clearly as possible: we are seeing actions in a dictatorial way, and this is not the way a world should discuss the future of humanity and the planet'. *Cuba:* 'What you call… "a group of representative leaders" is a gross violation of sovereign equality that is enshrined in the Charter of the United Nations and a mechanism that would impose decisions on the international community…'. However, support came from delegates from the African Union, the Maldives, the European Union, the United States, Australia, Japan, Russia, Norway, the Philippines, Bangladesh, Singapore, Barbados and Belize. The African Union arguably accepted the Accord because of the huge financial commitment involved. In the end the COP took note of the Copenhagen Accord and included it as part of its outcome document. Currently about 115 countries have formally endorsed the Accord. The Copenhagen events show how economic strength, powerful alignments, financial promises and managerialism could easily dictate the directions of future climate change negotiations and agreements. See R. Dimitrov, 'Inside UN Climate Change Negotiations: The Copenhagen Conference', *Review of Policy Research*, 27(6) (2010), 795–821.

Policy Dialogue); retain support for market mechanisms; and consolidate on progress made in efforts to combat climate change; or whether it would be read and interpreted as yet another attempt to smuggle human rights language into international climate change agreements through the back door. The former reading is, arguably, the better and sustainable one. Legal and institutional frameworks discussed earlier provide mutually reinforcing pillars through which human rights principles could be reflected in climate actions and projects. This way, many human rights infringements arising from carbon projects may be avoided and prevented in the first place.

While the intrigues and convolutions that have trailed ongoing negotiations supply evidence on how herculean a task it is to navigate through amendment procedures under the current international climate change regime, one could also point to the measure of compromise and success recorded at the Paris and Doha climate conferences to show that some level of global consensus is achievable, even if it is at the last hour.[12] Arguably, there are reasons to believe that developing countries affected by human rights impacts of carbon projects could build a consensus that might culminate in a rights-based reform or amendment of the international climate change regime. For example, the Nigerian delegation to the Doha Conference successfully spearheaded efforts to enlist the support of other African countries in the developing country bloc (G77 and China) to push for the operationalization of a green climate fund, the approval of new financing and technologies required to accelerate green transition in developing countries and a recommendation that the COP should take note of the findings of the CDM Policy Dialogue and to implement it.[13] Through such negotiation blocs, developing countries (especially African countries and the Alliance of Small Island States) that have been negatively affected the human rights impacts of carbon projects could present a formidable front to push

[12] See the Doha Amendment to the Kyoto Protocol, UNFCCC, Doha Amendment to the Kyoto Protocol (Adopted 8 December 2012, Not Yet in Force) C.N.718.2012. TREATIES-XXVII.7.c. Also the 2015 Paris climate conference has been described as the world's greatest diplomatic success for producing an agreement that is hailed as 'historic, durable and ambitious'. See F. Harvey, 'Paris Climate Change Agreement: The World's Greatest Diplomatic Success' (*The Guardian*, 14 December 2015), www.theguardian.com/environment/2015/dec/13/paris-climate-deal-cop-diplomacy-developing-united-nations, accessed 15 December 2015.

[13] Even though Nigerian delegates were arguably more interested in the CDM Policy Dialogue's recommendations on the need for more funding and the equitable distribution of CDM projects to African countries. See EnvironewsNigeria, 'Nigerian Climate Negotiators Talk Tough', www.environewsnigeria.com/tag/climate-change-negotiation/, accessed 12 January 2015; See also 'Developing Countries Firm up Common Doha Climate Talks Position' (All African News, 30 October 2012).

for the rights-based amendment of the international climate change regime to include more robust human rights obligations and procedures. To achieve requisite votes to push through such amendment, there might be a need to enlist the support of negotiators from countries in the developed country blocs. With a well-communicated proposal, perhaps support from some developed countries could be achieved.

9.2.3 Unresolved dilemmas and regime uncertainties

A major challenge to a rights-based reform of the current international climate change regime is the current uncertainties surrounding the future of the international climate change regime in general.[14] While parties seem to agree on the need for new emission reduction commitments, long-term finance and technology transfer, deep ideological disagreements and contentions remain on the historical responsibility for climate change actions, common but differentiated responsibilities, REDD+ mechanisms, the place of human rights in climate change negotiations and the governance of climate finance. Even among the negotiation blocks and groups, there are different ideological divisions. For example, while the eyes of the African Union has been fixed on financial and technological transfers and benefits through climate finance and the CDM, the Alliance of small Island States (AOSIS) Group has argued for human rights recognitions in the emerging regime to ensure the survival of small island states.[15] Despite recent progress in climate negotiations and governance, this fundamental conflict of values remains an inhibiting factor to consensus building within

[14] See S. Karcher and T. Forth, 'Carbon Markets: Which Way Forward: Essentials on Cooperation with Developing Countries', *Carbon Mechanisms Review* 1 (1), (2013), 4–6, where the authors suggest that the ability to form a consensus in designing the future outlook of carbon markets has reached its limits as far as climate change negotiations are concerned.

[15] For example, during the debates on the Copenhagen Accord, while the African Union was quick to support the Copenhagen Accord, negotiators from Tuvalu submitted that:

> There is no guarantee that the Kyoto Protocol will continue into the next commitment period in this text... And we see there is money put before us. Can I suggest, in biblical terms, it looks like we are being offered 30 pieces of silver to betray our people and our future. Mr. President, our future is not for sale. I regret to inform you that Tuvalu cannot accept this document.

See Dimitrov, 'Inside UN Climate Change Negotiations', (n 11); see also A. Averchenkova, *The Outcomes of Copenhagen: The Negotiations and the Accord* (New York: United Nations Development Programme Climate Policy Series, 2010).

the climate change multilateral decision-making process, especially with contentious topics such as human rights and climate justice. As such, while some countries argue for firm and binding commitments, including a reform of project-based mechanisms to make them more transparent, some argue that excluding projects with high emission potentials and opportunities for funding, on human rights grounds, runs counter to the ultimate objective of the international climate change regime – 'the stabilization of greenhouse gas concentrations in the atmosphere at a level that would prevent dangerous anthropogenic interference with the climate system'.[16] As discussed earlier in this book, a key contention is that any legal instrument that helps to address climate change would be consistent with this objective, while any proposal that could hinder or limit the abilities of countries to address climate change and to harness climate change funding runs counter to the ultimate objective of the Convention.

These differing standpoints cast doubt on the future outlook of the international climate change regime as a whole. They also underscore the relatively bleak long-term prospects of implementing a robust human rights threshold and project guidelines in the international climate change regime. These unresolved dilemmas and ideological differences have often inhibited reform processes required to truly address some of the threats posed by climate change through the consensus-based processes of the COP. They also underscore growing concerns on the limits of international governance structures to effectively respond to human rights risks in carbon projects and markets. These questions have in recent years prompted a growing shift away from the UN climate change processes to the possibilities of addressing climate change through national, bilateral and regional programmes.

Despite the uncertain prospects and outlook, the importance of integrating human rights standards into the design, approval, finance and implementation of carbon projects cannot be overemphasized.[17] For example, negative backlash and protests associated with CDM and REDD+ projects have increasingly affected market confidence and have resulted in sharp drops in demand for, and prices of, carbon credits.[18]

[16] See UNFCCC, (n 10), Article 2, (n 1).

[17] See A. Johl and S. Duyck, 'Promoting Human Rights in the Future Climate Regime', *Ethics, Policy and Environment*, 15 (3), (2012), 1–5.

[18] Renew Economy, 'Doha Dispatches: Carbon Market Creates another Fine Mess', (6 December 2012), www.reneweconomy.com.au/2012/doha-dispatches-carbon-market-collapse-creates-another-fine-mess-21440, accessed 12 January 2015.

The need to restore confidence in carbon markets should be an impetus for action in reforming project-based mechanisms in line with the threshold discussed in this book. The process-based framework proposed here could enhance public confidence in the effectiveness and efficiency of the carbon markets.

Lessons on the implementation challenges of the CDM should inform thoughts on how to shape new and emerging market mechanisms to anticipate and avoid human rights concerns that have trailed the CDM. If human rights problems and risks associated with carbon projects are not addressed and reflected in emerging rules for carbon projects, it could leave wide open the possibilities that the current trends in which national governments play active roles in the repression of human rights, land grabs, displacement of citizens and concentration of projects in poor communities may continue under new project-based portfolios.

To enhance the effectiveness of extant and emerging project-based mechanisms, a holistic legal and institutional framework, as discussed in this book, should be considered as part of emerging rules and modalities to avoid similar concerns on lack of accountability, land grabs without compensation, governmental secrecy and lack of access to project information, lack of participation and the absence of project review mechanisms. Addressing these concerns would significantly restore faith in carbon markets; it would also enhance the overall effectiveness of the international climate regime to mitigate climate change in a transparent, accountable and rights-based manner.

9.3 Conclusion

To ensure that participants in the international carbon markets respect, promote and fulfil human rights, there is a need to establish clear and mandatory guidelines and threshold that must be met prior to the approval, finance and registration of carbon projects. Parties to the international climate regime must look beyond political considerations and alignments, and understand that revitalizing legal and procedural frameworks on project implementation would not only promote coherence between climate change and human rights obligations, but in fact also restore faith in carbon markets, prevent future resistance to viable carbon projects and also promote the integrity of the international climate change regime in general. This expansive understanding could result in a rights-based international climate change regime, in which, rights of host communities and key stakeholders are considered and reflected in the design and execution of carbon projects.

The next and final chapter synthesizes the salient themes discussed and offers practical recommendations on how to overcome the practical barriers to a rights-based approach to carbon finance.

10

Implementing a human rights-based approach to carbon finance

Summary for policymakers

10.1 Summary

This book set out to examine the practicability of mainstreaming human rights principles into the design, approval, finance and implementation of carbon projects under the extant and emerging international climate change regime. This final chapter provides a review of the salient themes discussed, and then explores directions for future action and research.

Much of the literature on climate change and human rights has focused on how the direct impacts of climate change such as sea level rise, flooding and extreme weather events could directly affect human rights in different parts of the world. However, recent findings show that policy measures and projects aimed at combating climate change are in turn producing even more serious human rights concerns, especially in developing countries. These human rights issues include mass displacement of citizens from their homes to allow for climate change mitigation projects, lack of participation by citizens in project planning and implementation, siting and concentration of projects in poor and vulnerable communities, lack of governmental accountability on projects and the absence of review and complaint mechanisms for victims to obtain redress for these problems. These secondary human rights impacts of policy measures and projects aimed at mitigating climate change have not received sufficient attention in existing literature. This book has explored these secondary impacts of climate change on human rights.

This book has examined how carbon projects, particularly CDM and REDD+ projects, are producing human rights consequences in developing countries. This is a trend recognized by the 2015 Paris Climate Agreement and the 2011 Decision of the Conference of the Parties (COP) in Cancun, both of which declare that parties should, in all climate change related actions, fully respect human rights; and by the CDM

Policy Dialogue which called for the need to restructure the CDM EB towards making it a more transparent, accountable and rights-based organization.[1] However, despite these recognitions, lack of adequate guidelines and frameworks on how parties could effectively integrate human rights protection into measures and projects designed to combat climate change, specifically how the CDM EB could better reflect human rights considerations in its project approval processes and procedures, have obscured how this declaration could move from mere pious wishes to reality. This book has attempted to answer this question of 'how' – that is, the legal and institutional frameworks required to integrate human rights principles into the execution of carbon projects.

This book has discussed the need for a reform of the international climate change regime to address the growing trend of human rights violations and procedural impunity in the execution of carbon projects. It examined the theoretical basis and value of the human rights mainstream paradigm and also identified the legal justifications, basis and value of adopting the United Nations Human Rights-Based Approach (HRBA) as a framework for mainstreaming human rights standards into project execution and approval under the international climate change regime. Previous scholarly efforts have not devoted sufficient attention to examining the theoretical basis of the human rights mainstream paradigm, hence the charge that human rights mainstreaming is fuzzy, ambiguous, without legal foundation and would only devalue the human rights language. This book has examined how the principle of systemic integration of international law, inter-doctrinal legal transplant/cross-fertilization theory, participatory development theory and the full belly thesis provide legal and theoretical anchor for the human rights mainstream paradigm. These theoretical perspectives underpin the need for a holistic and mutually reinforcing approach, based on integrating human rights obligations and institutions with climate change obligations and institutions, to produce a coherent, anticipatory and participatory framework for international climate governance.

[1] See Para. 8, The Cancun Agreements: Outcome of the work of the Ad Hoc Working Group on Long-term Cooperative Action under the Convention which noted Resolution 10/4 of the United Nations Human Rights Council on human rights and climate change on the adverse direct and indirect effects of climate change on the effective enjoyment of human rights. See also CDM Policy Dialogue, 'Climate Change, Carbon Markets and the CDM: A Call to Action: Recommendations of the High Level Panel on the CDM Policy Dialogue', (2012), www.cdmpolicydialogue.org/report/ues_en.pdf, accessed 10 January 2015.

While linking climate change and human rights issues is not a revolutionary suggestion, the potentials and requirements for adopting the HRBA framework for mainstreaming human rights principles into the international climate change regime have not been sufficiently explored and communicated. One reason for the lack of action on a HRBA to carbon finance is the lack of precedence or operational guidelines that will guide countries through such an approach.[2] Consequently, despite the rapid ascendancy of the HRBA approach and its emergence as a holistic approach for linking human rights with broad issues of development, questions on the legal and institutional framework required to apply it to the international climate change regime have not been considered in sufficient detail. This has been the main contribution of this book. It has examined the potentials and value of adopting the HRBA as a framework for integrating human rights obligations and standards into the international climate change regime, specifically to ensure that climate actions and projects do not produce human rights consequences. First, it identified and elaborated the normative contents and elements of a human rights-based approach: information, participation, accountability, equality and non-discrimination, and access to justice. It examined how these elements have been recognized and enshrined in core international human rights instruments. It also discussed practical/on-the-ground problems that a HRBA to carbon finance might face, specifically issues of resources, striking coordination between human rights and epistemic communities, lack of required capacity and training gaps. To address these problems and to identify how the elements of a HRBA could be reflected in the international climate change regime, Chapter 6 proposed a legal and institutional framework that could serve as the guiding principles and operational template on how a rights-based approach could be successfully implemented under the international climate change regime.

The proposed legal reforms include the need to entrench and adopt minimum legal thresholds that must be met and demonstrated before projects are approved and registered. This book has discussed a six-part legal threshold that could serve as the core content of a reformed rights-based international climate change regime. It includes project eligibility screen, sustainable development screen, human rights impact

[2] United Nations Environment Program (UNEP), *Climate Change and Human Rights* (UNEP, 2015), 14, stating that although the UNFCCC COP also formally recognized the linkages between climate change and human rights, the COP has done very little to elaborate on *how* countries should fulfill that goal.

assessment, stakeholder identification and consultation, information disclosure system and project review mechanism. These six elements emphasize the importance of an integrated assessment to identify the human rights, social and environmental impacts and risks of projects; the importance of effective community engagement and consultation with local communities on matters that directly affect them; as well as the need for proponents of a project to manage and disclose the human rights impact and the social and environmental performance of a project throughout its lifecycle.

Similarly, this book has examined the nature of the institutional reforms necessary to spearhead such a legal reform process. It identified four complementary institutional reforms that would go a long way to shaping how the legal threshold would be operationalized and enforced, how progress would be monitored and evaluated, and what further action should be taken in case of unsatisfactory progress. They are the creation of a specific human rights and climate change unit (HRCU) to be responsible for the overall mainstreaming strategy; the expansion of the terms of reference and powers of supervisory boards for carbon projects; empowering the enforcement branch of the compliance committee to ensure that after registration the execution of projects are in line with approved standards and do not create unanticipated problems; and the setting up of a public complaint branch that could hear appeals by stakeholders who may be adversely affected. This institutional framework mirrors the shift away from top-down planning towards more participatory forms of development and recognition for stakeholders in climate change mitigation. It provides a window of opportunity and a platform for stakeholders to bring their complaints about a project before an international supervisory body at every stage of project planning and execution – that is, prior to registration and post-registration. Mobilizing and empowering institutions would enable individuals, communities and NGOs to have a voice in the processes of approving climate change projects, and the current pervasive culture of approving emission reduction projects without adequate attention to their potential impact on human rights could be addressed in a balanced and transparent way.

The framework discussed and developed in this book provides a preventive step-by-step process through which human rights are systematically integrated into project design and approval so as to anticipate and avoid human rights abuses in the first instance. The legal and institutional frameworks proposed could also help reduce opposition, protests and litigation arising from emission reduction projects as it allows citizens to take part in the processes and procedures for project approval and execution. By providing the public with an opportunity to highlight the environmental and human rights threats posed by projects, project proponents and national authorities can better anticipate and respond to these threats and take necessary precautions to avoid them. This would go a long way towards preventing violations, thereby reducing the need for litigation.

Finally, this book examined practical limitations to such a proposal. One important question is whether policymakers at national and international levels would be able to muster the political will and courage to replace business-as-usual models with a more rights-based and transformative one. This book has demonstrated that these challenges, though substantial, are not insurmountable; neither do they call for intellectual surrender. It has emphasized the need for negotiators and national authorities to understand that these reforms are not only essential but are also important to restore the integrity of carbon markets, address some of the human rights problems that bedevil project-based mechanisms such as the CDM as documented by the CDM Policy Dialogue, retain support for the Kyoto Protocol and its market mechanisms and consolidate on progress made in efforts to combat climate change. NGOs and interest groups have especially important roles to play in communicating and raising awareness on the importance of these rights-based reforms to the protection of human rights and fundamental freedoms while addressing the problem of climate change. Legal and institutional frameworks proposed in this book provide mutually reinforcing pillars through which human rights principles could be reflected in climate actions and projects. This way human rights infringements arising from carbon projects may be anticipated and prevented in the first place.

10.2 Recommendations

A first step in adopting a human rights-based approach to carbon finance is to introduce human rights language and framework into the international climate change regime. Countries will have to develop proposals aimed at integrating human rights threshold into climate change instruments. This section offers an outline of the essential contents of a detailed amendment proposal for mainstreaming human rights norms into the international climate change regime.

10.2.1 Integrate robust human rights provisions

First there is the need for international climate change agreements to clearly recognize that the direct and indirect linkages between climate change and human rights are real and concrete. Even though the UNFCCC COP and the United Nations Human Rights Council have already recognized these linkages in the 2015 Paris Agreement, and in several resolutions, there is a need to move beyond preambular recognition and adopt more robust human rights provisions and obligations in the international climate change regime.[3] There is the need for the COP to play its roles of reviewing the Convention, in the light of newly available information on the impacts of climate change on human rights, by establishing concrete human rights guidelines that mandate project participants and stakeholders to respect, protect and fulfil human rights when investing in carbon projects. Affirming these linkages in the context of COP negotiations, and within the international climate change regime, is the first important step towards implementing a human rights-based approach to carbon finance.

The UNFCCC and the Kyoto Protocol provide strong lessons on the importance of integrating and recognizing robust human rights provisions and norms as necessary first steps for enforcing a rights-based approach to carbon finance. Even though both instruments provide that parties shall minimize any adverse economic, social and environmental impact resulting from the implementation of measures taken to mitigate or adapt to climate change (response measures), implementation challenges arise from the fact that both instruments fail to establish mandatory human rights impact assessment procedures or guidelines that provide legal basis for parties to minimize the adverse *human rights* impact of climate change response measures.[4] Both instruments also fail to codify human rights provisions, establish detailed rights-based guidelines or address human rights impacts of climate change measures and actions. These omissions meant that supervisory bodies, such as the CDM EB, Compliance Committee or

[3] See UNHCR Res. 17/23 (2008) UN Doc E/CN.4/RES 2008; UNHCR Res. 10/4 (2009) UN Doc E/CN.4/RES/2009; and UNHCR Res. 18/22 (2011) UN Doc E/CN.4/RES 2011.

[4] See UNFCCC, Article 4(1) (h); Kyoto Protocol, Conference of the Parties to the Framework Convention on Climate Change, Kyoto Protocol, 10 December 1997, UN Doc FCCC/CP/1997/L.7/add.1, 37 ILM 22 (1998) [Kyoto Protocol], Articles 2(3) and 3(14).

the UNFCCC Secretariat, lacked legal mandates and authority to consider human rights questions.

To address these gaps, the international climate change regime must recognize and adopt more expansive language to include fundamental human rights. Specifically, the international climate regime must go beyond mere textual or preambular stipulations that 'Parties should, when taking action to address climate change, respect, promote and consider their respective obligations on human rights'.[5] The climate change regime should, as a next step, expressly stipulate adequate safeguards and procedures that parties must take to ensure and demonstrate that climate actions and projects are coordinated with the respect, protection and fulfilment of fundamental human rights, in an integrated manner. Respect for human rights in the implementation of climate actions cannot be achieved by mere preambular reference or textual acknowledgement, the climate regime must provide detailed guidance on how procedural human rights such as access to information, participation, accountability, equality and access to justice, can be mainstreamed in the design, approval, finance and implementation of carbon projects.

The step-wise approach discussed in this book provides key lessons and guidance on *how* the international climate architecture may be holistically reformed to better integrate human rights safeguards in the execution of carbon projects. Furthermore, the legal and governance structure discussed in this book unpacks the roles of key stakeholders such as governments, business enterprises, project proponents, financing institutions, project auditors, carbon verifiers, and law firms in developing and implementing human rights risk management frameworks to holistically anticipate and address impacts and concerns.

10.2.2 Redefine vulnerability

As already discussed, many of the direct and indirect effects of climate change on human rights are often exacerbated by inherent societal and non-climatic factors such as discrimination, unequal power relationships, lack of transparency and accountability by governments, lack of adequate data and government secrecy, lack of access to formal education, poverty, social marginalization, exclusion from policy and decision-making institutions

[5] See Preamble to the Paris Agreement, UNFCCC Decision 1/CP 21, Conference of the Parties Twenty-first session Paris, 30 November to 11 December 2015, FCCC/CP/2015/L.9/Rev.1.

and processes and lack of deliberative democracy. These are problems that are entrenched in the political norms of different countries with varying intensities and thereby make the impact of climate change on human rights more pronounced in some jurisdictions than in others. For example, in many developing countries, women are not allowed to take part in important town hall meetings where decisions on projects or policies are taken. Consequently at the Doha climate conference, parties agreed to achieve gender balance in all climate change programmes and in climate change governance bodies.[6]

To implement this decision, there is a need for the international climate change regime to acknowledge that causes of vulnerabilities extend beyond poverty and lack of infrastructure. Without international recognition of these skewed relations that worsen vulnerabilities, national approaches to climate change might continue to be shaped and patterned along the lines of these societal and non-climatic misgivings. The end result would be an uncoordinated, skewed and ineffective carbon market in which some stakeholders are excluded from playing active parts in project approval processes. To prevent policies that discriminate based on status or gender, there is the need to establish and maintain an effective database for assessing risks on the basis of income, gender roles and the constraints operating against their participation in decision-making processes. The international climate change regime can address the issue of discrimination by mandating participants in carbon projects to disaggregate project data and consider the specific impacts of projects on ethnic and indigenous groups, poor communities, youths, women, immigrant communities and other marginalized groups that typically bear disproportionate burdens of carbon projects. Governments, corporations and other participants in carbon projects must also incorporate vulnerability proofing as part of the wider human rights proofing of all climate investments, actions and projects. Gender proofing, for example, will help to assess the implications of any planned project on women, the representation afforded to them in decision-making and ways to protect their interests through policies, legislation and programmes.

A second discriminatory trend discussed is that in some countries, poor and low-income neighbourhoods are targeted as potential locations

[6] See UNFCCC, Decisions Adopted by the Conference of the Parties on its Eighteenth Session, held in Doha from 26 November to 8 December 2012 23/CP.18, Promoting Gender Balance and Improving the Participation of Women in UNFCCC Negotiations and in the Representation of Parties in Bodies Established Pursuant to the Convention of the Kyoto Protocol UNFCCC. Article 7 of the Paris Agreement also provides that 'adaptation action should follow a country-driven, gender-responsive, participatory and fully transparent approach, taking into consideration vulnerable groups, communities and ecosystems'.

for carbon projects; this worsens the already poor standard of living in these communities, making them even more vulnerable to the impact of climate change. This target selection and concentration of projects in poor communities must be addressed. A rights-based regime to carbon finance would ensure that the vulnerabilities of the poor and minorities are identified and protected in the design, approval, finance and implementation of carbon projects. Implementing non-discriminatory policies would include laying down clear criteria for selecting project locations, conducting environmental impact assessment on the effects of a project at that location, and providing public information on the outcome of the assessment. If the EIA reveals that residents of the specific location would be affected by that project, then the project should be re-designed to avoid the impact or as a last option discontinued, irrespective of the amount of emission reduction the project would generate.

To address discriminatory factors that worsen vulnerabilities, human rights standards and principles should inform and underpin project design and approval processes. The international climate change regime must establish detailed project requirements that would make it compulsory for project proponents to demonstrate that a particular section of the society is not disadvantaged by a project before it could be approved or certified for carbon finance. This would be achieved by developing concrete risk assessment procedures to ensure better characterization of risk across populations, communities or geographic areas. Measures would then be put in place to reduce high concentrations of risk among specific population groups. When making project decisions, project proponents should be required to demonstrate the distribution of projected risk. As part of the assessment, project proponents must identify individuals and groups that may be differentially or disproportionately affected by the project because of their status. Where groups are identified as disadvantaged or vulnerable, project proponents must propose and implement measures so that no adverse impact falls disproportionately on them.

10.2.3 Ensure equity and fairness

One of the lessons learned from the implementation of CDM and REDD+ projects is that their conceptualizations are themselves, arguably, premised on inequality. There have been general concerns about the distribution of carbon projects.[7] For example, CDM investments and projects have concentrated in some developing countries leaving others out. Over 80 per cent of current CDM projects are clustered in Asian countries, namely, India, China and Indonesia, while Africa accounts for less than 2 per cent of the entire CDM project portfolio, same as Latin American countries.[8] This inequitable distribution has led to intense competition among developing countries to host CDM projects. This lopsided contest at times even involves doing everything possible to secure the confidence of project proponents to invest in a country. This has led to a situation whereby national governments play active roles in the repression of human rights, land grabs, displacement of citizens and concentration of projects in poor communities.[9]

There is a need to address these equality concerns in the international climate change regime. To promote inclusiveness, the international climate architecture must foster balanced and geographical spread of carbon projects and investments. Without addressing the competitive nature of carbon markets, countries might continue to adopt such extreme measures aimed at winning the patronage of project proponents at the expense of the human rights of the public. In reforming the international climate change regime, there is the need to create a level playing field among countries. An approach would be to create more incentives for developed countries to invest in the least developed countries (LDCs) that are perceived as unattractive locations for carbon projects. For example, doubling the worth of emission credits generated from projects carried out in least attractive countries could encourage developed countries to consider investing in these least attractive destinations. Another policy approach could be to provide tailored financial assistance to LDCs to reduce local risks that stifle their competitiveness in the carbon market. This could be through clean funds, low interest loans and development assistance that would allow LDCs to reduce their infrastructural gaps and be able to compete in attracting projects.

[7] See D. Olawuyi, 'Towards a Transparent and Accountable Clean Development Mechanism: Legal and Institutional Imperatives', *Nordic Journal of Environmental Law*, 2, (2012), 33–35.

[8] These Asian countries are now referred to as the CDM giants. For a breakdown of current CDM projects see The UNFCCC, 'Statistics of CDM Projects' cdm.unfccc.int/Statistics/Registration/AmountOfReductRegisteredProjP, accessed 20 March 2015.

[9] See Olawuyi, 'Towards a Transparent and Accountable Clean Development Mechanism', (n 7), 33.

10.2.4 *Clarify the roles of responsible business cultures and corporate human rights risk management policies*

One of the lessons learnt from the implementation of CDM and REDD+ projects is that human rights risks in carbon projects are not properly evaluated and considered by project participants during project design and planning stages. In most cases, human rights considerations only come to the table when business enterprises and project participants are faced with protests and petitions about their carbon projects or investments.

While much has been written and said on the need for corporate responsibility in climate actions, the international climate regime has not done enough to establish robust requirements for business enterprises to adopt holistic internal human rights risk management frameworks that will enhance their abilities to anticipate and manage human rights risks. Consequently, while several companies have internal corporate policies that articulate their carbon footprints and corporate actions to reduce greenhouse gases, only very few have internal human rights risk management policies in place to address the impacts of their carbon investments, projects or credit sourcing activities on fundamental human rights. One reason for this lack of action is that the roles of responsible business cultures when investing in or financing carbon projects, or sourcing for carbon credits, have not been clearly articulated and clarified in the international climate regime.

There is a need for the international climate regime to evolve coherent and consistent standards policies and practices expected of responsible corporations when investing in or financing carbon projects. Highly important is the need to provide concrete and mandatory guidelines for businesses to develop internal corporate policies that clearly articulate a human rights risk management framework aimed at minimizing human rights risks when investing in carbon projects.

It is important for the climate change regime to introduce more stringent requirements that emphasise responsible business culture in carbon investments, and mandate participants in the carbon markets to demonstrate that human rights issues have been considered in the design, financing and planning of carbon projects. Participants in the carbon markets must, as a prerequisite for participation and project registration, be required to demonstrate compliance with the guiding principles of human rights impact assessment as discussed in this book. Conducting human rights diligence and disseminating its results would provide the public, and the carbon market in general, with immediate notice and information that a project is being planned and that its potential human rights impacts have been evaluated and appropriately mitigated.

The international climate regime must provide guidance on the roles of responsible business culture in carbon investments, by developing model internal risk management policies to guide business on how to effectively integrate human rights safeguards in carbon investments and projects.[10] However, business enterprises themselves must take the lead in developing human rights risks management frameworks to ensure compliance with human rights obligations under the climate regime. With the increased recognition of human rights in the climate regime, and their mainstreaming into project requirements, the expectations of carbon investors and business enterprises to protect, respect and fulfil human rights in carbon investments and projects is no longer an option but a necessity. Business enterprises and organizations investing in carbon projects can avoid the backlash associated with carbon projects, as learnt from CDM and REDD+ projects, by incorporating human rights due diligence into corporate risk management processes.

There is a strong business case, in terms of cost, reputation and effectiveness, for participants in the carbon markets to manage human rights risks responsibly. Failure to effectively manage human rights issues associated with carbon projects carries significant financial, legal and reputational risks. Such risks may manifest themselves in community protests due to health and safety concerns, disruption of project construction and potential harm to employees due to such protests. There may be legal risks related to the court and regulatory actions to challenge projects, and reputational risks related to loss of life or injuries from actions of security forces in responding to protest situations. Failure to effectively address human rights risks can lead to significant costs in terms of the management time required to respond to crises, costly suspensions or closures of projects by supervisory bodies and may impact a company's ability to receive project funding from carbon finance institutions or export credit agencies. Project proponents will be evaluated and judged by stakeholders (including financial institutions and supervisory institutions) based on the strength of the internal procedures, practices and approaches put in place to anticipate and manage

[10] See International Bar Association, *Climate Change Justice and Human Rights Task Force Report, Achieving Justice and Human Rights in an Era of Climate Disruption* (London: International Bar Association, 2014), 148, calling on the OHCHR to develop a model internal corporate policy on how corporations can address human rights impacts of climate actions.

human rights risks associated with its carbon investment and project. Furthermore, adopting sound internal human rights screening processes as part of corporate risk management frameworks can help business enterprises to build and maintain strong social licence to operate and to foster relationships with communities, customers and other stakeholders based on mutual respect, acceptance and trust.

As for the role of lawyers representing companies and investors in carbon projects and investments, it is important to accentuate the significance of human rights risk management when providing counsel. The growing insistence of courts and tribunals on the rights of information, public participation and benefit-sharing for those persons potentially affected by proposed projects should lead lawyers to advise clients of the wisdom of early compliance with human rights requirements in carbon investments and projects. Nearly all cases of conflict over CDM and REDD+ projects arise from projects executed without proper consultations with the local communities. These communities are not generally opposed to carbon projects or investment, but they know and insist upon their rights, even if governments and supervisory bodies are unwilling or unable to protect them. Lawyers must provide detailed guidance to clients on the procedures and processes for integrating human rights due diligence into general corporate risk management processes, specifically in the design and implementation of carbon projects.

10.3 Further research

This book has developed a legal and institutional framework that could serve as a process-based guideline or template for mainstreaming human rights principles into project execution and approval procedures under the international climate change regime. However, even though establishment of legal and institutional mechanisms for mainstreaming human rights into climate change mitigation is facilitated at the international level, strengthening democratic and accountability processes at the national level remains a prerequisite for overall effectiveness. Arguably, human rights mainstreaming would be more effective if it takes root at all levels of decision-making, especially at the local levels, particularly for national mitigation projects or plans that are not executed within the scope of the international climate change regime; for example, large hydroelectric power plant projects or biofuel projects executed under a national programme.

There is therefore a need for further research on the extent to which the human rights-based approach to carbon finance could be embedded in legal frameworks for climate change mitigation at the national level. Specifically, there is a need to consider the potentials and paradoxes of adopting a human rights-based approach to carbon finance at the national level. What are the web of laws, regulations and policies that could be adopted at the national level to coordinate national mitigation plans with human rights laws and institutions; what national institutions, departments and ministries would foster the linkages between human rights and climate change communities; what are the potentials for success; and what are the practical limitations and challenges of adopting a rights-based approach to mitigation at the national level. Answers to these questions would further clarify how national governments can take ownership of the HRBA framework and how national authorities could tailor the rights-based threshold discussed in this book, to guide national mitigation plans and policies, most especially project-based actions and measures.

INDEX

Lightning Source UK Ltd.
Milton Keynes UK
UKOW06n2205290516

275190UK00004B/138/P

9 781107 105515